Modern Quantum Mechanics

Modern Quantum Mechanics

Edited by
Kendrick Porter

⊟ Larsen & Keller
www.larsen-keller.com

Modern Quantum Mechanics
Edited by Kendrick Porter
ISBN: 978-1-63549-245-3 (Hardback)

☰ Larsen & Keller

Published by Larsen and Keller Education,
5 Penn Plaza,
19th Floor,
New York, NY 10001, USA

Cataloging-in-Publication Data

Modern quantum mechanics / edited by Kendrick Porter.
    p. cm.
Includes bibliographical references and index.
ISBN 978-1-63549-245-3
1. Quantum theory. 2. Mechanics.
3. Physics. I. Porter, Kendrick.
QC174.12 .M63 2017
530.12--dc23

The publisher's policy is to use permanent paper from mills that operate a sustainable forestry policy. Furthermore, the publisher ensures that the text paper and cover boards used have met acceptable environmental accreditation standards.

Printed and bound in the United States of America.

For more information regarding Larsen and Keller Education and its products, please visit the publisher's website www.larsen-keller.com

Table of Contents

Preface

This book is a compilation of chapters that discuss the most vital concepts in the field of modern quantum mechanics. It is designed to provide students with the basic concepts and applications of this field. Modern quantum mechanics refers to the study of processes of photons and atoms. It is based on quantum field theory. This text attempts to understand the multiple branches that fall under the discipline of modern quantum mechanics and how such concepts have practical applications. Such selected concepts that redefine the field of modern quantum mechanics have been presented in this book. It studies, analyses and upholds the pillars of the subject and its utmost significance in modern times. This textbook attempts to assist those with a goal of delving into this field.

To facilitate a deeper understanding of the contents of this book a short introduction of every chapter is written below:

Chapter 1- Quantum mechanics is a fundamental branch of physics. This branch is concerned with processes that involve atoms and photons. This chapter is an overview of the subject matter incorporating all the major aspects of quantum mechanics.

Chapter 2- The important concepts of quantum mechanics discussed in the chapter are Stern-Gerlach experiment, bra-ket notation, uncertainty principle, wave function, old quantum theory, double-slit experiment etc. Bra-ket is a standard that is used in describing quantum states. This section elucidates the crucial theories and concepts of quantum mechanics.

Chapter 3- Schrödinger equation explains how the quantum state of a quantum system changes, and how it changes with time. A Schrödinger field is a quantum field which goes as per the Schrödinger equation. The section on Schrödinger equation offers an insightful focus, keeping in mind the complex subject matter.

Chapter 4- Path integral formulation is the description of quantum theory that helps in the generalization of the action principle of classical mechanics. The topics elucidated in this chapter are relation between Schrödinger's equation and the path integral formulation of quantum mechanics, propagator and the Feynman diagram. The chapter serves as a source to understand the path integrals in quantum mechanics.

Chapter 5- In physics, symmetry is the physical feature of the system that remains unchanged under some transformation. Spacetime symmetries, supersymmetry, Noether's theorem and parity are some of the aspects of symmetry. This section helps the reader in understanding the features of symmetry in quantum mechanics.

Chapter 6- Quantum field theories include theories such as common integrals in quantum field theory, first quantization and second quantization. Quantum field theory is the theoretical framework; this framework is used in constructing quantum mechanical models. The topics discussed in the chapter are of great importance to broaden the existing knowledge on quantum field theories.

I would like to share the credit of this book with my editorial team who worked tirelessly on this book. I owe the completion of this book to the never-ending support of my family, who supported me throughout the project.

Editor

Introduction to Quantum Mechanics

Quantum mechanics is a fundamental branch of physics. This branch is concerned with processes that involve atoms and photons. This chapter is an overview of the subject matter incorporating all the major aspects of quantum mechanics.

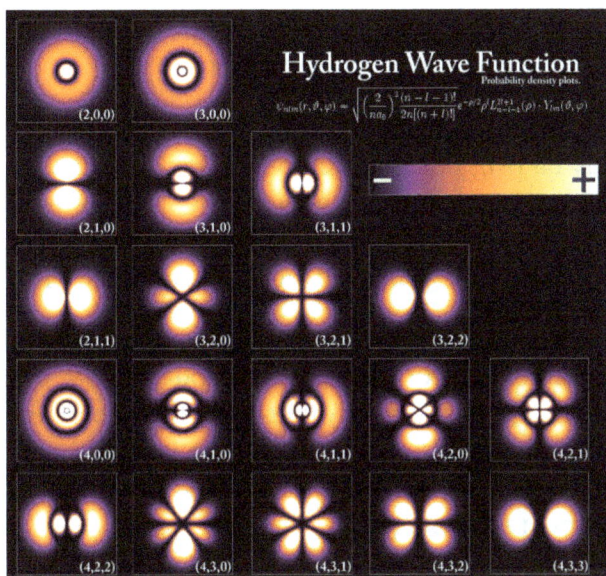

Solution to Schrödinger's equation for the hydrogen atom at different energy levels. The brighter areas represent a higher probability of finding an electron.

Quantum mechanics (QM; also known as quantum physics or quantum theory), including quantum field theory, is a fundamental branch of physics concerned with processes involving, for example, atoms and photons. Systems such as these which obey quantum mechanics can be in a quantum superposition of different states, unlike in classical physics.

Quantum mechanics gradually arose from Max Planck's solution in 1900 to the black-body radiation problem (reported 1859) and Albert Einstein's 1905 paper which offered a quantum-based theory to explain the photoelectric effect (reported 1887). Early quantum theory was profoundly reconceived in the mid-1920s.

The reconceived theory is formulated in various specially developed mathematical formalisms. In one of them, a mathematical function, the wave function, provides information about the probability amplitude of position, momentum, and other physical properties of a particle.

Important applications of quantum theory include superconducting magnets, light-emitting diodes, and the laser, the transistor and semiconductors such as the microprocessor, medical and research imaging such as magnetic resonance imaging and electron microscopy, and explanations for many biological and physical phenomena.

History

Scientific inquiry into the wave nature of light began in the 17th and 18th centuries, when scientists such as Robert Hooke, Christiaan Huygens and Leonhard Euler proposed a wave theory of light based on experimental observations. In 1803, Thomas Young, an English polymath, performed the famous double-slit experiment that he later described in a paper titled *On the nature of light and colours*. This experiment played a major role in the general acceptance of the wave theory of light.

In 1838, Michael Faraday discovered cathode rays. These studies were followed by the 1859 statement of the black-body radiation problem by Gustav Kirchhoff, the 1877 suggestion by Ludwig Boltzmann that the energy states of a physical system can be discrete, and the 1900 quantum hypothesis of Max Planck. Planck's hypothesis that energy is radiated and absorbed in discrete "quanta" (or energy packets) precisely matched the observed patterns of black-body radiation.

In 1896, Wilhelm Wien empirically determined a distribution law of black-body radiation, known as Wien's law in his honor. Ludwig Boltzmann independently arrived at this result by considerations of Maxwell's equations. However, it was valid only at high frequencies and underestimated the radiance at low frequencies. Later, Planck corrected this model using Boltzmann's statistical interpretation of thermodynamics and proposed what is now called Planck's law, which led to the development of quantum mechanics.

Following Max Planck's solution in 1900 to the black-body radiation problem (reported 1859), Albert Einstein offered a quantum-based theory to explain the photoelectric effect (1905, reported 1887). Around 1900-1910, the atomic theory and the corpuscular theory of light first came to be widely accepted as scientific fact; these latter theories can be viewed as quantum theories of matter and electromagnetic radiation, respectively.

Among the first to study quantum phenomena in nature were Arthur Compton, C. V. Raman, and Pieter Zeeman, each of whom has a quantum effect named after him. Robert Andrews Millikan studied the photoelectric effect experimentally, and Albert Einstein developed a theory for it. At the same time, Ernest Rutherford experimentally discovered the nuclear model of the atom, for which Niels Bohr developed his theory of the atomic structure, which was later confirmed by the experiments of Henry Moseley. In 1913, Peter Debye extended Niels Bohr's theory of atomic structure, introducing elliptical orbits, a concept also introduced by Arnold Sommerfeld. This phase is known as old quantum theory.

According to Planck, each energy element (E) is proportional to its frequency (v):

$$E = hv$$

where h is Planck's constant.

Planck cautiously insisted that this was simply an aspect of the *processes* of absorption and emission of radiation and had nothing to do with the *physical reality* of the radiation itself. In fact, he considered his quantum hypothesis a mathematical trick to get the right answer rather than a sizable discovery. However, in 1905 Albert Einstein interpreted Planck's quantum hypothesis realistically and used it to explain the photoelectric effect, in which shining light on certain materials can eject electrons from the material. He won the 1921 Nobel Prize in Physics for this work.

Max Planck is considered the father of the quantum theory.

Einstein further developed this idea to show that an electromagnetic wave such as light could also be described as a particle (later called the photon), with a discrete quantum of energy that was dependent on its frequency.

The 1927 Solvay Conference in Brussels.

The foundations of quantum mechanics were established during the first half of the 20th century by Max Planck, Niels Bohr, Werner Heisenberg, Louis de Broglie, Arthur Compton, Albert Einstein, Erwin Schrödinger, Max Born, John von Neumann, Paul Dirac, Enrico Fermi, Wolfgang Pauli, Max von Laue, Freeman Dyson, David Hilbert, Wilhelm Wien, Satyendra Nath Bose, Arnold Sommerfeld, and others. The Copenhagen interpretation of Niels Bohr became widely accepted.

In the mid-1920s, developments in quantum mechanics led to its becoming the standard formulation for atomic physics. In the summer of 1925, Bohr and Heisenberg published results that closed the old quantum theory. Out of deference to their particle-like behavior in certain processes and measurements, light quanta came to be called photons (1926). From Einstein's simple postulation was born a flurry of debating, theorizing, and testing. Thus, the entire field of quantum physics emerged, leading to its wider acceptance at the Fifth Solvay Conference in 1927.

It was found that subatomic particles and electromagnetic waves are neither simply particle nor wave but have certain properties of each. This originated the concept of wave–particle duality.

By 1930, quantum mechanics had been further unified and formalized by the work of David Hilbert, Paul Dirac and John von Neumann with greater emphasis on measurement, the statistical nature of our knowledge of reality, and philosophical speculation about the 'observer'. It has since permeated many disciplines including quantum chemistry, quantum electronics, quantum optics, and quantum information science. Its speculative modern developments include string theory and quantum gravity theories. It also provides a useful framework for many features of the modern periodic table of elements, and describes the behaviors of atoms during chemical bonding and the flow of electrons in computer semiconductors, and therefore plays a crucial role in many modern technologies.

While quantum mechanics was constructed to describe the world of the very small, it is also needed to explain some macroscopic phenomena such as superconductors, and superfluids.

The word *quantum* derives from the Latin, meaning "how great" or "how much". In quantum mechanics, it refers to a discrete unit assigned to certain physical quantities such as the energy of an atom at rest. The discovery that particles are discrete packets of energy with wave-like properties led to the branch of physics dealing with atomic and subatomic systems which is today called quantum mechanics. It underlies the mathematical framework of many fields of physics and chemistry, including condensed matter physics, solid-state physics, atomic physics, molecular physics, computational physics, computational chemistry, quantum chemistry, particle physics, nuclear chemistry, and nuclear physics. Some fundamental aspects of the theory are still actively studied.

Quantum mechanics is essential to understanding the behavior of systems at atomic length scales and smaller. If the physical nature of an atom were solely described by classical mechanics, electrons would not *orbit* the nucleus, since orbiting electrons emit radiation (due to circular motion) and would eventually collide with the nucleus due to this loss of energy. This framework was unable to explain the stability of atoms. Instead, electrons remain in an uncertain, non-deterministic, *smeared*, probabilistic wave–particle orbital about the nucleus, defying the traditional assumptions of classical mechanics and electromagnetism.

Quantum mechanics was initially developed to provide a better explanation and description of the atom, especially the differences in the spectra of light emitted by different isotopes of the same chemical element, as well as subatomic particles. In short, the quantum-mechanical atomic model has succeeded spectacularly in the realm where classical mechanics and electromagnetism falter.

Broadly speaking, quantum mechanics incorporates four classes of phenomena for which classical physics cannot account:

- quantization of certain physical properties
- quantum entanglement
- principle of uncertainty
- wave–particle duality

Mathematical Formulations

In the mathematically rigorous formulation of quantum mechanics developed by Paul Dirac, David Hilbert, John von Neumann, and Hermann Weyl, the possible states of a quantum mechanical system are symbolized as unit vectors (called *state vectors*). Formally, these reside in a complex separable Hilbert space—variously called the *state space* or the *associated Hilbert space* of the system—that is well defined up to a complex number of norm 1 (the phase factor). In other words, the possible states are points in the projective space of a Hilbert space, usually called the complex projective space. The exact nature of this Hilbert space is dependent on the system—for example, the state space for position and momentum states is the space of square-integrable functions, while the state space for the spin of a single proton is just the product of two complex planes. Each observable is represented by a maximally Hermitian (precisely: by a self-adjoint) linear operator acting on the state space. Each eigenstate of an observable corresponds to an eigenvector of the operator, and the associated eigenvalue corresponds to the value of the observable in that eigenstate. If the operator's spectrum is discrete, the observable can attain only those discrete eigenvalues.

In the formalism of quantum mechanics, the state of a system at a given time is described by a complex wave function, also referred to as state vector in a complex vector space. This abstract mathematical object allows for the calculation of probabilities of outcomes of concrete experiments. For example, it allows one to compute the probability of finding an electron in a particular region around the nucleus at a particular time. Contrary to classical mechanics, one can never make simultaneous predictions of conjugate variables, such as position and momentum, to arbitrary precision. For instance, electrons may be considered (to a certain probability) to be located somewhere within a given region of space, but with their exact positions unknown. Contours of constant probability, often referred to as "clouds", may be drawn around the nucleus of an atom to conceptualize where the electron might be located with the most probability. Heisenberg's uncertainty principle quantifies the inability to precisely locate the particle given its conjugate momentum.

According to one interpretation, as the result of a measurement the wave function containing the probability information for a system collapses from a given initial state to a particular eigenstate. The possible results of a measurement are the eigenvalues of the operator representing the observable—which explains the choice of *Hermitian* operators, for which all the eigenvalues are real. The probability distribution of an observable in a given state can be found by computing the spectral decomposition of the corresponding operator. Heisenberg's uncertainty principle is represented by the statement that the operators corresponding to certain observables do not commute.

The probabilistic nature of quantum mechanics thus stems from the act of measurement. This is one of the most difficult aspects of quantum systems to understand. It was the central topic in the famous Bohr–Einstein debates, in which the two scientists attempted to clarify these fundamental principles by way of thought experiments. In the decades after the formulation of quantum mechanics, the question of what constitutes a "measurement" has been extensively studied. Newer interpretations of quantum mechanics have been formulated that do away with the concept of "wave function collapse" (for example, the relative state interpretation). The basic idea is that when a quantum system interacts with a measuring apparatus, their respective wave functions become entangled, so that the original quantum system ceases to exist as an independent entity.

Generally, quantum mechanics does not assign definite values. Instead, it makes a prediction using a probability distribution; that is, it describes the probability of obtaining the possible outcomes from measuring an observable. Often these results are skewed by many causes, such as dense probability clouds. Probability clouds are approximate (but better than the Bohr model) whereby electron location is given by a probability function, the wave function eigenvalue, such that the probability is the squared modulus of the complex amplitude, or quantum state nuclear attraction. Naturally, these probabilities will depend on the quantum state at the "instant" of the measurement. Hence, uncertainty is involved in the value. There are, however, certain states that are associated with a definite value of a particular observable. These are known as eigenstates of the observable ("eigen" can be translated from German as meaning "inherent" or "characteristic").

In the everyday world, it is natural and intuitive to think of everything (every observable) as being in an eigenstate. Everything appears to have a definite position, a definite momentum, a definite energy, and a definite time of occurrence. However, quantum mechanics does not pinpoint the exact values of a particle's position and momentum (since they are conjugate pairs) or its energy and time (since they too are conjugate pairs); rather, it provides only a range of probabilities in which that particle might be given its momentum and momentum probability. Therefore, it is helpful to use different words to describe states having *uncertain* values and states having *definite* values (eigenstates). Usually, a system will not be in an eigenstate of the observable (particle) we are interested in. However, if one measures the observable, the wave function will instantaneously be an eigenstate (or "generalized" eigenstate) of that observable. This process is known as wave function collapse, a controversial and much-debated process that involves expanding the system under study to include the measurement device. If one knows the corresponding wave function at the instant before the measurement, one will be able to compute the probability of the wave function collapsing into each of the possible eigenstates. For example, the free particle in the previous example will usually have a wave function that is a wave packet centered around some mean position x_0 (neither an eigenstate of position nor of momentum). When one measures the position of the particle, it is impossible to predict with certainty the result. It is probable, but not certain, that it will be near x_0, where the amplitude of the wave function is large. After the measurement is performed, having obtained some result x, the wave function collapses into a position eigenstate centered at x.

The time evolution of a quantum state is described by the Schrödinger equation, in which the Hamiltonian (the operator corresponding to the total energy of the system) generates the time evolution. The time evolution of wave functions is deterministic in the sense that - given a wave function at an *initial* time - it makes a definite prediction of what the wave function will be at any *later* time.

During a measurement, on the other hand, the change of the initial wave function into another, later wave function is not deterministic, it is unpredictable (i.e., random). A time-evolution simulation can be seen here.

Wave functions change as time progresses. The Schrödinger equation describes how wave functions change in time, playing a role similar to Newton's second law in classical mechanics. The Schrödinger equation, applied to the aforementioned example of the free particle, predicts that the center of a wave packet will move through space at a constant velocity (like a classical particle with no forces acting on it). However, the wave packet will also spread out as time progresses, which means that the position becomes more uncertain with time. This also has the effect of turning a position eigenstate (which can be thought of as an infinitely sharp wave packet) into a broadened wave packet that no longer represents a (definite, certain) position eigenstate.

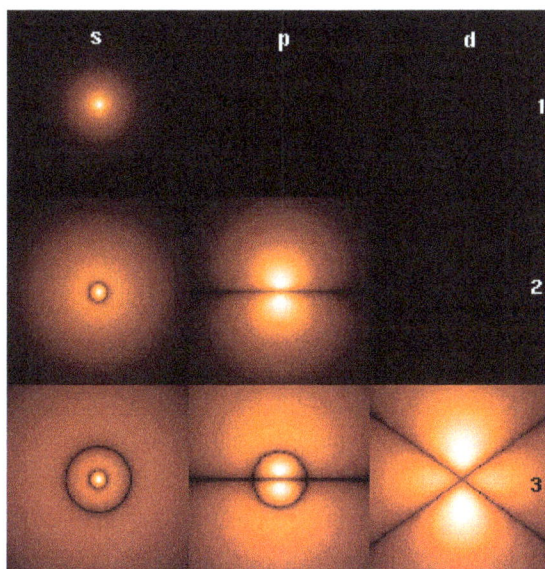

Fig. 1: Probability densities corresponding to the wave functions of an electron in a hydrogen atom possessing definite energy levels (increasing from the top of the image to the bottom: $n = 1, 2, 3, ...$) and angular momenta (increasing across from left to right: s, p, d, ...). Brighter areas correspond to higher probability density in a position measurement. Such wave functions are directly comparable to Chladni's figures of acoustic modes of vibration in classical physics, and are modes of oscillation as well, possessing a sharp energy and, thus, a definite frequency. The angular momentum and energy are quantized, and take **only** discrete values like those shown (as is the case for resonant frequencies in acoustics)

Some wave functions produce probability distributions that are constant, or independent of time—such as when in a stationary state of constant energy, time vanishes in the absolute square of the wave function. Many systems that are treated dynamically in classical mechanics are described by such "static" wave functions. For example, a single electron in an unexcited atom is pictured classically as a particle moving in a circular trajectory around the atomic nucleus, whereas in quantum mechanics it is described by a static, spherically symmetric wave function surrounding the nucleus (Fig. 1) (note, however, that only the lowest angular momentum states, labeled s, are spherically symmetric).

The Schrödinger equation acts on the *entire* probability amplitude, not merely its absolute value. Whereas the absolute value of the probability amplitude encodes information about probabilities, its phase encodes information about the interference between quantum states. This gives rise to the "wave-like" behavior of quantum states. As it turns out, analytic solutions of the Schrödinger equation are available for only a very small number of relatively simple model Hamiltonians, of which the quantum harmonic oscillator, the particle in a box, the dihydrogen cation, and the hydrogen atom are the most important representatives. Even the helium atom—which contains just one more electron than does the hydrogen atom—has defied all attempts at a fully analytic treatment.

There exist several techniques for generating approximate solutions, however. In the important method known as perturbation theory, one uses the analytic result for a simple quantum mechanical model to generate a result for a more complicated model that is related to the simpler model by (for one example) the addition of a weak potential energy. Another method is the "semi-classical equation of motion" approach, which applies to systems for which quantum mechanics produces only weak (small) deviations from classical behavior. These deviations can then be computed based on the classical motion. This approach is particularly important in the field of quantum chaos.

Mathematically Equivalent Formulations of Quantum Mechanics

There are numerous mathematically equivalent formulations of quantum mechanics. One of the oldest and most commonly used formulations is the "transformation theory" proposed by Paul Dirac, which unifies and generalizes the two earliest formulations of quantum mechanics - matrix mechanics (invented by Werner Heisenberg) and wave mechanics (invented by Erwin Schrödinger).

Especially since Werner Heisenberg was awarded the Nobel Prize in Physics in 1932 for the creation of quantum mechanics, the role of Max Born in the development of QM was overlooked until the 1954 Nobel award. The role is noted in a 2005 biography of Born, which recounts his role in the matrix formulation of quantum mechanics, and the use of probability amplitudes. Heisenberg himself acknowledges having learned matrices from Born, as published in a 1940 *festschrift* honoring Max Planck. In the matrix formulation, the instantaneous state of a quantum system encodes the probabilities of its measurable properties, or "observables". Examples of observables include energy, position, momentum, and angular momentum. Observables can be either continuous (e.g., the position of a particle) or discrete (e.g., the energy of an electron bound to a hydrogen atom). An alternative formulation of quantum mechanics is Feynman's path integral formulation, in which a quantum-mechanical amplitude is considered as a sum over all possible classical and non-classical paths between the initial and final states. This is the quantum-mechanical counterpart of the action principle in classical mechanics.

Interactions with other Scientific Theories

The rules of quantum mechanics are fundamental. They assert that the state space of a system is a Hilbert space and that observables of that system are Hermitian operators acting on that space—although they do not tell us which Hilbert space or which operators. These can be chosen appropriately in order to obtain a quantitative description of a quantum system. An important guide for making these choices is the correspondence principle, which states that the predictions of quantum mechanics reduce to those of classical mechanics when a system moves to higher energies or, equivalently, larger quantum numbers, i.e. whereas a single particle exhibits a degree of randomness, in systems incorporating millions of particles averaging takes over and, at the high energy limit, the statistical probability of random behaviour approaches zero. In other words, classical mechanics is simply a quantum mechanics of large systems. This "high energy" limit is known as the *classical* or *correspondence limit*. One can even start from an established classical model of a particular system, then attempt to guess the underlying quantum model that would give rise to the classical model in the correspondence limit.

When quantum mechanics was originally formulated, it was applied to models whose correspondence limit was non-relativistic classical mechanics. For instance, the well-known model of the quantum harmonic oscillator uses an explicitly non-relativistic expression for the kinetic energy of the oscillator, and is thus a quantum version of the classical harmonic oscillator.

Early attempts to merge quantum mechanics with special relativity involved the replacement of the Schrödinger equation with a covariant equation such as the Klein–Gordon equation or the Dirac equation. While these theories were successful in explaining many experimental results, they had certain unsatisfactory qualities stemming from their neglect of the relativistic creation and annihilation of particles. A fully relativistic quantum theory required the development of quantum field theory, which applies quantization to a field (rather than a fixed set of particles). The first complete quantum field theory, quantum electrodynamics, provides a fully quantum description of the electromagnetic interaction. The full apparatus of quantum field theory is often unnecessary for describing electrodynamic systems. A simpler approach, one that has been employed since the inception of quantum mechanics, is to treat charged particles as quantum mechanical objects being acted on by a classical electromagnetic field. For example, the elementary quantum model of the hydrogen atom describes the electric field of the hydrogen atom using a classical $-e^2 / (4\pi \epsilon_0 \, r)$ Coulomb potential. This "semi-classical" approach fails if quantum fluctuations in the electromagnetic field play an important role, such as in the emission of photons by charged particles .

Quantum field theories for the strong nuclear force and the weak nuclear force have also been developed. The quantum field theory of the strong nuclear force is called quantum chromodynamics, and describes the interactions of subnuclear particles such as quarks and gluons. The weak nuclear force and the electromagnetic force were unified, in their quantized forms, into a single quantum field theory (known as electroweak theory), by the physicists Abdus Salam, Sheldon Glashow and Steven Weinberg. These three men shared the Nobel Prize in Physics in 1979 for this work.

It has proven difficult to construct quantum models of gravity, the remaining fundamental force. Semi-classical approximations are workable, and have led to predictions such as Hawking radiation. However, the formulation of a complete theory of quantum gravity is hindered by apparent incompatibilities between general relativity (the most accurate theory of gravity currently known) and some of the fundamental assumptions of quantum theory. The resolution of these incompatibilities is an area of active research, and theories such as string theory are among the possible candidates for a future theory of quantum gravity.

Classical mechanics has also been extended into the complex domain, with complex classical mechanics exhibiting behaviors similar to quantum mechanics.

Quantum Mechanics and Classical Physics

Predictions of quantum mechanics have been verified experimentally to an extremely high degree of accuracy. According to the correspondence principle between classical and quantum mechanics, all objects obey the laws of quantum mechanics, and classical mechanics is just an approximation for large systems of objects (or a statistical quantum mechanics of a large collection of particles). The laws of classical mechanics thus follow from the laws of quantum mechanics as a statistical average at the limit of large systems or large quantum numbers. However, chaotic systems do not have good quantum numbers, and quantum chaos studies the relationship between classical and quantum descriptions in these systems.

Quantum coherence is an essential difference between classical and quantum theories as illustrated by the Einstein–Podolsky–Rosen (EPR) paradox — an attack on a certain philosophical interpretation of quantum mechanics by an appeal to local realism. Quantum interference involves adding together *probability amplitudes*, whereas classical "waves" infer that there is an adding together of *intensities*. For microscopic bodies, the extension of the system is much smaller than the coherence length, which gives rise to long-range entanglement and other nonlocal phenomena characteristic of quantum systems. Quantum coherence is not typically evident at macroscopic scales, though an exception to this rule may occur at extremely low temperatures (i.e. approaching absolute zero) at which quantum behavior may manifest itself macroscopically. This is in accordance with the following observations:

- Many macroscopic properties of a classical system are a direct consequence of the quantum behavior of its parts. For example, the stability of bulk matter (consisting of atoms and molecules which would quickly collapse under electric forces alone), the rigidity of solids, and the mechanical, thermal, chemical, optical and magnetic properties of matter are all results of the interaction of electric charges under the rules of quantum mechanics.

- While the seemingly "exotic" behavior of matter posited by quantum mechanics and relativity theory become more apparent when dealing with particles of extremely small size or velocities approaching the speed of light, the laws of classical, often considered "Newtonian", physics remain accurate in predicting the behavior of the vast majority of "large" objects (on the order of the size of large molecules or bigger) at velocities much smaller than the velocity of light.

Copenhagen Interpretation of Quantum Versus Classical Kinematics

A big difference between classical and quantum mechanics is that they use very different kinematic descriptions.

In Niels Bohr's mature view, quantum mechanical phenomena are required to be experiments, with complete descriptions of all the devices for the system, preparative, intermediary, and finally measuring. The descriptions are in macroscopic terms, expressed in ordinary language, supplemented with the concepts of classical mechanics. The initial condition and the final condition of the system are respectively described by values in a configuration space, for example a position space, or some equivalent space such as a momentum space. Quantum mechanics does not admit a completely precise description, in terms of both position and momentum, of an initial condition or "state" (in the classical sense of the word) that would support a precisely deterministic and causal prediction of a final condition. In this sense, advocated by Bohr in his mature writings, a quantum phenomenon is a process, a passage from initial to final condition, not an instantaneous "state" in the classical sense of that word. Thus there are two kinds of processes in quantum mechanics: stationary and transitional. For a stationary process, the initial and final condition are the same. For a transition, they are different. Obviously by definition, if only the initial condition is given, the process is not determined. Given its initial condition, prediction of its final condition is possible, causally but only probabilistically, because the Schrödinger equation is deterministic for wave function evolution, but the wave function describes the system only probabilistically.

For many experiments, it is possible to think of the initial and final conditions of the system as being a particle. In some cases it appears that there are potentially several spatially distinct pathways or trajectories by which a particle might pass from initial to final condition. It is an important feature of the quantum kinematic description that it does not permit a unique definite statement of which of those pathways is actually followed. Only the initial and final conditions are definite, and, as stated in the foregoing paragraph, they are defined only as precisely as allowed by the configuration space description or its equivalent. In every case for which a quantum kinematic description is needed, there is always a compelling reason for this restriction of kinematic precision. An example of such a reason is that for a particle to be experimentally found in a definite position, it must be held motionless; for it to be experimentally found to have a definite momentum, it must have free motion; these two are logically incompatible.

Classical kinematics does not primarily demand experimental description of its phenomena. It allows completely precise description of an instantaneous state by a value in phase space, the Cartesian product of configuration and momentum spaces. This description simply assumes or imagines a state as a physically existing entity without concern about its experimental measurability. Such a description of an initial condition, together with Newton's laws of motion, allows a precise deterministic and causal prediction of a final condition, with a definite trajectory of passage. Hamiltonian dynamics can be used for this. Classical kinematics also allows the description of a process analogous to the initial and final condition description used by quantum mechanics. Lagrangian mechanics applies to this. For processes that need account to be taken of actions of a small number of Planck constants, classical kinematics is not adequate; quantum mechanics is needed.

General Relativity and Quantum Mechanics

Even with the defining postulates of both Einstein's theory of general relativity and quantum theory being indisputably supported by rigorous and repeated empirical evidence, and while they do not directly contradict each other theoretically (at least with regard to their primary claims), they have proven extremely difficult to incorporate into one consistent, cohesive model.

Gravity is negligible in many areas of particle physics, so that unification between general relativity and quantum mechanics is not an urgent issue in those particular applications. However, the lack of a correct theory of quantum gravity is an important issue in cosmology and the search by physicists for an elegant "Theory of Everything" (TOE). Consequently, resolving the inconsistencies between both theories has been a major goal of 20th and 21st century physics. Many prominent physicists, including Stephen Hawking, have labored for many years in the attempt to discover a theory underlying *everything*. This TOE would combine not only the different models of subatomic physics, but also derive the four fundamental forces of nature - the strong force, electromagnetism, the weak force, and gravity - from a single force or phenomenon. While Stephen Hawking was initially a believer in the Theory of Everything, after considering Gödel's Incompleteness Theorem, he has concluded that one is not obtainable, and has stated so publicly in his lecture "Gödel and the End of Physics" (2002).

Attempts at a Unified Field Theory

The quest to unify the fundamental forces through quantum mechanics is still ongoing. Quantum electrodynamics (or "quantum electromagnetism"), which is currently (in the perturbative regime at least) the most accurately tested physical theory in competition with general relativity, has been successfully merged with the weak nuclear force into the electroweak force and work is currently being done to merge the electroweak and strong force into the electrostrong force. Current predictions state that at around 10^{14} GeV the three aforementioned forces are fused into a single unified field. Beyond this "grand unification", it is speculated that it may be possible to merge gravity with the other three gauge symmetries, expected to occur at roughly 10^{19} GeV. However — and while special relativity is parsimoniously incorporated into quantum electrodynamics — the expanded general relativity, currently the best theory describing the gravitation force, has not been fully incorporated into quantum theory. One of those searching for a coherent TOE is Edward Witten, a theoretical physicist who formulated the M-theory, which is an attempt at describing the supersymmetrical based string theory. M-theory posits that our apparent 4-dimensional spacetime is, in reality, actually an 11-dimensional spacetime containing 10 spatial dimensions and 1 time dimension, although 7 of the spatial dimensions are - at lower energies - completely "compactified" (or infinitely curved) and not readily amenable to measurement or probing.

Another popular theory is Loop quantum gravity (LQG), a theory first proposed by Carlo Rovelli that describes the quantum properties of gravity. It is also a theory of quantum space and quantum time, because in general relativity the geometry of spacetime is a manifestation of gravity. LQG is an attempt to merge and adapt standard quantum mechanics and standard general relativity. The main output of the theory is a physical picture of space where space is granular. The granularity is a direct consequence of the quantization. It has the same nature of the granularity of the photons in the quantum theory of electromagnetism or the discrete levels of the energy of the atoms. But here it is space itself which is discrete. More precisely, space can be viewed as an extremely fine fabric or network "woven" of finite loops. These networks of loops are called spin networks. The evolution of a spin network over time is called a spin foam. The predicted size of this structure is the Planck length, which is approximately 1.616×10^{-35} m. According to theory, there is no meaning to length shorter than this (cf. Planck scale energy). Therefore, LQG predicts that not just matter, but also space itself, has an atomic structure.

Philosophical Implications

Since its inception, the many counter-intuitive aspects and results of quantum mechanics have provoked strong philosophical debates and many interpretations. Even fundamental issues, such as Max Born's basic rules concerning probability amplitudes and probability distributions, took decades to be appreciated by society and many leading scientists. Richard Feynman once said, "I think I can safely say that nobody understands quantum mechanics." According to Steven Weinberg, "There is now in my opinion no entirely satisfactory interpretation of quantum mechanics."

The Copenhagen interpretation — due largely to Niels Bohr and Werner Heisenberg — remains most widely accepted amongst physicists, some 75 years after its enunciation. According to this interpretation, the probabilistic nature of quantum mechanics is not a *temporary* feature which will eventually be replaced by a deterministic theory, but instead must be considered a *final* renunciation of the classical idea of "causality." It is also believed therein that any well-defined application of the quantum mechanical formalism must always make reference to the experimental arrangement, due to the conjugate nature of evidence obtained under different experimental situations.

Albert Einstein, himself one of the founders of quantum theory, did not accept some of the more philosophical or metaphysical interpretations of quantum mechanics, such as rejection of determinism and of causality. He is famously quoted as saying, in response to this aspect, "God does not play with dice". He rejected the concept that the state of a physical system depends on the experimental arrangement for its measurement. He held that a state of nature occurs in its own right, regardless of whether or how it might be observed. In that view, he is supported by the currently accepted definition of a quantum state, which remains invariant under arbitrary choice of configuration space for its representation, that is to say, manner of observation. He also held that underlying quantum mechanics there should be a theory that thoroughly and directly expresses the rule against action at a distance; in other words, he insisted on the principle of locality. He considered, but rejected on theoretical grounds, a particular proposal for hidden variables to obviate the indeterminism or acausality of quantum mechanical measurement. He considered that quantum mechanics was a currently valid but not a permanently definitive theory for quantum phenomena. He thought its future replacement would require profound conceptual advances, and would not come quickly or easily. The Bohr-Einstein debates provide a vibrant critique of the Copenhagen Interpretation from an epistemological point of view. In arguing for his views, he produced a series of objections, the most famous of which has become known as the Einstein–Podolsky–Rosen paradox.

John Bell showed that this "EPR" paradox led to experimentally testable differences between quantum mechanics and theories that rely on added hidden variables. Experiments have been performed confirming the accuracy of quantum mechanics, thereby demonstrating that quantum mechanics cannot be improved upon by addition of hidden variables. Alain Aspect's initial experiments in 1982, and many subsequent experiments since, have definitively verified quantum entanglement.

Entanglement, as demonstrated in Bell-type experiments, does not, however, violate causality, since no transfer of information happens. Quantum entanglement forms the basis of quantum cryptography, which is proposed for use in high-security commercial applications in banking and government.

The Everett many-worlds interpretation, formulated in 1956, holds that *all* the possibilities described by quantum theory *simultaneously* occur in a multiverse composed of mostly independent parallel universes. This is not accomplished by introducing some "new axiom" to quantum mechanics, but on the contrary, by *removing* the axiom of the collapse of the wave packet. *All* of the possible consistent states of the measured system and the measuring apparatus (including the observer) are present in a *real* physical - not just formally mathematical, as in other interpretations - quantum superposition. Such a superposition of consistent state combinations of different systems is called an entangled state. While the multiverse is deterministic, we perceive non-deterministic behavior governed by probabilities, because we can only observe the universe (i.e., the consistent state contribution to the aforementioned superposition) that we, as observers, inhabit. Everett's interpretation is perfectly consistent with John Bell's experiments and makes them intuitively understandable. However, according to the theory of quantum decoherence, these "parallel universes" will never be accessible to us. The inaccessibility can be understood as follows: once a measurement is done, the measured system becomes entangled with *both* the physicist who measured it *and* a huge number of other particles, some of which are photons flying away at the speed of light towards the other end of the universe. In order to prove that the wave function did not collapse, one would have to bring *all* these particles back and measure them again, together with the system that was originally measured. Not only is this completely impractical, but even if one *could* theoretically do this, it would have to destroy any evidence that the original measurement took place (including the physicist's memory). In light of these Bell tests, Cramer (1986) formulated his transactional interpretation. Relational quantum mechanics appeared in the late 1990s as the modern derivative of the Copenhagen Interpretation.

Applications

Quantum mechanics has had enormous success in explaining many of the features of our universe. Quantum mechanics is often the only tool available that can reveal the individual behaviors of the subatomic particles that make up all forms of matter (electrons, protons, neutrons, photons, and others). Quantum mechanics has strongly influenced string theories, candidates for a Theory of Everything.

Quantum mechanics is also critically important for understanding how individual atoms combine covalently to form molecules. The application of quantum mechanics to chemistry is known as quantum chemistry. Relativistic quantum mechanics can, in principle, mathematically describe most of chemistry. Quantum mechanics can also provide quantitative insight into ionic and covalent bonding processes by explicitly showing which molecules are energetically favorable to which others and the magnitudes of the energies involved. Furthermore, most of the calculations performed in modern computational chemistry rely on quantum mechanics.

In many aspects modern technology operates at a scale where quantum effects are significant.

Electronics

Many modern electronic devices are designed using quantum mechanics. Examples include the laser, the transistor (and thus the microchip), the electron microscope, and magnetic resonance imaging (MRI). The study of semiconductors led to the invention of the diode and the transistor, which are indispensable parts of modern electronics systems, computer and telecommunication devices. Another application is the light emitting diode which is a high-efficiency source of light.

Many electronic devices operate under effect of Quantum tunneling. It even exists in the simple light switch. The switch would not work if electrons could not quantum tunnel through the layer of oxidation on the metal contact surfaces. Flash memory chips found in USB drives use quantum tunneling to erase their memory cells. Some negative differential resistance devices also utilizes quantum tunneling effect, such as resonant tunneling diode. Unlike classical diodes, its current is carried by resonant tunneling through two potential barriers. Its negative resistance behavior can only be understood with quantum mechanics: As the confined state moves close to Fermi level, tunnel current increases. As it moves away, current decreases. Quantum mechanics is vital to understanding and designing such electronic devices.

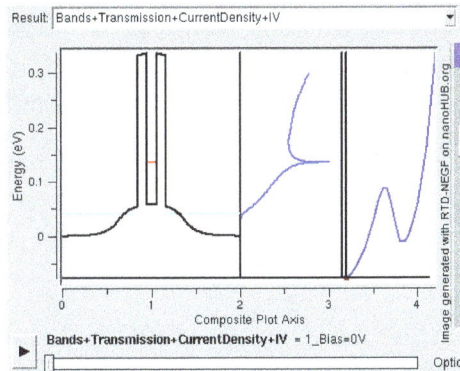

A working mechanism of a resonant tunneling diode device, based on the phenomenon of quantum tunneling through potential barriers. (Left: band diagram; Center: transmission coefficient; Right: current-voltage characteristics) As shown in the band diagram(left), although there are two barriers, electrons still tunnel through via the confined states between two barriers(center), conducting current.

Cryptography

Researchers are currently seeking robust methods of directly manipulating quantum states. Efforts are being made to more fully develop quantum cryptography, which will theoretically allow guaranteed secure transmission of information.

Quantum Computing

A more distant goal is the development of quantum computers, which are expected to perform certain computational tasks exponentially faster than classical computers. Instead of using classical bits, quantum computers use qubits, which can be in superpositions of states. Another active research topic is quantum teleportation, which deals with techniques to transmit quantum information over arbitrary distances.

Macroscale Quantum Effects

While quantum mechanics primarily applies to the smaller atomic regimes of matter and energy, some systems exhibit quantum mechanical effects on a large scale. Superfluidity, the frictionless flow of a liquid at temperatures near absolute zero, is one well-known example. So is the closely related phenomenon of superconductivity, the frictionless flow of an electron gas in a conducting material (an electric current) at sufficiently low temperatures. The fractional quantum hall effect is a topological ordered state which corresponds to patterns of long-range quantum entanglement. States with different topological orders (or different patterns of long range entanglements) cannot change into each other without a phase transition.

Quantum Theory

Quantum theory also provides accurate descriptions for many previously unexplained phenomena, such as black-body radiation and the stability of the orbitals of electrons in atoms. It has also given insight into the workings of many different biological systems, including smell receptors and protein structures. Recent work on photosynthesis has provided evidence that quantum correlations play an essential role in this fundamental process of plants and many other organisms. Even so, classical physics can often provide good approximations to results otherwise obtained by quantum physics, typically in circumstances with large numbers of particles or large quantum numbers. Since classical formulas are much simpler and easier to compute than quantum formulas, classical approximations are used and preferred when the system is large enough to render the effects of quantum mechanics insignificant.

Examples

Free Particle

For example, consider a free particle. In quantum mechanics, there is wave–particle duality, so the properties of the particle can be described as the properties of a wave. Therefore, its quantum state can be represented as a wave of arbitrary shape and extending over space as a wave function. The position and momentum of the particle are observables. The Uncertainty Principle states that both the position and the momentum cannot simultaneously be measured with complete precision. However, one *can* measure the position (alone) of a moving free particle, creating an eigenstate of position with a wave function that is very large (a Dirac delta) at a particular position x, and zero everywhere else. If one performs a position measurement on such a wave function, the resultant x will be obtained with 100% probability (i.e., with full certainty, or complete precision). This is called an eigenstate of position—or, stated in mathematical terms, a *generalized position eigenstate (eigendistribution)*. If the particle is in an eigenstate of position, then its momentum is completely unknown. On the other hand, if the particle is in an eigenstate of momentum, then its position is completely unknown. In an eigenstate of momentum having a plane wave form, it can be shown that the wavelength is equal to h/p, where h is Planck's constant and p is the momentum of the eigenstate.

Step Potential

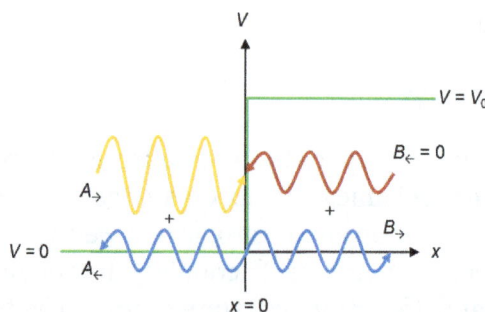

Scattering at a finite potential step of height V_0, shown in green. The amplitudes and direction of left- and right-moving waves are indicated. Yellow is the incident wave, blue are reflected and transmitted waves, red does not occur. $E > V_0$ for this figure.

The potential in this case is given by

$$V(x) = \begin{cases} 0, & x < 0, \\ V_0, & x \geq 0. \end{cases}$$

The solutions are superpositions of left- and right-moving waves:

$$\psi_1(x) = \frac{1}{\sqrt{k_1}}\left(A_\rightarrow e^{ik_1 x} + A_\leftarrow e^{-ik_1 x}\right) \quad x < 0$$

$$\psi_2(x) = \frac{1}{\sqrt{k_2}}\left(B_\rightarrow e^{ik_2 x} + B_\leftarrow e^{-ik_2 x}\right) \quad x > 0$$

where the wave vectors are related to the energy via

$$k_1 = \sqrt{2mE/\hbar^2} \text{ , and}$$

$$k_2 = \sqrt{2m(E - V_0)/\hbar^2}$$

with coefficients A and B determined from the boundary conditions and by imposing a continuous derivative on the solution.

Each term of the solution can be interpreted as an incident, reflected, or transmitted component of the wave, allowing the calculation of transmission and reflection coefficients. Notably, in contrast to classical mechanics, incident particles with energies greater than the potential step are partially reflected.

Rectangular Potential Barrier

This is a model for the quantum tunneling effect which plays an important role in the performance of modern technologies such as flash memory and scanning tunneling microscopy. Quantum tunneling is central to physical phenomena involved in superlattices.

Particle in a Box

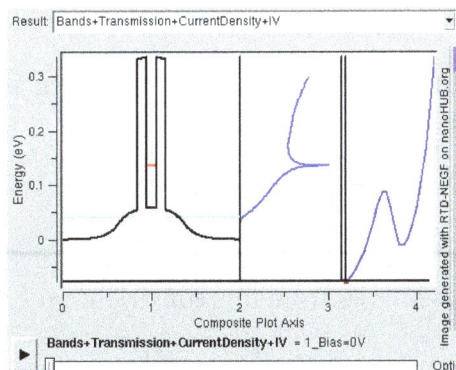

1-dimensional potential energy box (or infinite potential well)

The particle in a one-dimensional potential energy box is the most mathematically simple example where restraints lead to the quantization of energy levels. The box is defined as having zero potential energy everywhere *inside* a certain region, and infinite potential energy everywhere *outside* that region. For the one-dimensional case in the x direction, the time-independent Schrödinger equation may be written

$$-\frac{\hbar^2}{2m}\frac{d^2\psi}{dx^2} = E\psi.$$

With the differential operator defined by

$$\hat{p}_x = -i\hbar\frac{d}{dx}$$

the previous equation is evocative of the classic kinetic energy analogue,

$$\frac{1}{2m}\hat{p}_x^2 = E,$$

with state ψ in this case having energy E coincident with the kinetic energy of the particle.

The general solutions of the Schrödinger equation for the particle in a box are

$$\psi(x) = Ae^{ikx} + Be^{-ikx} \qquad E = \frac{\hbar^2 k^2}{2m}$$

or, from Euler's formula,

$$\psi(x) = C\sin kx + D\cos kx.$$

The infinite potential walls of the box determine the values of C, D, and k at $x = 0$ and $x = L$ where ψ must be zero. Thus, at $x = 0$,

$$\psi(0) = 0 = C\sin 0 + D\cos 0 = D$$

and $D = 0$. At $x = L$,

$$\psi(L) = 0 = C\sin kL.$$

in which C cannot be zero as this would conflict with the Born interpretation. Therefore, since $\sin(kL) = 0$, kL must be an integer multiple of π,

$$k = \frac{n\pi}{L} \qquad n = 1, 2, 3, \ldots.$$

The quantization of energy levels follows from this constraint on k, since

$$E = \frac{\hbar^2\pi^2 n^2}{2mL^2} = \frac{n^2 h^2}{8mL^2}.$$

Finite Potential Well

A finite potential well is the generalization of the infinite potential well problem to potential wells having finite depth.

The finite potential well problem is mathematically more complicated than the infinite particle-in-a-box problem as the wave function is not pinned to zero at the walls of the well. Instead, the wave function must satisfy more complicated mathematical boundary conditions as it is nonzero in regions outside the well.

Harmonic Oscillator

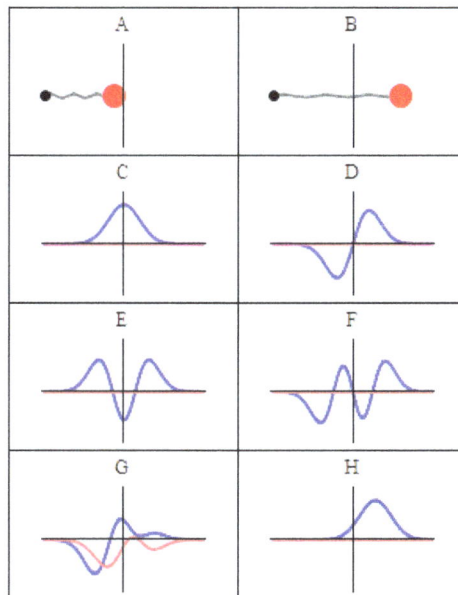

Some trajectories of a harmonic oscillator (i.e. a ball attached to a spring) in classical mechanics (A-B) and quantum mechanics (C-H). In quantum mechanics, the position of the ball is represented by a wave (called the wave function), with the real part shown in blue and the imaginary part shown in red. Some of the trajectories (such as C,D,E,and F) are standing waves (or "stationary states"). Each standing-wave frequency is proportional to a possible energy level of the oscillator. This "energy quantization" does not occur in classical physics, where the oscillator can have *any* energy.

As in the classical case, the potential for the quantum harmonic oscillator is given by

$$V(x) = \frac{1}{2} m\omega^2 x^2$$

This problem can either be treated by directly solving the Schrödinger equation, which is not trivial, or by using the more elegant "ladder method" first proposed by Paul Dirac. The eigenstates are given by

$$\psi_n(x) = \sqrt{\frac{1}{2^n n!}} \cdot \left(\frac{m\omega}{\pi\hbar}\right)^{1/4} \cdot e^{-\frac{m\omega x^2}{2\hbar}} \cdot H_n\left(\sqrt{\frac{m\omega}{\hbar}} x\right),$$

$$n = 0, 1, 2, \ldots.$$

where H_n are the Hermite polynomials,

$$H_n(x) = (-1)^n e^{x^2} \frac{d^n}{dx^n}\left(e^{-x^2}\right)$$

and the corresponding energy levels are

$$E_n = \hbar\omega\left(n + \frac{1}{2}\right).$$

This is another example illustrating the quantization of energy for bound states.

References

- The following titles, all by working physicists, attempt to communicate quantum theory to lay people, using a minimum of technical apparatus.

- Cox, Brian; Forshaw, Jeff (2011). The Quantum Universe: Everything That Can Happen Does Happen:. Allen Lane. ISBN 1-84614-432-9.

- N. David Mermin, 1990, "Spooky actions at a distance: mysteries of the QT" in his Boojums all the way through. Cambridge University Press: 110-76.

- Bryce DeWitt, R. Neill Graham, eds., 1973. The Many-Worlds Interpretation of Quantum Mechanics, Princeton Series in Physics, Princeton University Press. ISBN 0-691-08131-X

- Dirac, P. A. M. (1930). The Principles of Quantum Mechanics. ISBN 0-19-852011-5. The beginning chapters make up a very clear and comprehensible introduction.

- Feynman, Richard P.; Leighton, Robert B.; Sands, Matthew (1965). The Feynman Lectures on Physics. 1–3. Addison-Wesley. ISBN 0-7382-0008-5.

- Griffiths, David J. (2004). Introduction to Quantum Mechanics (2nd ed.). Prentice Hall. ISBN 0-13-111892-7. OCLC 40251748. A standard undergraduate text.

- Hagen Kleinert, 2004. Path Integrals in Quantum Mechanics, Statistics, Polymer Physics, and Financial Markets, 3rd ed. Singapore: World Scientific. Draft of 4th edition.

- George Mackey (2004). The mathematical foundations of quantum mechanics. Dover Publications. ISBN 0-486-43517-2.

- Albert Messiah, 1966. Quantum Mechanics (Vol. I), English translation from French by G. M. Temmer. North Holland, John Wiley & Sons. Cf. chpt. IV, section III.

- Omnès, Roland (1999). Understanding Quantum Mechanics. Princeton University Press. ISBN 0-691-00435-8. OCLC 39849482.

Essential Concepts of Quantum Mechanics

The important concepts of quantum mechanics discussed in the chapter are Stern-Gerlach experiment, bra-ket notation, uncertainty principle, wave function, old quantum theory, double-slit experiment etc. Bra-ket is a standard that is used in describing quantum states. This section elucidates the crucial theories and concepts of quantum mechanics.

Stern–Gerlach Experiment

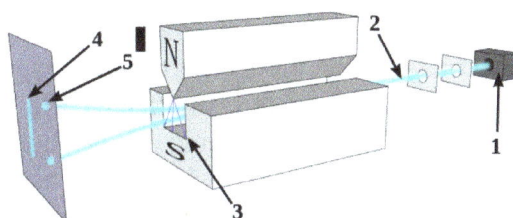

Stern–Gerlach experiment: silver atoms travel through an inhomogeneous magnetic field and are deflected up or down depending on their spin.

The Stern–Gerlach experiment showed that the spatial orientation of angular momentum is quantized. It demonstrated that atomic-scale systems have intrinsically quantum properties, and that measurement in quantum mechanics affects the system being measured. In the original experiment, silver atoms were sent through a non-uniform magnetic field, which deflected them before they struck a detector screen. Other kinds of particles can be used. If the particles have a magnetic moment related to their spin angular momentum, the magnetic field gradient deflects them from a straight path. The screen reveals discrete points of accumulation rather than a continuous distribution, owing to the quantum nature of spin. Historically, this experiment was decisive in convincing physicists of the reality of angular momentum quantization in all atomic-scale systems.

The experiment was first conducted by the German physicists Otto Stern and Walther Gerlach, in 1922.

Basic Theory and Description

The Stern–Gerlach experiment involves sending a beam of particles through an inhomogeneous magnetic field and observing their deflection. The results show that particles possess an intrinsic angular momentum that is closely analogous to the angular momentum of a classically spinning object, but that takes only certain quantized values. Another important result is that only one component of a particle's spin can be measured at one time, meaning that the measurement of the spin along the z-axis destroys information about a particle's spin along the x and y axis.

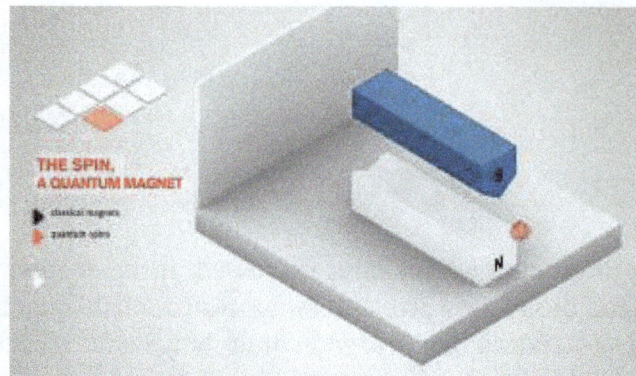

Video explaining quantum spin versus classical magnet in the Stern–Gerlach experiment

The experiment is normally conducted using electrically neutral particles or atoms. This avoids the large deflection to the orbit of a charged particle moving through a magnetic field and allows spin-dependent effects to dominate. For example, observation of the Stern-Gerlach effect with free electrons is infeasible. If the particle is treated as a classical spinning dipole, it will precess in a magnetic field because of the torque that the magnetic field exerts on the dipole. If it moves through a homogeneous magnetic field, the forces exerted on opposite ends of the dipole cancel each other out and the trajectory of the particle is unaffected. However, if the magnetic field is inhomogeneous then the force on one end of the dipole will be slightly greater than the opposing force on the other end, so that there is a net force which deflects the particle's trajectory. If the particles were classical spinning objects, one would expect the distribution of their spin angular momentum vectors to be random and continuous. Each particle would be deflected by a different amount, producing some density distribution on the detector screen. Instead, the particles passing through the Stern–Gerlach apparatus are deflected either up or down by a specific amount. This was a measurement of the quantum observable now known as spin angular momentum, which demonstrated possible outcomes of a measurement where the observable has a discrete set of values or point spectrum. Although some discrete quantum phenomena, such as atomic spectra, were observed much earlier, the Stern–Gerlach experiment allowed scientists to observe separation between discrete quantum states for the first time in the history of science.

By now it is known theoretically that quantum angular momentum *of any kind* has a discrete spectrum, which is sometimes imprecisely expressed as "angular momentum is quantized".

If the experiment is conducted using charged particles like electrons, there will be a Lorentz force that tends to bend the trajectory in a circle. This force can be cancelled by an electric field of appropriate magnitude oriented transverse to the charged particle's path.

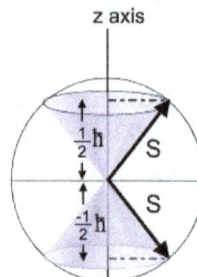

Spin values for fermions.

Electrons are spin-$\frac{1}{2}$ particles. These have only two possible spin angular momentum values measured along any axis, $+\hbar/2$ or $-\hbar/2$, a purely quantum mechanical phenomenon. Because its value is always the same, it is regarded as an intrinsic property of electrons, and is sometimes known as "intrinsic angular momentum" (to distinguish it from orbital angular momentum, which can vary and depends on the presence of other particles).

To describe the experiment with spin $+\frac{1}{2}$ particles mathematically, it is easiest to use Dirac's bra–ket notation. As the particles pass through the Stern–Gerlach device, they are being observed by the detector which resolves to either spin up or spin down. These are described by the angular momentum quantum number j, which can take on one of the two possible allowed values, either $+\hbar/2$ or $-\hbar/2$. The act of observing (measuring) the momentum along the z axis corresponds to the operator J_z. In mathematical terms, the initial state of the particles is

$$|\psi\rangle = c_1 \left|\psi_{j=+\frac{\hbar}{2}}\right\rangle + c_2 \left|\psi_{j=-\frac{\hbar}{2}}\right\rangle,$$

where constants c_1 and c_2 are complex numbers. This initial state spin can in fact point in any direction. The squares of the absolute values ($|c_1|^2$ and $|c_2|^2$) determine the probabilities that for a system in the initial state $|\psi\rangle$ one of the two possible values of j is found after the measurement is made. The constants must also be normalized in order that the probability of finding either one of the values be unity. However, this information is not sufficient to determine the values of c_1 and c_2, because they are complex numbers. Therefore, the measurement yields only the squared magnitudes of the constants, which are interpreted as probabilities.

Sequential Experiments

If we link multiple Stern–Gerlach apparatuses, we can clearly see that they do not act as simple selectors, but alter the states observed (as in light polarization), according to quantum mechanical law:

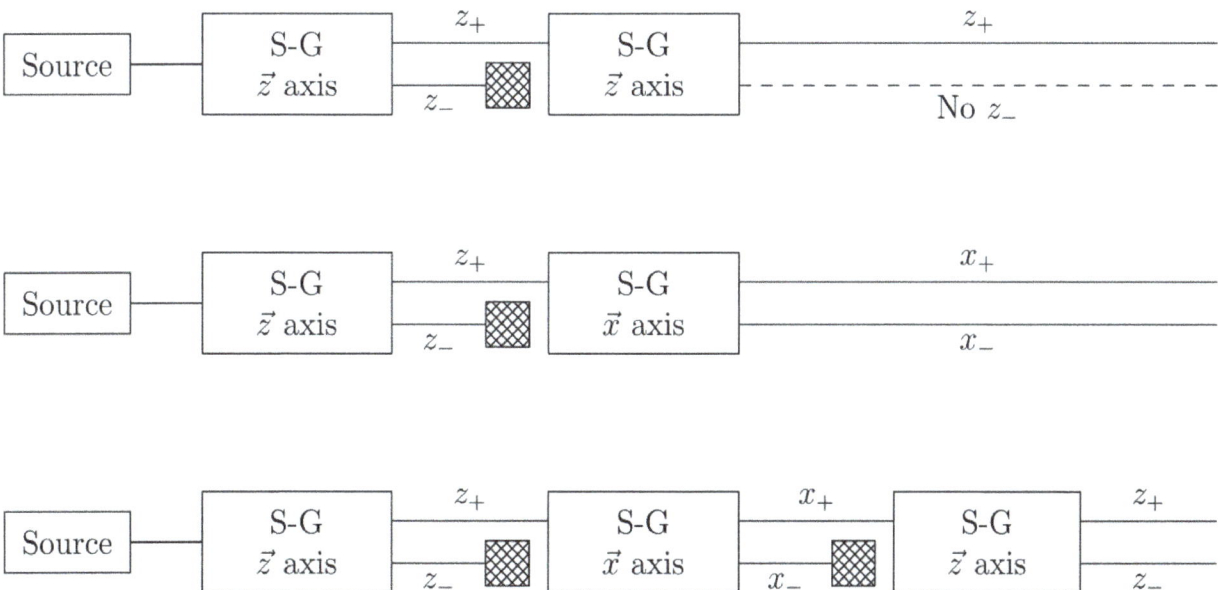

History

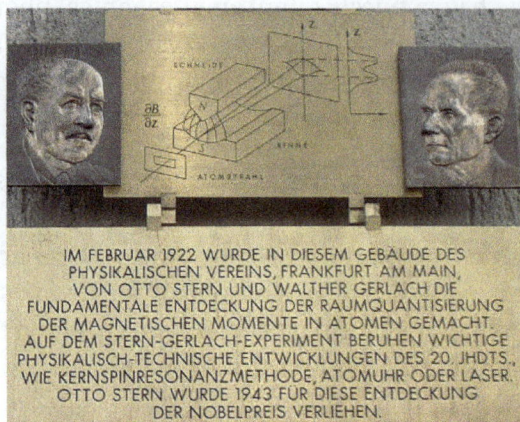

IM FEBRUAR 1922 WURDE IN DIESEM GEBÄUDE DES
PHYSIKALISCHEN VEREINS, FRANKFURT AM MAIN,
VON OTTO STERN UND WALTHER GERLACH DIE
FUNDAMENTALE ENTDECKUNG DER RAUMQUANTISIERUNG
DER MAGNETISCHEN MOMENTE IN ATOMEN GEMACHT.
AUF DEM STERN-GERLACH-EXPERIMENT BERUHEN WICHTIGE
PHYSIKALISCH-TECHNISCHE ENTWICKLUNGEN DES 20. JHDTS.,
WIE KERNSPINRESONANZMETHODE, ATOMUHR ODER LASER.
OTTO STERN WURDE 1943 FÜR DIESE ENTDECKUNG
DER NOBELPREIS VERLIEHEN.

A plaque at the Frankfurt institute commemorating the experiment

The Stern–Gerlach experiment was performed in Frankfurt, Germany in 1922 by Otto Stern and Walther Gerlach. At the time, Stern was an assistant to Max Born at the University of Frankfurt's Institute for Theoretical Physics, and Gerlach was an assistant at the same university's Institute for Experimental Physics.

At the time of the experiment, the most prevalent model for describing the atom was the Bohr model, which described electrons as going around the positively charged nucleus only in certain discrete atomic orbitals or energy levels. Since the electron was quantized to be only in certain positions in space, the separation into distinct orbits was referred to as space quantization. The Stern–Gerlach experiment was meant to test the Bohr–Sommerfeld hypothesis that the direction of the angular momentum of a silver atom is quantized.

Note that the experiment was performed several years before Uhlenbeck and Goudsmit formulated their hypothesis of the existence of the electron spin. Even though the result of the Stern–Gerlach experiment has later turned out to be in agreement with the predictions of quantum mechanics for a spin-$\frac{1}{2}$ particle, the experiment should be seen as a corroboration of the Bohr–Sommerfeld theory.

In 1927, T.E. Phipps and J.B. Taylor reproduced the effect using hydrogen atoms in their ground state, thereby eliminating any doubts that may have been caused by the use of silver atoms. (In 1926 the non-relativistic Schrödinger equation had incorrectly predicted the magnetic moment of hydrogen to be zero in its ground state. To correct this problem Wolfgang Pauli introduced "by hand", so to speak, the 3 Pauli matrices which now bear his name, but which were later shown by Paul Dirac in 1928 to be intrinsic in his relativistic equation.)

The experiment was first performed with an electromagnet that allowed the non-uniform magnetic field to be turned on gradually from a null value. When the field was null, the silver atoms were deposited as a single band on the detecting glass slide. When the field was made stronger, the middle of the band began to widen and eventually to split into two, so that the glass-slide image looked like a lip-print, with an opening in the middle, and closure at either end. In the middle, where the magnetic field was strong enough to split the beam into two, statistically half of the silver atoms had been deflected by the non-uniformity of the field.

Importance

The Stern–Gerlach experiment strongly influenced later developments in modern physics:

- In the decade that followed, scientists showed using similar techniques, that the nuclei of some atoms also have quantized angular momentum. It is the interaction of this nuclear angular momentum with the spin of the electron that is responsible for the hyperfine structure of the spectroscopic lines.

- In the 1930s, using an extended version of the Stern–Gerlach apparatus, Isidor Rabi and colleagues showed that by using a varying magnetic field, one can force the magnetic moment to go from one state to the other. The series of experiments culminated in 1937 when they discovered that state transitions could be induced using time varying fields or RF fields. The so-called Rabi oscillation is the working mechanism for the Magnetic Resonance Imaging equipment found in hospitals.

- Norman F. Ramsey later modified the Rabi apparatus to increase the interaction time with the field. The extreme sensitivity due to the frequency of the radiation makes this very useful for keeping accurate time, and it is still used today in atomic clocks.

- In the early sixties, Ramsey and Daniel Kleppner used a Stern–Gerlach system to produce a beam of polarized hydrogen as the source of energy for the hydrogen Maser, which is still one of the most popular atomic clocks.

- The direct observation of the spin is the most direct evidence of quantization in quantum mechanics.

- The Stern–Gerlach experiment has become a paradigm of quantum measurement.

Bra–ket Notation

In quantum mechanics, bra–ket notation is a standard notation for describing quantum states. The notation uses angle brackets ("$\langle$", and "$\rangle$") and vertical bars ("$|$"). It can also be used to denote abstract vectors and linear functionals in mathematics. In such terms, the scalar product, or action of a linear functional on a vector in a complex vector space, is denoted by

$$\langle \phi | \psi \rangle,$$

consisting of a left part,

$$\langle \phi |$$

called the bra , and a right part,

$$| \psi \rangle,$$

called the ket . The notation was introduced in 1939 by Paul Dirac and is also known as the Dirac notation, though the notation has precursors in Hermann Grassmann's use of the notation

$$[\phi\,|\,\psi]$$

for his inner products nearly 100 years earlier. The relevant quantity is actually

$$|\langle\phi\,|\,\psi\rangle|^2 = |\langle\psi\,|\,\phi\rangle|^2$$

and is interpreted according to the fundamental Born rule.

Bra–ket notation is widespread in quantum mechanics. Many phenomena that are explained using quantum mechanics are usually most clearly demonstrated with the help of the bra-ket notation. Part of the appeal of the notation is the abstract representation-independence it encodes, together with its versatility in producing a specific representation (e.g. x, or p, or eigenfunction base) without much ado, or excessive reliance on the nature of the linear spaces involved.

The standard mathematical notation for the inner product, preferred as well by some physicists, expresses exactly the same thing as the bra-ket notation,

$$(\phi,\psi) = \langle\phi\,|\,\psi\rangle = \langle\phi\,|\,(|\,\psi\rangle),$$

where the far right gives the value of the linear functional (the bra) on its argument (the ket), and one can choose freely to interpret the left argument of $\langle\cdot\,|\,\cdot\rangle$ as either a bra or a ket according to whether the far right or far left expression is the one thought of. It is only when a bra appears unpaired in an expression, such as,

$$|\,\psi\rangle\langle\phi\,|$$

that one should be aware that the bra technically is an element of the dual space of Hilbert space. It is in handling expressions containing these that the Dirac notation comes into its own. The notation does not introduce or imply any new physics.

Vector Spaces

Background: Vector Spaces

In physics, basis vectors allow any Euclidean vector to be represented geometrically using angles and lengths, in different directions, i.e. in terms of the spatial orientations. It is simpler to see the notational equivalences between ordinary notation and bra-ket notation; so, for now, consider a vector A starting at the origin and ending at an element of 3-d Euclidean space; the vector then is specified by this end-point, a triplet of elements in the field of real numbers, symbolically dubbed as $A \in R^3$.

The vector A can be written using any set of basis vectors and corresponding coordinate system. Informally, basis vectors are like "building blocks of a vector": they are added together to compose a vector, and the coordinates are the numerical coefficients of basis vectors in each direction. Two useful representations of a vector are simply a linear combination of basis vectors, and column matrices. Using the familiar Cartesian basis, a vector A may be written as

$$\mathbf{A} \doteq A_x \mathbf{e}_x + A_y \mathbf{e}_y + A_z \mathbf{e}_z = A_x \begin{pmatrix} 1 \\ 0 \\ 0 \end{pmatrix} + A_y \begin{pmatrix} 0 \\ 1 \\ 0 \end{pmatrix} + A_z \begin{pmatrix} 0 \\ 0 \\ 1 \end{pmatrix}$$

$$= \begin{pmatrix} A_x \\ 0 \\ 0 \end{pmatrix} + \begin{pmatrix} 0 \\ A_y \\ 0 \end{pmatrix} + \begin{pmatrix} 0 \\ 0 \\ A_z \end{pmatrix} = \begin{pmatrix} A_x \\ A_y \\ A_z \end{pmatrix}$$

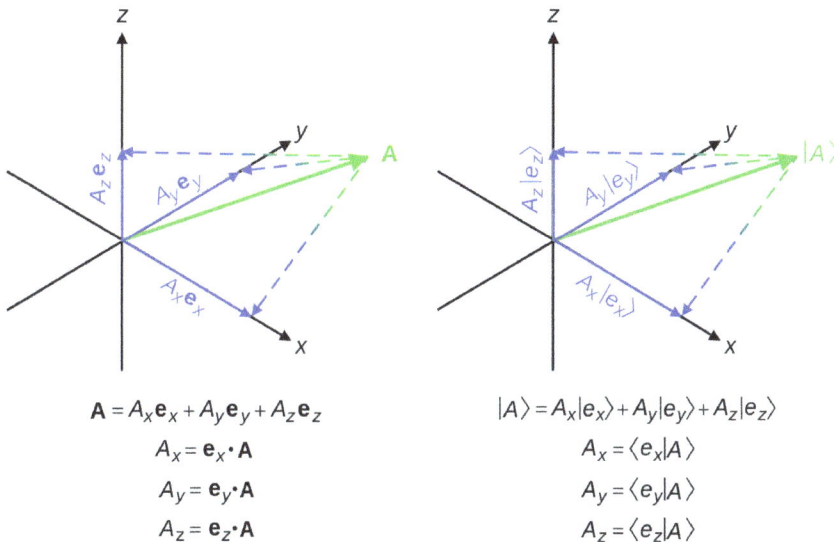

$$\mathbf{A} = A_x \mathbf{e}_x + A_y \mathbf{e}_y + A_z \mathbf{e}_z$$
$$A_x = \mathbf{e}_x \cdot \mathbf{A}$$
$$A_y = \mathbf{e}_y \cdot \mathbf{A}$$
$$A_z = \mathbf{e}_z \cdot \mathbf{A}$$

$$|A\rangle = A_x |e_x\rangle + A_y |e_y\rangle + A_z |e_z\rangle$$
$$A_x = \langle e_x | A \rangle$$
$$A_y = \langle e_y | A \rangle$$
$$A_z = \langle e_z | A \rangle$$

3d real vector components and bases projection; similarities between vector calculus notation and Dirac notation. Projection is an important feature of the Dirac notation.

respectively, where e_x, e_y, e_z denote the Cartesian basis vectors (all are orthogonal unit vectors) and A_x, A_y, A_z are the corresponding coordinates, in the x, y, z directions. In a more general notation, for any basis in 3-d space one writes

$$\mathbf{A} \doteq A_1 \mathbf{e}_1 + A_2 \mathbf{e}_2 + A_3 \mathbf{e}_3 = \begin{pmatrix} A_1 \\ A_2 \\ A_3 \end{pmatrix}$$

Generalizing further, consider a vector A in an N-dimensional vector space over the field of complex numbers C, symbolically stated as A $\in$ C^N. The vector A is still conventionally represented by a linear combination of basis vectors or a column matrix:

$$\mathbf{A} \doteq \sum_{n=1}^{N} A_n \mathbf{e}_n = \begin{pmatrix} A_1 \\ A_2 \\ \vdots \\ A_N \end{pmatrix}$$

though the coordinates are now all complex-valued.

Even more generally, A can be a vector in a complex Hilbert space. Some Hilbert spaces, like C^N, have finite dimension, while others have infinite dimension. In an infinite-dimensional space, the column-vector representation of A would be a list of infinitely many complex numbers.

Ket Notation for Vectors

Rather than such conventions as bold type, over arrows, and underscores (i.e., $\mathbf{A}, \vec{A}, \underline{A}$), which are used to describe vectors elsewhere in applicable fields, Dirac's notation for a vector uses vertical bars and angular brackets: $|A\rangle$. When this notation is used, these vectors are called "kets," and $|A\rangle$ is read as "ket-A". This applies to all vectors, the resultant vector and the basis. The previous vectors are now written

$$|A\rangle = A_x |e_x\rangle + A_y |e_y\rangle + A_z |e_z\rangle \doteq \begin{pmatrix} A_x \\ A_y \\ A_z \end{pmatrix},$$

or in a more easily generalized notation,

$$|A\rangle = A_1 |e_1\rangle + A_2 |e_2\rangle + A_3 |e_3\rangle \doteq \begin{pmatrix} A_1 \\ A_2 \\ A_3 \end{pmatrix}$$

The last one may be written in short as

$$|A\rangle = A_1 |1\rangle + A_2 |2\rangle + A_3 |3\rangle .$$

Note how any symbols, letters, numbers, or even words—whatever serves as a convenient label— can be used as the label inside a ket. In other words, the symbol "$|A\rangle$" has a specific and universal mathematical meaning, while just the "A" by itself does not. Nevertheless, for convenience, there is usually some logical scheme behind the labels inside kets, such as the common practice of labeling energy eigenkets in quantum mechanics through a listing of their quantum numbers. Further note that a ket and its representation by a coordinate vector are not the same mathematical object: a ket does not require specification of a basis, whereas the coordinate vector needs a basis in order to be well defined (the same holds for an operator and its representation by a matrix). In this context, one should best use a symbol different than the equal sign, for example the symbol, read as "is represented by".

Inner Products and Bras

An inner product is a generalization of the dot product. The inner product of two vectors is a scalar. bra-ket notation uses a specific notation for inner products:

$\langle A | B \rangle$ = the inner product of ket $|A\rangle$ with ket $|B\rangle$

For example, in three-dimensional complex Euclidean space,

$$\langle A | B \rangle \doteq A_x^* B_x + A_y^* B_y + A_z^* B_z$$

where A_i^* denotes the complex conjugate of A_i. A special case is the inner product of a vector with itself, which is the square of its norm (magnitude):

$$\langle A | A \rangle \doteq |A_x|^2 + |A_y|^2 + |A_z|^2$$

Bra-ket notation splits this inner product (also called a "bracket") into two pieces, the "bra" and the "ket":

$$\langle A | B \rangle = \left(\langle A | \right)\left(| B \rangle \right)$$

where $\langle A |$ is called a bra, read as "bra-A", and $| B \rangle$ is a ket as above.

The purpose of "splitting" the inner product into a bra and a ket is that *both* the bra $\langle \quad$ and the ket $| B \rangle$ are meaningful *on their own*, and can be used in other contexts besides within an inner product. There are two main ways to think about the meanings of separate bras and kets:

Bras and Kets as Row and Column Vectors

For a finite-dimensional vector space, using a fixed orthonormal basis, the inner product can be written as a matrix multiplication of a row vector with a column vector:

$$\langle A | B \rangle \doteq A_1^* B_1 + A_2^* B_2 + \cdots + A_N^* B_N = \begin{pmatrix} A_1^* & A_2^* & \cdots & A_N^* \end{pmatrix} \begin{pmatrix} B_1 \\ B_2 \\ \vdots \\ B_N \end{pmatrix}$$

Based on this, the bras and kets can be defined as:

$$\langle A | \doteq \begin{pmatrix} A_1^* & A_2^* & \cdots & A_N^* \end{pmatrix}$$

$$| B \rangle \doteq \begin{pmatrix} B_1 \\ B_2 \\ \vdots \\ B_N \end{pmatrix}$$

and then it is understood that a bra next to a ket implies matrix multiplication.

The conjugate transpose (also called *Hermitian conjugate*) of a bra is the corresponding ket and vice versa:

$$\langle A|^\dagger = | A\rangle, \quad | A\rangle^\dagger = \langle A|$$

because if one starts with the bra

$$\begin{pmatrix} A_1^* & A_2^* & \cdots & A_N^* \end{pmatrix},$$

then performs a complex conjugation, and then a matrix transpose, one ends up with the ket

$$\begin{pmatrix} A_1 \\ A_2 \\ \vdots \\ A_N \end{pmatrix}$$

Bras as Linear Transformation on Kets

A more abstract definition, which is equivalent but more easily generalized to infinite-dimensional spaces, is to say that bras are linear functionals on kets, i.e. linear transformations that input a ket and output a complex number. The bra linear functionals are defined to be consistent with the inner product.

In mathematics terminology, the vector space of bras is the dual space to the vector space of kets, and corresponding bras and kets are related by the Riesz representation theorem.

Non-normalizable States and Non-Hilbert Spaces

bra-ket notation can be used even if the vector space is not a Hilbert space.

In quantum mechanics, it is common practice to write down kets which have infinite norm, i.e. non-normalisable wavefunctions. Examples include states whose wavefunctions are Dirac delta functions or infinite plane waves. These do not, technically, belong to the Hilbert space itself. However, the definition of "Hilbert space" can be broadened to accommodate these states. The bra-ket notation continues to work in an analogous way in this broader context.

For a rigorous treatment of the Dirac inner product of non-normalizable states, see the definition given by D. Carfi. For a rigorous definition of basis with a continuous set of indices and consequently for a rigorous definition of position and momentum basis. For a rigorous statement of the expansion of an S-diagonalizable operator, or observable, in its eigenbasis or in another basis.

Banach spaces are a different generalization of Hilbert spaces. In a Banach space B, the vectors may be notated by kets and the continuous linear functionals by bras. Over any vector space without topology, we may also notate the vectors by kets and the linear functionals by bras. In these more general contexts, the bracket does not have the meaning of an inner product, because the Riesz representation theorem does not apply.

Usage in Quantum Mechanics

The mathematical structure of quantum mechanics is based in large part on linear algebra:

- Wave functions and other quantum states can be represented as vectors in a complex Hilbert space. (The exact structure of this Hilbert space depends on the situation.) In bra-ket notation, for example, an electron might be in the "state" $|\psi\rangle$. (Technically, the quantum states are *rays* of vectors in the Hilbert space, as $c|\psi\rangle$ corresponds to the same state for any nonzero complex number c.)

- Quantum superpositions can be described as vector sums of the constituent states. For example, an electron in the state $|1\rangle + i\,|2\rangle$ is in a quantum superposition of the states $|1\rangle$ and $|2\rangle$.

- Measurements are associated with linear operators (called observables) on the Hilbert space of quantum states.

- Dynamics are also described by linear operators on the Hilbert space. For example, in the Schrödinger picture, there is a linear time evolution operator U with the property that if an electron is in state $|\psi\rangle$ right now, at a later time it will be in the state $U|\psi\rangle$, the same U for every possible $|\psi\rangle$.

- Wave function normalization is scaling a wave function so that its norm is 1.

Since virtually every calculation in quantum mechanics involves vectors and linear operators, it can involve, and often *does* involve, bra-ket notation. A few examples follow:

Spinless Position–space Wave Function

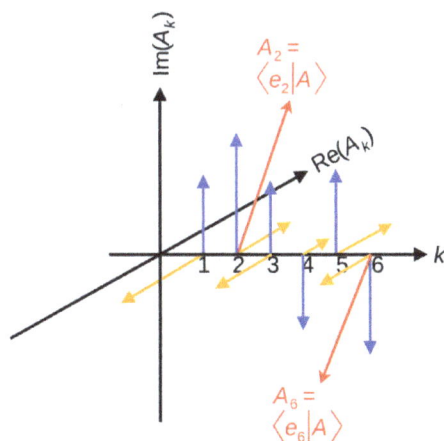

Discrete components A_k of a complex vector $|A\rangle = \sum_k A_k|e_k\rangle$, which belongs to a *countably infinite*-dimensional Hilbert space; there are countably infinitely many k values and basis vectors $|e_k\rangle$.

Components of complex vectors plotted against index number; discrete k and continuous x. Two particular components out of infinitely many are highlighted.

The Hilbert space of a spin-0 point particle is spanned by a "position basis" $\{\,|r\rangle\,\}$, where the label r extends over the set of all points in position space. Since there are an uncountably infinite number of vector components in the basis, this is an uncountably infinite-dimensional Hilbert space. The dimensions of the Hilbert space (usually infinite) and position space (usually 1, 2 or 3) are not to be conflated.

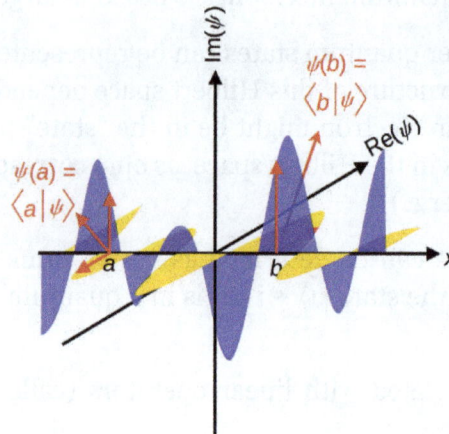

Continuous components $\psi(x)$ of a complex vector $|\psi\rangle = \int dx\, \psi(x)|x\rangle$, which belongs to an *uncountably infinite-dimensional* Hilbert space; there are infinitely many x values and basis vectors $|x\rangle$.

Starting from any ket $|\Psi\rangle$ in this Hilbert space, we can *define* a complex scalar function of r, known as a wavefunction:

$$\Psi(\mathbf{r}) \overset{\text{def}}{=} \langle \mathbf{r} | \Psi \rangle.$$

On the left side, $\Psi(\mathbf{r})$ is a function mapping any point in space to a complex number; on the right side, $|\Psi\rangle = \int d^3\mathbf{r}\, \Psi(\mathbf{r}) |\mathbf{r}\rangle$ is a ket.

It is then customary to define linear operators acting on wavefunctions in terms of linear operators acting on kets, by

$$A\Psi(\mathbf{r}) \overset{\text{def}}{=} \langle \mathbf{r} | A | \Psi \rangle.$$

For instance, the momentum operator p has the following form,

$$\mathbf{p}\Psi(\mathbf{r}) \overset{\text{def}}{=} \langle \mathbf{r} | \mathbf{p} | \Psi \rangle = -i\hbar \nabla \Psi(\mathbf{r}).$$

One occasionally encounters an expression as

$$\nabla | \Psi \rangle,$$

though this is something of an abuse of notation. The differential operator must be understood to be an abstract operator, acting on kets, that has the effect of differentiating wavefunctions once the expression is projected into the position basis,

$$\nabla \langle \mathbf{r} | \Psi \rangle,$$

even though, in the momentum basis, the operator amounts to a mere multiplication operator (by $i\hbar p$).

Overlap of States

In quantum mechanics the expression $\langle \phi | \psi \rangle$ is typically interpreted as the probability amplitude for the state ψ to collapse into the state ϕ. Mathematically, this means the coefficient for the projection of ψ onto ϕ. It is also described as the projection of state ψ onto state ϕ.

Changing Basis for a Spin-1/2 Particle

A stationary spin-½ particle has a two-dimensional Hilbert space. One orthonormal basis is:

$$|\uparrow_z\rangle, |\downarrow_z\rangle$$

where $|\uparrow_z\rangle$ is the state with a definite value of the spin operator S_z equal to $+1/2$ and $|\downarrow_z\rangle$ is the state with a definite value of the spin operator S_z equal to $-1/2$.

Since these are a basis, *any* quantum state of the particle can be expressed as a linear combination (i.e., quantum superposition) of these two states:

$$|\psi\rangle = a_\psi |\uparrow_z\rangle + b_\psi |\downarrow_z\rangle$$

where a_ψ, b_ψ are complex numbers.

A *different* basis for the same Hilbert space is:

$$|\uparrow_x\rangle, |\downarrow_x\rangle$$

defined in terms of S_x rather than S_z.

Again, *any* state of the particle can be expressed as a linear combination of these two:

$$|\psi\rangle = c_\psi |\uparrow_x\rangle + d_\psi |\downarrow_x\rangle$$

In vector form, you might write

$$|\psi\rangle \doteq \begin{pmatrix} a_\psi \\ b_\psi \end{pmatrix}, \text{ OR } |\psi\rangle \doteq \begin{pmatrix} c_\psi \\ d_\psi \end{pmatrix}$$

depending on which basis you are using. In other words, the "coordinates" of a vector depend on the basis used.

There is a mathematical relationship between a_ψ, b_ψ, c_ψ, d_ψ.

Misleading Uses

There are a few conventions and abuses of notation that are generally accepted by the physics community, but which might confuse the non-initiated.

It is common to use the same symbol for *labels* and *constants* in the same equation. For example,

$\hat{\alpha} |\alpha\rangle = \alpha|\alpha\rangle$, where the symbol α is used simultaneously as the *name of the operator* $\hat{\alpha}$, its *eigen-vector* $|\alpha\rangle$ and the associated *eigenvalue* α.

Something similar occurs in component notation of vectors. While Ψ (uppercase) is traditionally associated with wavefunctions, ψ (lowercase) may be used to denote a *label*, a *wave function* or *complex constant* in the same context, usually differentiated only by a subscript.

The main abuses are including operations inside the vector labels. This is usually done for a fast notation of scaling vectors. E.g. if the vector $|\alpha\rangle$ is scaled by $\sqrt{2}$, it might be denoted by $|\alpha/\sqrt{2}\rangle$, which makes no sense since α is a label, not a function or a number, so you can't perform operations on it.

This is especially common when denoting vectors as tensor products, where part of the labels are moved outside the designed slot, e.g. $|\alpha\rangle = |\alpha/\sqrt{2}_1\rangle \otimes |\alpha/\sqrt{2}_2\rangle$. Here part of the labeling that should state that all three vectors are different was moved outside the kets, as subscripts 1 and 2. And a further abuse occurs, since α is meant to refer to the norm of the first vector—which is a *label* denoting a *value*.

Linear Operators

Linear Operators Acting on Kets

A linear operator is a map that inputs a ket and outputs a ket. (In order to be called "linear", it is required to have certain properties.) In other words, if A is a linear operator and $|\psi\rangle$ is a ket, then $A|\psi\rangle$ is another ket.

In an N-dimensional Hilbert space, $|\psi\rangle$ can be written as an $N\times 1$ column vector, and then A is an $N\times N$ matrix with complex entries. The ket $A|\psi\rangle$ can be computed by normal matrix multiplication.

Linear operators are ubiquitous in the theory of quantum mechanics. For example, observable physical quantities are represented by self-adjoint operators, such as energy or momentum, whereas transformative processes are represented by unitary linear operators such as rotation or the progression of time.

Linear Operators Acting on Bras

Operators can also be viewed as acting on bras *from the right hand side*. Specifically, if A is a linear operator and $\langle\varphi|$ is a bra, then $\langle\varphi|A$ is another bra defined by the rule

$$\left(\langle\phi| A\right)|\psi\rangle = \langle\phi|\left(A|\psi\rangle\right),$$

(in other words, a function composition). This expression is commonly written as (cf. energy inner product)

$$\langle\phi| A |\psi\rangle.$$

In an N-dimensional Hilbert space, $\langle\varphi|$ can be written as a $1\times N$ row vector, and A (as in the previous section) is an $N\times N$ matrix. Then the bra $\langle\varphi|A$ can be computed by normal matrix multiplication.

If the same state vector appears on both bra and ket side,

$$\langle \psi \,|\, A \,|\, \psi \rangle,$$

then this expression gives the expectation value, or mean or average value, of the observable represented by operator A for the physical system in the state $|\psi\rangle$.

Outer Products

A convenient way to define linear operators on Hilbert space H is given by the outer product: if $\langle\varphi|$ is a bra and $|\psi\rangle$ is a ket, the outer product

$$|\phi\rangle\langle\psi|$$

denotes the rank-one operator with the rule

$$(|\phi\rangle\langle\psi|)(x) = \langle\psi\,|\,x\rangle\,|\,\phi\rangle.$$

For a finite-dimensional vector space, the outer product can be understood as simple matrix multiplication:

$$|\phi\rangle\langle\psi| \doteq \begin{pmatrix} \phi_1 \\ \phi_2 \\ \vdots \\ \phi_N \end{pmatrix} \begin{pmatrix} \psi_1^* & \psi_2^* & \cdots & \psi_N^* \end{pmatrix} = \begin{pmatrix} \phi_1\psi_1^* & \phi_1\psi_2^* & \cdots & \phi_1\psi_N^* \\ \phi_2\psi_1^* & \phi_2\psi_2^* & \cdots & \phi_2\psi_N^* \\ \vdots & \vdots & \ddots & \vdots \\ \phi_N\psi_1^* & \phi_N\psi_2^* & \cdots & \phi_N\psi_N^* \end{pmatrix}$$

The outer product is an N×N matrix, as expected for a linear operator.

One of the uses of the outer product is to construct projection operators. Given a ket $|\psi\rangle$ of norm 1, the orthogonal projection onto the subspace spanned by $|\psi\rangle$ is

$$|\psi\rangle\langle\psi|.$$

Hermitian Conjugate Operator

Just as kets and bras can be transformed into each other (making $|\psi\rangle$ into $\langle\psi|$), the element from the dual space corresponding to $A|\psi\rangle$ is $\langle\psi|A^\dagger$, where $A^\dagger$ denotes the Hermitian conjugate (or adjoint) of the operator A. In other words,

$$|\phi\rangle = A\,|\psi\rangle \quad \text{if and only if} \quad \langle\phi| = \langle\psi\,|\,A^\dagger.$$

If A is expressed as an N×N matrix, then $A^\dagger$ is its conjugate transpose.

Self-adjoint operators, where $A = A^\dagger$, play an important role in quantum mechanics; for example, an observable is always described by a self-adjoint operator. If A is a self-adjoint operator, then $\langle\psi|A|\psi\rangle$ is always a real number (not complex). This implies that expectation values of observables are real.

Properties

bra-ket notation was designed to facilitate the formal manipulation of linear-algebraic expressions. Some of the properties that allow this manipulation are listed herein. In what follows, c_1 and c_2 denote arbitrary complex numbers, c^* denotes the complex conjugate of c, A and B denote arbitrary linear operators, and these properties are to hold for any choice of bras and kets.

Linearity

- Since bras are linear functionals,

$$\langle \phi | \big(c_1 | \psi_1 \rangle + c_2 | \psi_2 \rangle \big) = c_1 \langle \phi | \psi_1 \rangle + c_2 \langle \phi | \psi_2 \rangle.$$

- By the definition of addition and scalar multiplication of linear functionals in the dual space,

$$\big(c_1 \langle \phi_1 | + c_2 \langle \phi_2 | \big) | \psi \rangle = c_1 \langle \phi_1 | \psi \rangle + c_2 \langle \phi_2 | \psi \rangle.$$

Associativity

Given any expression involving complex numbers, bras, kets, inner products, outer products, and/or linear operators (but not addition), written in bra-ket notation, the parenthetical groupings do not matter (i.e., the associative property holds). For example:

$$\langle \psi | (A | \phi \rangle) = (\langle \psi | A) | \phi \rangle \overset{\text{def}}{=} \langle \psi | A | \phi \rangle$$

$$(A | \psi \rangle) \langle \phi | = A(| \psi \rangle \langle \phi |) \overset{\text{def}}{=} A | \psi \rangle \langle \phi |$$

and so forth. The expressions on the right (with no parentheses whatsoever) are allowed to be written unambiguously *because* of the equalities on the left. Note that the associative property does *not* hold for expressions that include non-linear operators, such as the antilinear time reversal operator in physics.

Hermitian Conjugation

bra-ket notation makes it particularly easy to compute the Hermitian conjugate (also called *dagger*, and denoted †) of expressions. The formal rules are:

- The Hermitian conjugate of a bra is the corresponding ket, and vice versa.

- The Hermitian conjugate of a complex number is its complex conjugate.

- The Hermitian conjugate of the Hermitian conjugate of anything (linear operators, bras, kets, numbers) is itself—i.e.,

$$(x^\dagger)^\dagger = x.$$

- Given any combination of complex numbers, bras, kets, inner products, outer products, and/or linear operators, written in bra-ket notation, its Hermitian conjugate can be computed by reversing the order of the components, and taking the Hermitian conjugate of each.

These rules are sufficient to formally write the Hermitian conjugate of any such expression; some examples are as follows:

- Kets:

$$\left(c_1 |\psi_1\rangle + c_2 |\psi_2\rangle\right)^\dagger = c_1^* \langle\psi_1| + c_2^* \langle\psi_2|.$$

- Inner products:

$$\langle\phi|\psi\rangle^* = \langle\psi|\phi\rangle.$$

(Note that $\langle\phi|\psi\rangle$ is a scalar, so the Hermitian conjugate is just the complex conjugate i.e. $\langle\phi|\psi\rangle^* = \overline{\langle\phi|\psi\rangle}$)

- Matrix elements:

$$\langle\phi|A|\psi\rangle^* = \langle\psi|A^\dagger|\phi\rangle$$

$$\langle\phi|A^\dagger B^\dagger|\psi\rangle^* = \langle\psi|BA|\phi\rangle.$$

- Outer products:

$$\left((c_1 |\phi_1\rangle\langle\psi_1|) + (c_2 |\phi_2\rangle\langle\psi_2|)\right)^\dagger = (c_1^* |\psi_1\rangle\langle\phi_1|) + (c_2^* |\psi_2\rangle\langle\phi_2|).$$

Composite Bras and Kets

Two Hilbert spaces V and W may form a third space $V \otimes W$ by a tensor product. In quantum mechanics, this is used for describing composite systems. If a system is composed of two subsystems described in V and W respectively, then the Hilbert space of the entire system is the tensor product of the two spaces. (The exception to this is if the subsystems are actually identical particles. In that case, the situation is a little more complicated.)

If $|\psi\rangle$ is a ket in V and $|\varphi\rangle$ is a ket in W, the direct product of the two kets is a ket in $V \otimes W$. This is written in various notations:

$$|\psi\rangle|\phi\rangle, \quad |\psi\rangle\otimes|\phi\rangle, \quad |\psi\phi\rangle, \quad |\psi,\phi\rangle.$$

The Unit Operator

Consider a complete orthonormal system (*basis*), $\{e_i | i \in \mathbb{N}\}$, for a Hilbert space H, with respect to the norm from an inner product $\langle\cdot,\cdot\rangle$. From basic functional analysis we know that any ket $|\psi\rangle$ can also be written as

$$|\psi\rangle = \sum_{i\in\mathbb{N}} \langle e_i | \psi\rangle | e_i\rangle,$$

with $\langle\cdot|\cdot\rangle$ the inner product on the Hilbert space.

From the commutativity of kets with (complex) scalars now follows that

$$\sum_{i\in\mathbb{N}} |e_i\rangle\langle e_i| = \hat{1}$$

must be the identity operator, which sends each vector to itself. This can be inserted in any expression without affecting its value, for example

$$\langle v | w\rangle = \langle v | \sum_{i\in\mathbb{N}} |e_i\rangle\langle e_i | w\rangle = \langle v | \sum_{i\in\mathbb{N}} |e_i\rangle\langle e_i | \sum_{j\in\mathbb{N}} |e_j\rangle\langle e_j | w\rangle = \langle v | e_i\rangle\langle e_i | e_j\rangle\langle e_j | w\rangle,$$

where, in the last identity, the Einstein summation convention has been used.

In quantum mechanics, it often occurs that little or no information about the inner product $\langle\psi|\phi\rangle$ of two arbitrary (state) kets is present, while it is still possible to say something about the expansion coefficients $\langle\psi|e_i\rangle = \langle e_i|\psi\rangle^*$ and $\langle e_i|\phi\rangle$ of those vectors with respect to a specific (orthonormalized) basis. In this case, it is particularly useful to insert the unit operator into the bracket one time or more.

Resolution of the identity, $1 = \int dx\, |x\rangle\langle x| = \int dp\, |p\rangle\langle p|$, where $|p\rangle = \int dx\, e^{ixp/\hbar}|x\rangle/\sqrt{2\pi\hbar}$; since $\langle x'|x\rangle = \delta(x - x')$, plane waves follow, $\langle x|p\rangle = \exp(ixp/\hbar)/\sqrt{2\pi\hbar}$.

Notation used by Mathematicians

The object physicists are considering when using the "bra-ket" notation is a Hilbert space (a complete inner product space).

Let $\mathcal{H}$ be a Hilbert space and $h \in \mathcal{H}$ is a vector in $\mathcal{H}$. What physicists would denote as $|h\rangle$ is the vector itself. That is

$$|h\rangle \in \mathcal{H}.$$

Let $\mathcal{H}^*$ be the dual space of $\mathcal{H}$. This is the space of linear functionals on $\mathcal{H}$. The isomorphism $\Phi : \mathcal{H} \to \mathcal{H}^*$ is defined by $\Phi(h) = \phi_h$ where for all $g \in \mathcal{H}$ we have

$$\phi_h(g) = \mathrm{IP}(h, g) = (h, g) = \langle h, g\rangle = \langle h | g\rangle,$$

where $\mathrm{IP}(\cdot,\cdot), (\cdot,\cdot), \langle\cdot,\cdot\rangle$ and $\langle\cdot|\cdot\rangle$ are just different notations for expressing an inner product between two elements in a Hilbert space (or for the first three, in *any* inner product space). Notational confusion arises when identifying ϕ_h and g with $\langle h|$ and $|g\rangle$ respectively. This is because of literal symbolic substitutions. Let $\phi_h = H = \langle h|$ and let $g = G = |g\rangle$. This gives

$$\phi_h(g) = H(g) = H(G) = \langle h|(G) = \langle h|(|g\rangle).$$

One ignores the parentheses and removes the double bars. Some properties of this notation are convenient since we are dealing with linear operators and composition acts like a ring multiplication.

Moreover, mathematicians usually write the dual entity not at the first place, as the physicists do, but at the second one, and they don't use the *-symbol, but an overline (which the physicists reserve for averages and Dirac conjugation) to denote conjugate-complex numbers, i.e. for scalar products mathematicians usually write

$$(\phi, \psi) = \int \phi(x) \cdot \overline{\psi(x)} \mathrm{d}x,$$

whereas physicists would write for the same quantity

$$\langle \psi \mid \phi \rangle = \int \mathrm{d}x \psi^*(x) \cdot \phi(x).$$

Uncertainty Principle

In quantum mechanics, the uncertainty principle, also known as Heisenberg's uncertainty principle, is any of a variety of mathematical inequalities asserting a fundamental limit to the precision with which certain pairs of physical properties of a particle, known as complementary variables, such as position x and momentum p, can be known.

Introduced first in 1927, by the German physicist Werner Heisenberg, it states that the more precisely the position of some particle is determined, the less precisely its momentum can be known, and vice versa. The formal inequality relating the standard deviation of position σ_x and the standard deviation of momentum σ_p was derived by Earle Hesse Kennard later that year and by Hermann Weyl in 1928:

$$\sigma_x \sigma_p \geq \frac{\hbar}{2}$$

($\hbar$ is the reduced Planck constant, $h / 2\pi$).

Historically, the uncertainty principle has been confused with a somewhat similar effect in physics, called the observer effect, which notes that measurements of certain systems cannot be made without affecting the systems, that is, without changing something in a system. Heisenberg offered such an observer effect at the quantum level as a physical "explanation" of quantum uncertainty. It has since become clear, however, that the uncertainty principle is inherent in the properties of all wave-like systems, and that it arises in quantum mechanics simply due to the matter wave nature of all quantum objects. Thus, *the uncertainty principle actually states a fundamental property of quantum systems, and is not a statement about the observational success of current technology.* It must be emphasized that *measurement* does not mean only a process in which a physicist-observer takes part, but rather any interaction between classical and quantum objects regardless of any observer.

Since the uncertainty principle is such a basic result in quantum mechanics, typical experiments in quantum mechanics routinely observe aspects of it. Certain experiments, however, may deliberately test a particular form of the uncertainty principle as part of their main research program. These include, for example, tests of number–phase uncertainty relations in superconducting or quantum optics systems. Applications dependent on the uncertainty principle for their operation include extremely low-noise technology such as that required in gravitational wave interferometers.

Introduction

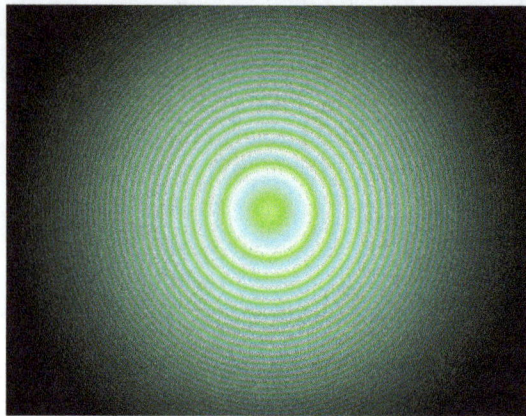

The evolution of an initially very localized gaussian wave function of a free particle in two-dimensional space, with colour and intensity indicating phase and amplitude. The spreading of the wave function in all directions shows that the initial momentum has a spread of values, unmodified in time; while the spread in position increases in time: as a result, the uncertainty $\Delta x \, \Delta p$ increases in time.

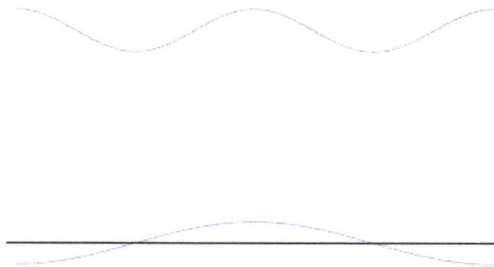

The superposition of several plane waves to form a wave packet. This wave packet becomes increasingly localized with the addition of many waves. The Fourier transform is a mathematical operation that separates a wave packet into its individual plane waves. Note that the waves shown here are real for illustrative purposes only, whereas in quantum mechanics the wave function is generally complex.

The uncertainty principle is not readily apparent on the macroscopic scales of everyday experience. So it is helpful to demonstrate how it applies to more easily understood physical situations. Two alternative frameworks for quantum physics offer different explanations for the uncertainty principle. The wave mechanics picture of the uncertainty principle is more visually intuitive, but the more abstract matrix mechanics picture formulates it in a way that generalizes more easily.

Mathematically, in wave mechanics, the uncertainty relation between position and momentum arises because the expressions of the wavefunction in the two corresponding orthonormal bases in

Hilbert space are Fourier transforms of one another (i.e., position and momentum are conjugate variables). A nonzero function and its Fourier transform cannot both be sharply localized. A similar tradeoff between the variances of Fourier conjugates arises in all systems underlain by Fourier analysis, for example in sound waves: A pure tone is a sharp spike at a single frequency, while its Fourier transform gives the shape of the sound wave in the time domain, which is a completely delocalized sine wave. In quantum mechanics, the two key points are that the position of the particle takes the form of a matter wave, and momentum is its Fourier conjugate, assured by the de Broglie relation $p = \hbar k$, where k is the wavenumber.

In matrix mechanics, the mathematical formulation of quantum mechanics, any pair of non-commuting self-adjoint operators representing observables are subject to similar uncertainty limits. An eigenstate of an observable represents the state of the wavefunction for a certain measurement value (the eigenvalue). For example, if a measurement of an observable A is performed, then the system is in a particular eigenstate Ψ of that observable. However, the particular eigenstate of the observable A need not be an eigenstate of another observable B: If so, then it does not have a unique associated measurement for it, as the system is not in an eigenstate of that observable.

Wave Mechanics Interpretation

$$\Psi = Ae^{i(px - \omega t)}$$

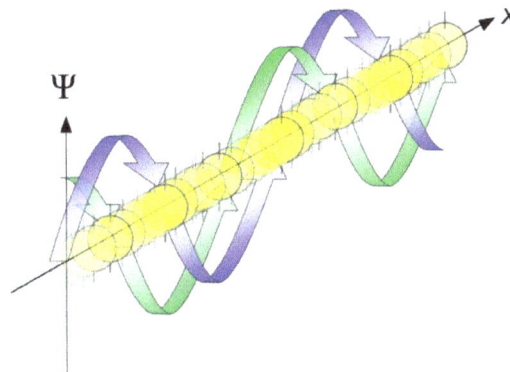

Plane wave

$$\Psi = \sum_n A_n e^{i(p_n x - \omega_n t)}$$

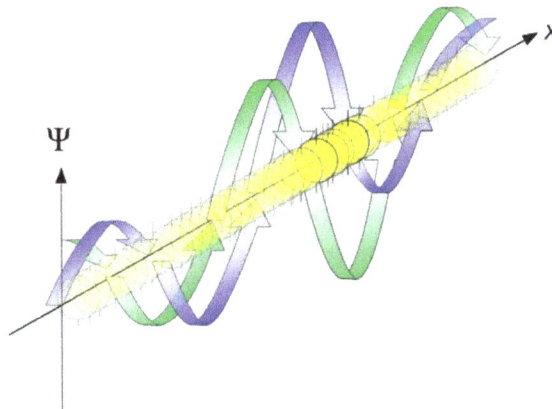

Wave packet

Propagation of de Broglie waves in 1d—real part of the complex amplitude is blue, imaginary part is green. The probability (shown as the colour opacity) of finding the particle at a given point x is spread out like a waveform, there is no definite position of the particle. As the amplitude increases above zero the curvature reverses sign, so the amplitude begins to decrease again, and vice versa— the result is an alternating amplitude: a wave.

According to the de Broglie hypothesis, every object in the universe is a wave, i.e., a situation which gives rise to this phenomenon. The position of the particle is described by a wave function $\Psi(x,t)$. The time-independent wave function of a single-moded plane wave of wavenumber k_0 or momentum p_0 is

$$\psi(x) \propto e^{ik_0 x} = e^{ip_0 x/\hbar} \ .$$

The Born rule states that this should be interpreted as a probability density amplitude function in the sense that the probability of finding the particle between a and b is

$$P[a \leq X \leq b] = \int_a^b |\psi(x)|^2 \, dx.$$

In the case of the single-moded plane wave, $|\psi(x)|^2$ is a uniform distribution. In other words, the particle position is extremely uncertain in the sense that it could be essentially anywhere along the wave packet. Consider a wave function that is a sum of many waves, however, we may write this as

$$\psi(x) \propto \sum_n A_n e^{ip_n x/\hbar} \ ,$$

where A_n represents the relative contribution of the mode p_n to the overall total. The figures to the right show how with the addition of many plane waves, the wave packet can become more localized. We may take this a step further to the continuum limit, where the wave function is an integral over all possible modes

$$\psi(x) = \frac{1}{\sqrt{2\pi\hbar}} \int_{-\infty}^{\infty} \phi(p) \cdot e^{ipx/\hbar} \, dp \ ,$$

with $\phi(p)$ representing the amplitude of these modes and is called the wave function in momentum space. In mathematical terms, we say that $\phi(p)$ is the *Fourier transform* of $\psi(x)$ and that x and p are conjugate variables. Adding together all of these plane waves comes at a cost, namely the momentum has become less precise, having become a mixture of waves of many different momenta.

One way to quantify the precision of the position and momentum is the standard deviation σ. Since $|\psi(x)|^2$ is a probability density function for position, we calculate its standard deviation.

The precision of the position is improved, i.e. reduced σ_x, by using many plane waves, thereby weakening the precision of the momentum, i.e. increased σ_p. Another way of stating this is that σ_x and σ_p have an inverse relationship or are at least bounded from below. This is the uncertainty principle, the exact limit of which is the Kennard bound.

Matrix Mechanics Interpretation

In matrix mechanics, observables such as position and momentum are represented by self-adjoint operators. When considering pairs of observables, an important quantity is the *commutator*. For a pair of operators $\hat{A}$ and $\hat{B}$, one defines their commutator as

$$[\hat{A}, \hat{B}] = \hat{A}\hat{B} - \hat{B}\hat{A}.$$

In the case of position and momentum, the commutator is the canonical commutation relation

$$[\hat{x}, \hat{p}] = i\hbar.$$

The physical meaning of the non-commutativity can be understood by considering the effect of the commutator on position and momentum eigenstates. Let $|\psi\rangle$ be a right eigenstate of position with a constant eigenvalue x_0. By definition, this means that $\hat{x}|\psi\rangle = x_0|\psi\rangle$. Applying the commutator to $|\psi\rangle$ yields

$$[\hat{x}, \hat{p}]|\psi\rangle = (\hat{x}\hat{p} - \hat{p}\hat{x})|\psi\rangle = (\hat{x} - x_0\hat{I})\hat{p}|\psi\rangle = i\hbar|\psi\rangle,$$

where $\hat{I}$ is the identity operator.

Suppose, for the sake of proof by contradiction, that $|\psi\rangle$ is also a right eigenstate of momentum, with constant eigenvalue p_0. If this were true, then one could write

$$(\hat{x} - x_0\hat{I})\hat{p}|\psi\rangle = (\hat{x} - x_0\hat{I})p_0|\psi\rangle = (x_0\hat{I} - x_0\hat{I})p_0|\psi\rangle = 0.$$

On the other hand, the above canonical commutation relation requires that

$$[\hat{x}, \hat{p}]|\psi\rangle = i\hbar|\psi\rangle \neq 0.$$

This implies that no quantum state can simultaneously be both a position and a momentum eigenstate.

When a state is measured, it is projected onto an eigenstate in the basis of the relevant observable. For example, if a particle's position is measured, then the state amounts to a position eigenstate. This means that the state is *not* a momentum eigenstate, however, but rather it can be represented as a sum of multiple momentum basis eigenstates. In other words, the momentum must be less precise. This precision may be quantified by the standard deviations,

$$\sigma_x = \sqrt{\langle \hat{x}^2 \rangle - \langle \hat{x} \rangle^2}$$

$$\sigma_p = \sqrt{\langle \hat{p}^2 \rangle - \langle \hat{p} \rangle^2}.$$

As in the wave mechanics interpretation above, one sees a tradeoff between the respective precisions of the two, quantified by the uncertainty principle.

Robertson–Schrödinger Uncertainty Relations

The most common general form of the uncertainty principle is the *Robertson uncertainty relation*.

For an arbitrary Hermitian operator $\hat{O}$ we can associate a standard deviation

$$\sigma_O = \sqrt{\langle \hat{O}^2 \rangle - \langle \hat{O} \rangle^2},$$

where the brackets $\langle O \rangle$ indicate an expectation value. For a pair of operators $\hat{A}$ and $\hat{B}$, we may define their *commutator* as

$$[\hat{A}, \hat{B}] = \hat{A}\hat{B} - \hat{B}\hat{A},$$

In this notation, the Robertson uncertainty relation is given by

$$\sigma_A \sigma_B \geq \left| \frac{1}{2i} \langle [\hat{A}, \hat{B}] \rangle \right| = \frac{1}{2} \left| \langle [\hat{A}, \hat{B}] \rangle \right|,$$

The Robertson uncertainty relation immediately follows from a slightly stronger inequality, the *Schrödinger uncertainty relation*,

$$\sigma_A^2 \sigma_B^2 \geq \left| \frac{1}{2} \langle \{\hat{A}, \hat{B}\} \rangle - \langle \hat{A} \rangle \langle \hat{B} \rangle \right|^2 + \left| \frac{1}{2i} \langle [\hat{A}, \hat{B}] \rangle \right|^2,$$

where we have introduced the *anticommutator*,

$$\{\hat{A}, \hat{B}\} = \hat{A}\hat{B} + \hat{B}\hat{A}.$$

Since the Robertson and Schrödinger relations are for general operators, the relations can be applied to any two observables to obtain specific uncertainty relations. A few of the most common relations found in the literature are given below.

- For position and linear momentum, the canonical commutation relation $[\hat{x}, \hat{p}] = i\hbar$ implies the Kennard inequality from above:

$$\sigma_x \sigma_p \geq \frac{\hbar}{2}$$

- For two orthogonal components of the total angular momentum operator of an object:

$$\sigma_{J_i} \sigma_{J_j} \geq \frac{\hbar}{2} \left| \langle J_k \rangle \right|,$$

where i, j, k are distinct and J_i denotes angular momentum along the x_i axis. This relation implies that unless all three components vanish together, only a single component of a system's angular momentum can be defined with arbitrary precision, normally the component parallel to an external (magnetic or electric) field. Moreover, for $[J_x, J_y] = i\hbar \epsilon_{xyz} J_z$, a choice $\hat{A} = J_x$, $\hat{B} = J_y$,, in angular momentum multiplets, $\psi = |j, m\rangle$, bounds the Casimir invariant (angular momentum squared, $\langle J_x^2 + J_y^2 + J_z^2 \rangle$) from below and thus yields useful constraints such as $j(j + 1) \geq m(m + 1)$, and hence $j \geq m$, among others.

- In non-relativistic mechanics, time is privileged as an independent variable. Nevertheless, in 1945, L. I. Mandelshtam and I. E. Tamm derived a non-relativistic *time–energy uncertainty relation*, as follows. For a quantum system in a non-stationary state ψ and an observable B represented by a self-adjoint operator $\hat{B}$, the following formula holds:

$$\sigma_E \frac{\sigma_B}{\left| \dfrac{d\langle \hat{B} \rangle}{dt} \right|} \geq \frac{\hbar}{2}.$$

where σ_E is the standard deviation of the energy operator (Hamiltonian) in the state ψ, σ_B stands for the standard deviation of B. Although the second factor in the left-hand side has dimension of time, it is different from the time parameter that enters the Schrödinger equation. It is a *lifetime* of the state ψ with respect to the observable B: In other words, this is the *time interval* (Δt) after which the expectation value $\langle \hat{B} \rangle$ changes appreciably.

An informal, heuristic meaning of the principle is the following: A state that only exists for a short time cannot have a definite energy. To have a definite energy, the frequency of the state must be defined accurately, and this requires the state to hang around for many cycles, the reciprocal of the required accuracy. For example, in spectroscopy, excited states have a finite lifetime. By the time–energy uncertainty principle, they do not have a definite energy, and, each time they decay, the energy they release is slightly different. The average energy of the outgoing photon has a peak at the theoretical energy of the state, but the distribution has a finite width called the *natural linewidth*. Fast-decaying states have a broad linewidth, while slow decaying states have a narrow linewidth.

The same linewidth effect also makes it difficult to specify the rest mass of unstable, fast-decaying particles in particle physics. The faster the particle decays (the shorter its lifetime), the less certain is its mass (the larger the particle's width).

- For the number of electrons in a superconductor and the phase of its Ginzburg–Landau order parameter

$$\Delta N \Delta \phi \geq 1.$$

Examples

Quantum Harmonic Oscillator Stationary States

Consider a one-dimensional quantum harmonic oscillator (QHO). It is possible to express the position and momentum operators in terms of the creation and annihilation operators:

$$\hat{x} = \sqrt{\frac{\hbar}{2m\omega}} (a + a^\dagger)$$

$$\hat{p} = i\sqrt{\frac{m\omega\hbar}{2}}(a^\dagger - a).$$

Using the standard rules for creation and annihilation operators on the eigenstates of the QHO,

$$a^\dagger \mid n \rangle = \sqrt{n+1} \mid n+1 \rangle$$

$$a \mid n \rangle = \sqrt{n} \mid n-1 \rangle \,,,$$

the variances may be computed directly,

$$\sigma_x^2 = \frac{\hbar}{m\omega}\left(n + \frac{1}{2}\right)$$

$$\sigma_p^2 = \hbar m\omega\left(n + \frac{1}{2}\right).$$

The product of these standard deviations is then

$$\sigma_x \sigma_p = \hbar\left(n + \frac{1}{2}\right) \geq \frac{\hbar}{2}.$$

In particular, the above Kennard bound is saturated for the ground state $n=0$, for which the probability density is just the normal distribution.

Quantum Harmonic Oscillator with Gaussian Initial Condition

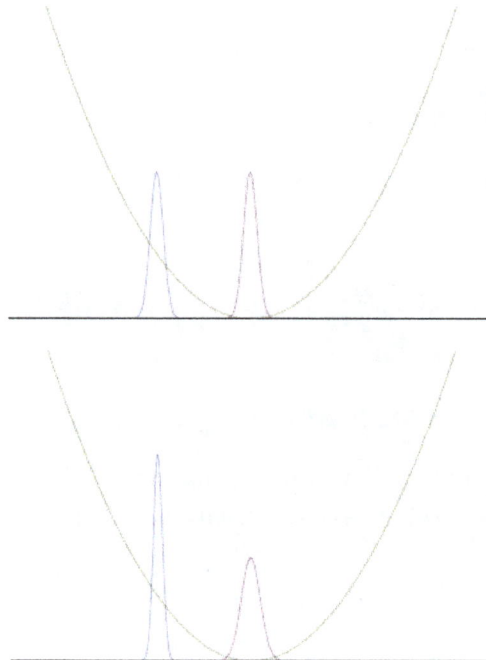

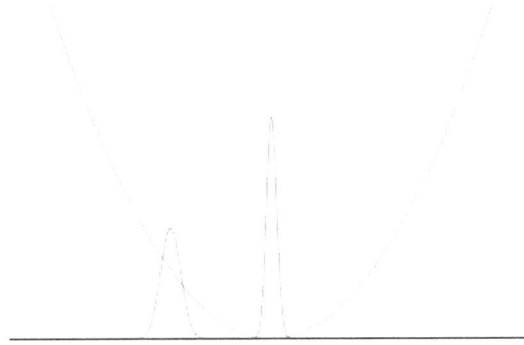

Position (blue) and momentum (red) probability densities for an initially Gaussian distribution. From top to bottom, the animations show the cases $\Omega=\omega$, $\Omega=2\omega$, and $\Omega=\omega/2$. Note the tradeoff between the widths of the distributions.

In a quantum harmonic oscillator of characteristic angular frequency ω, place a state that is offset from the bottom of the potential by some displacement x_0 as

$$\psi(x) = \left(\frac{m\Omega}{\pi\hbar}\right)^{1/4} \exp\left(-\frac{m\Omega(x-x_0)^2}{2\hbar}\right),$$

where Ω describes the width of the initial state but need not be the same as ω. Through integration over the propagator, we can solve for the full time-dependent solution. After many cancelations, the probability densities reduce to

$$|\Psi(x,t)|^2 \sim \mathcal{N}\left(x_0\cos(\omega t), \frac{\hbar}{2m\Omega}\left(\cos^2(\omega t)+\frac{\Omega^2}{\omega^2}\sin^2(\omega t)\right)\right)$$

$$|\Phi(p,t)|^2 \sim \mathcal{N}\left(-mx_0\omega\sin(\omega t), \frac{\hbar m\Omega}{2}\left(\cos^2(\omega t)+\frac{\omega^2}{\Omega^2}\sin^2(\omega t)\right)\right),$$

where we have used the notation $\mathcal{N}(\mu,\sigma^2)$ to denote a normal distribution of mean μ and variance σ^2. Copying the variances above and applying trigonometric identities, we can write the product of the standard deviations as

$$\sigma_x\sigma_p = \frac{\hbar}{2}\sqrt{\left(\cos^2(\omega t)+\frac{\Omega^2}{\omega^2}\sin^2(\omega t)\right)\left(\cos^2(\omega t)+\frac{\omega^2}{\Omega^2}\sin^2(\omega t)\right)}$$

$$= \frac{\hbar}{4}\sqrt{3+\frac{1}{2}\left(\frac{\Omega^2}{\omega^2}+\frac{\omega^2}{\Omega^2}\right)-\left(\frac{1}{2}\left(\frac{\Omega^2}{\omega^2}+\frac{\omega^2}{\Omega^2}\right)-1\right)\cos(4\omega t)}$$

From the relations

$$\frac{\Omega^2}{\omega^2}+\frac{\omega^2}{\Omega^2} \geq 2, |\cos(4\omega t)| \leq 1, ,$$

we can conclude

$$\sigma_x\sigma_p \geq \frac{\hbar}{4}\sqrt{3+\frac{1}{2}\left(\frac{\Omega^2}{\omega^2}+\frac{\omega^2}{\Omega^2}\right)-\left(\frac{1}{2}\left(\frac{\Omega^2}{\omega^2}+\frac{\omega^2}{\Omega^2}\right)-1\right)} = \frac{\hbar}{2}$$

Coherent States

A coherent state is a right eigenstate of the annihilation operator,

$$\hat{a}\,|\,\alpha\rangle = \alpha\,|\,\alpha\rangle,$$

which may be represented in terms of Fock states as

$$|\,\alpha\rangle = e^{-\frac{|\alpha|^2}{2}}\sum_{n=0}^{\infty}\frac{\alpha^n}{\sqrt{n!}}\,|\,n\rangle$$

In the picture where the coherent state is a massive particle in a QHO, the position and momentum operators may be expressed in terms of the annihilation operators in the same formulas above and used to calculate the variances,

$$\sigma_x^2 = \frac{\hbar}{2m\omega}$$

$$\sigma_p^2 = \frac{\hbar m\omega}{2}.$$

Therefore, every coherent state saturates the Kennard bound

$$\sigma_x\sigma_p = \sqrt{\frac{\hbar}{2m\omega}}\sqrt{\frac{\hbar m\omega}{2}} = \frac{\hbar}{2}.$$

with position and momentum each contributing an amount $\sqrt{\hbar/2}$ in a "balanced" way. Moreover, every squeezed coherent state also saturates the Kennard bound although the individual contributions of position and momentum need not be balanced in general.

Particle in a Box

Consider a particle in a one-dimensional box of length L. The eigenfunctions in position and momentum space are

$$\psi_n(x,t) = \begin{cases} A\sin(k_n x)e^{-i\omega_n t}, & 0 < x < L, \\ 0, & \text{otherwise,} \end{cases}$$

and

$$\phi_n(p,t) = \sqrt{\frac{\pi L}{\hbar}}\frac{n\left(1-(-1)^n e^{-ikL}\right)e^{-i\omega_n t}}{\pi^2 n^2 - k^2 L^2},$$

where $\omega_n = \dfrac{\pi^2 \hbar n^2}{8L^2 m}$ and we have used the de Broglie relation $p = \hbar k$. The variances of x and p can be calculated explicitly:

$$\sigma_x^2 = \frac{L^2}{12}\left(1 - \frac{6}{n^2 \pi^2}\right)$$

$$\sigma_p^2 = \left(\frac{\hbar n \pi}{L}\right)^2.$$

The product of the standard deviations is therefore

$$\sigma_x \sigma_p = \frac{\hbar}{2}\sqrt{\frac{n^2 \pi^2}{3} - 2}.$$

For all $n = 1,2,3...$, the quantity $\sqrt{\dfrac{n^2 \pi^2}{3} - 2}$ is greater than 1, so the uncertainty principle is never violated. For numerical concreteness, the smallest value occurs when $n = 1,$, in which case

$$\sigma_x \sigma_p = \frac{\hbar}{2}\sqrt{\frac{\pi^2}{3} - 2} \approx 0.568\hbar > \frac{\hbar}{2}..$$

Constant Momentum

Position space probability density of an initially Gaussian state moving at minimally uncertain, constant momentum in free space

Assume a particle initially has a momentum space wave function described by a normal distribution around some constant momentum p_0 according to

$$\phi(p) = \left(\frac{x_0}{\hbar\sqrt{\pi}}\right)^{1/2} \cdot \exp\left(\frac{-x_0^2(p - p_0)^2}{2\hbar^2}\right),$$

where we have introduced a reference scale $x_0 = \sqrt{\hbar/m\omega_0}$, with $\omega_0 > 0$ describing the width of the distribution--cf. nondimensionalization. If the state is allowed to evolve in free space, then the time-dependent momentum and position space wave functions are

$$\Phi(p,t) = \left(\frac{x_0}{\hbar\sqrt{\pi}}\right)^{1/2} \cdot \exp\left(\frac{-x_0^2(p-p_0)^2}{2\hbar^2} - \frac{ip^2 t}{2m\hbar}\right),$$

$$\Psi(x,t) = \left(\frac{1}{x_0\sqrt{\pi}}\right)^{1/2} \cdot \frac{e^{-x_0^2 p_0^2/2\hbar^2}}{\sqrt{1+i\omega_0 t}} \cdot \exp\left(-\frac{(x - ix_0^2 p_0/\hbar)^2}{2x_0^2(1+i\omega_0 t)}\right).$$

Since $\langle p(t) \rangle = p_0$ and $\sigma_p(t) = \hbar / x_0\sqrt{2}$, this can be interpreted as a particle moving along with constant momentum at arbitrarily high precision. On the other hand, the standard deviation of the position is

$$\sigma_x = \frac{x_0}{\sqrt{2}}\sqrt{1+\omega_0^2 t^2}$$

such that the uncertainty product can only increase with time as

$$\sigma_x(t)\sigma_p(t) = \frac{\hbar}{2}\sqrt{1+\omega_0^2 t^2}$$

Additional Uncertainty Relations

Mixed States

The Robertson–Schrödinger uncertainty relation may be generalized in a straightforward way to describe mixed states.

$$\sigma_A^2 \sigma_B^2 \geq \left(\frac{1}{2}\operatorname{tr}(\rho\{A,B\}) - \operatorname{tr}(\rho A)\operatorname{tr}(\rho B)\right)^2 + \left(\frac{1}{2i}\operatorname{tr}(\rho[A,B])\right)^2$$

Phase Space

In the phase space formulation of quantum mechanics, the Robertson–Schrödinger relation follows from a positivity condition on a real star-square function. Given a Wigner function $W(x,p)$ with star product $\star$ and a function f, the following is generally true:

$$\langle f^* \star f \rangle = \int (f^* \star f)W(x,p)dxdp \geq 0.$$

Choosing $f = a + bx + cp$, we arrive at

$$\langle f^* \star f \rangle = \begin{bmatrix} a^* & b^* & c^* \end{bmatrix} \begin{bmatrix} 1 & \langle x \rangle & \langle p \rangle \\ \langle x \rangle & \langle x \star x \rangle & \langle x \star p \rangle \\ \langle p \rangle & \langle p \star x \rangle & \langle p \star p \rangle \end{bmatrix} \begin{bmatrix} a \\ b \\ c \end{bmatrix} \geq 0.$$

Since this positivity condition is true for all a, b, and c, it follows that all the eigenvalues of the matrix are positive. The positive eigenvalues then imply a corresponding positivity condition on the determinant:

$$\det \begin{bmatrix} 1 & \langle x \rangle & \langle p \rangle \\ \langle x \rangle & \langle x \star x \rangle & \langle x \star p \rangle \\ \langle p \rangle & \langle p \star x \rangle & \langle p \star p \rangle \end{bmatrix} = \det \begin{bmatrix} 1 & \langle x \rangle & \langle p \rangle \\ \langle x \rangle & \langle x^2 \rangle & \left\langle xp + \frac{i\hbar}{2} \right\rangle \\ \langle p \rangle & \left\langle xp - \frac{i\hbar}{2} \right\rangle & \langle p^2 \rangle \end{bmatrix} \geq 0,$$

or, explicitly, after algebraic manipulation,

$$\sigma_x^2 \sigma_p^2 = \left(\langle x^2 \rangle - \langle x \rangle^2 \right)\left(\langle p^2 \rangle - \langle p \rangle^2 \right) \geq \left(\langle xp \rangle - \langle x \rangle \langle p \rangle \right)^2 + \frac{\hbar^2}{4}.$$

Systematic and Statistical Errors

The inequalities above focus on the *statistical imprecision* of observables as quantified by the standard deviation σ. Heisenberg's original version, however, was dealing with the *systematic error*, a disturbance of the quantum system produced by the measuring apparatus, i.e., an observer effect.

If we let ϵ_A represent the error (i.e., inaccuracy) of a measurement of an observable A and η_B the disturbance produced on a subsequent measurement of the conjugate variable B by the former measurement of A, then the inequality proposed by Ozawa — encompassing both systematic and statistical errors — holds:

$$\epsilon_A \eta_B + \epsilon_A \sigma_B + \sigma_A \eta_B \geq \frac{1}{2} \left| \langle [\hat{A}, \hat{B}] \rangle \right|$$

Heisenberg uncertainty principle, as originally described in the 1927 formulation, mentions only the first term of Ozawa inequality, regarding the *systematic error*. Using the notation above to describe the *error/disturbance* effect of *sequential measurements* (first A, then B), it could be written as

$$\epsilon_A \eta_B \geq \frac{1}{2} \left| \langle [\hat{A}, \hat{B}] \rangle \right|$$

The formal derivation of Heisenberg relation is possible but far from intuitive. It was *not* proposed by Heisenberg, but formulated in a mathematically consistent way only in recent years. Also, it must be stressed that the Heisenberg formulation is not taking into account the intrinsic statistical errors σ_A and σ_B.. There is increasing experimental evidence that the total quantum uncertainty cannot be described by the Heisenberg term alone, but requires the presence of all the three terms of the Ozawa inequality.

Using the same formalism, it is also possible to introduce the other kind of physical situation, often confused with the previous one, namely the case of *simultaneous measurements* (A and B at the same time):

$$\epsilon_A \epsilon_B \geq \frac{1}{2}\left|\langle[\hat{A},\hat{B}]\rangle\right|$$

The two simultaneous measurements on A and B are necessarily *unsharp* or *weak*.

It is also possible to derive an uncertainty relation that, as the Ozawa's one, combines both the statistical and systematic error components, but keeps a form very close to the Heisenberg original inequality. By adding Robertson

$$\sigma_A \sigma_B \geq \frac{1}{2}\left|\langle[\hat{A},\hat{B}]\rangle\right|$$

and Ozawa relations we obtain

$$\epsilon_A \eta_B + \epsilon_A \sigma_B + \sigma_A \eta_B + \sigma_A \sigma_B \geq \left|\langle[\hat{A},\hat{B}]\rangle\right|.$$

The four terms can be written as:

$$(\epsilon_A + \sigma_A)(\eta_B + \sigma_B) \geq \left|\langle[\hat{A},\hat{B}]\rangle\right|.$$

Defining:

$$\bar{\epsilon}_A \equiv (\epsilon_A + \sigma_A)$$

as the *inaccuracy* in the measured values of the variable A and

$$\bar{\eta}_B \equiv (\eta_B + \sigma_B)$$

as the *resulting fluctuation* in the conjugate variable B, Fujikawa established an uncertainty relation similar to the Heisenberg original one, but valid both for *systematic and statistical errors*:

$$\bar{\epsilon}_A \bar{\eta}_B \geq \left|\langle[\hat{A},\hat{B}]\rangle\right|$$

Quantum Entropic Uncertainty Principle

For many distributions, the standard deviation is not a particularly natural way of quantifying the structure. For example, uncertainty relations in which one of the observables is an angle has little physical meaning for fluctuations larger than one period. Other examples include highly bimodal distributions, or unimodal distributions with divergent variance.

A solution that overcomes these issues is an uncertainty based on entropic uncertainty instead of the product of variances. While formulating the many-worlds interpretation of quantum mechanics in 1957, Hugh Everett III conjectured a stronger extension of the uncertainty principle based on entropic certainty. This conjecture, also studied by Hirschman and proven in 1975 by Beckner and by Iwo Bialynicki-Birula and Jerzy Mycielski is that, for two normalized, dimensionless Fourier transform pairs $f(a)$ and $g(b)$ where

$$f(a) = \int_{-\infty}^{\infty} g(b) \, e^{2\pi i a b} \, dx \quad \text{and} \quad g(b) = \int_{-\infty}^{\infty} f(a) \, e^{-2\pi i a b} \, dx$$

the Shannon information entropies

$$H_a = \int_{-\infty}^{\infty} f(a) \log(f(a)) dx,$$

and

$$H_b = \int_{-\infty}^{\infty} g(b) \log(g(b)) dy$$

are subject to the following constraint,

$$\boxed{H_a + H_b \geq \log(e/2)}$$

where the logarithms may be in any base.

The probability distribution functions associated with the position wave function $\psi(x)$ and the momentum wave function $\varphi(x)$ have dimensions of inverse length and momentum respectively, but the entropies may be rendered dimensionless by

$$H_x = -\int |\psi(x)|^2 \, \ln(x_0 |\psi(x)|^2) dx = -\left\langle \ln(x_0 |\psi(x)|^2) \right\rangle$$

$$H_p = -\int |\phi(p)|^2 \, \ln(p_0 |\phi(p)|^2) dp = -\left\langle \ln(p_0 |\phi(p)|^2) \right\rangle$$

where x_o and p_o are some arbitrarily chosen length and momentum respectively, which render the arguments of the logarithms dimensionless. Note that the entropies will be functions of these chosen parameters. Due to the Fourier transform relation between the position wave function $\psi(x)$ and the momentum wavefuction $\varphi(p)$, the above constraint can be written for the corresponding entropies as

$$\boxed{H_x + H_p \geq \log\left(\frac{eh}{2x_0 p_0}\right)}$$

where h is Planck's constant.

Depending on one's choice of the $x_o p_o$ product, the expression may be written in many ways. If $x_o p_o$ is chosen to be h, then

$$H_x + H_p \geq \log\left(\frac{e}{2}\right)$$

If, instead, $x_o p_o$ is chosen to be $\hbar$, then

$$H_x + H_p \geq \log(e\pi)$$

If x_o and p_o are chosen to be unity in whatever system of units are being used, then

$$H_x + H_p \geq \log\left(\frac{eh}{2}\right)$$

where h is interpreted as a dimensionless number equal to the value of Planck's constant in the chosen system of units.

The quantum entropic uncertainty principle is more restrictive than the Heisenberg uncertainty principle. From the inverse logarithmic Sobolev inequalities

$$H_x \leq \frac{1}{2}\log(2e\pi\sigma_x^2 / x_0^2),$$

$$H_p \leq \frac{1}{2}\log(2e\pi\sigma_p^2 / p_0^2),$$

(equivalently, from the fact that normal distributions maximize the entropy of all such with a given variance), it readily follows that this entropic uncertainty principle is *stronger than the one based on standard deviations*, because

$$\sigma_x \sigma_p \geq \frac{\hbar}{2}\exp\left(H_x + H_p - \log\left(\frac{eh}{2x_0 p_0}\right)\right) \geq \frac{\hbar}{2}.$$

In other words, the Heisenberg uncertainty principle, is a consequence of the quantum entropic uncertainty principle, but not vice versa. A few remarks on these inequalities. First, the choice of base e is a matter of popular convention in physics. The logarithm can alternatively be in any base, provided that it be consistent on both sides of the inequality. Second, recall the Shannon entropy has been used, *not* the quantum von Neumann entropy. Finally, the normal distribution saturates the inequality, and it is the only distribution with this property, because it is the maximum entropy probability distribution among those with fixed variance.

Harmonic Analysis

In the context of harmonic analysis, a branch of mathematics, the uncertainty principle implies that one cannot at the same time localize the value of a function and its Fourier transform. To wit, the following inequality holds,

$$\left(\int_{-\infty}^{\infty} x^2 \mid f(x) \mid^2 dx\right)\left(\int_{-\infty}^{\infty} \xi^2 \mid \hat{f}(\xi) \mid^2 d\xi\right) \geq \frac{\|f\|_2^4}{16\pi^2}.$$

Further mathematical uncertainty inequalities, including the above entropic uncertainty, hold between a function f and its Fourier transform $\hat{f}$:

$$H_x + H_\xi \geq \log(e/2)$$

Signal Processing

In the context of signal processing, and in particular time–frequency analysis, uncertainty principles are referred to as the Gabor limit, after Dennis Gabor, or sometimes the *Heisenberg–Gabor limit*. The basic result, which follows from "Benedicks's theorem", below, is that a function cannot be both time limited and band limited (a function and its Fourier transform cannot both have bounded domain).

Stated alternatively, "One cannot simultaneously sharply localize a signal (function f) in both the time domain and frequency domain ($\hat{f}$, its Fourier transform)".

When applied to filters, the result implies that one cannot achieve high temporal resolution and frequency resolution at the same time; a concrete example are the resolution issues of the short-time Fourier transform—if one uses a wide window, one achieves good frequency resolution at the cost of temporal resolution, while a narrow window has the opposite trade-off.

Alternate theorems give more precise quantitative results, and, in time–frequency analysis, rather than interpreting the (1-dimensional) time and frequency domains separately, one instead interprets the limit as a lower limit on the support of a function in the (2-dimensional) time–frequency plane. In practice, the Gabor limit limits the *simultaneous* time–frequency resolution one can achieve without interference; it is possible to achieve higher resolution, but at the cost of different components of the signal interfering with each other.

Benedicks's Theorem

Amrein-Berthier and Benedicks's theorem intuitively says that the set of points where f is non-zero and the set of points where $\hat{f}$ is nonzero cannot both be small.

Specifically, it is impossible for a function f in $L^2(\mathbf{R})$ and its Fourier transform $\hat{f}$ to both be supported on sets of finite Lebesgue measure. A more quantitative version is

$$\|f\|_{L^2(\mathbf{R}^d)} \leq Ce^{C|S||\Sigma|}\left(\|f\|_{L^2(S^c)} + \|\hat{f}\|_{L^2(\Sigma^c)}\right).$$

One expects that the factor $Ce^{C|S||\Sigma|}$ may be replaced by $Ce^{C(|S||\Sigma|)^{1/d}}$, which is only known if either S or Σ is convex.

Hardy's Uncertainty Principle

The mathematician G. H. Hardy formulated the following uncertainty principle: it is not possible for f and $\hat{f}$ to both be "very rapidly decreasing." Specifically, if f in $L^2(R)$ is such that

$$| f(x) |\le C(1+| x |)^N e^{-a\pi x^2}$$

and

$$| \hat{f}(\xi) |\le C(1+| \xi |)^N e^{-b\pi\xi^2} \; (C > 0, N \text{ an integer}),$$

then, if $ab > 1, f = 0$, while if $ab = 1$, then there is a polynomial P of degree $\le N$ such that

$$f(x) = P(x)e^{-a\pi x^2}.$$

This was later improved as follows: if $f \in L^2(R^d)$ is such that

$$\int_{R^d} \int_{R^d} |f(x)| \| \hat{f}(\xi) | \frac{e^{\pi|\langle x,\xi\rangle|}}{(1+| x |+| \xi |)^N}dxd\xi < +\infty ,$$

then

$$f(x) = P(x)e^{-\pi\langle Ax,x\rangle} ,$$

where P is a polynomial of degree $(N - d)/2$ and A is a real $d\times d$ positive definite matrix.

This result was stated in Beurling's complete works without proof and proved in Hörmander (the case $d = 1, N = 0$) and Bonami, Demange, and Jaming for the general case. Note that Hörmander–Beurling's version implies the case $ab > 1$ in Hardy's Theorem while the version by Bonami–Demange–Jaming covers the full strength of Hardy's Theorem. A different proof of Beurling's theorem based on Liouville's theorem appeared in ref.

A full description of the case $ab < 1$ as well as the following extension to Schwartz class distributions appears in ref.

Theorem. If a tempered distribution $f \in S'(\mathbb{R}^d)$ is such that

$$e^{\pi|x|^2} f \in S'(\mathbb{R}^d)$$

and

$$e^{\pi|\xi|^2} \hat{f} \in S'(\mathbb{R}^d) ,$$

then

$$f(x) = P(x)e^{-\pi\langle Ax,x\rangle} ,$$

for some convenient polynomial P and real positive definite matrix A of type $d \times d$.

History

Werner Heisenberg formulated the Uncertainty Principle at Niels Bohr's institute in Copenhagen, while working on the mathematical foundations of quantum mechanics.

Werner Heisenberg and Niels Bohr

In 1925, following pioneering work with Hendrik Kramers, Heisenberg developed matrix mechanics, which replaced the ad hoc old quantum theory with modern quantum mechanics. The central premise was that the classical concept of motion does not fit at the quantum level, as electrons in an atom do not travel on sharply defined orbits. Rather, their motion is smeared out in a strange way: the Fourier transform of its time dependence only involves those frequencies that could be observed in the quantum jumps of their radiation.

Heisenberg's paper did not admit any unobservable quantities like the exact position of the electron in an orbit at any time; he only allowed the theorist to talk about the Fourier components of the motion. Since the Fourier components were not defined at the classical frequencies, they could not be used to construct an exact trajectory, so that the formalism could not answer certain overly precise questions about where the electron was or how fast it was going.

In March 1926, working in Bohr's institute, Heisenberg realized that the non-commutativity implies the uncertainty principle. This implication provided a clear physical interpretation for the non-commutativity, and it laid the foundation for what became known as the Copenhagen interpretation of quantum mechanics. Heisenberg showed that the commutation relation implies an uncertainty, or in Bohr's language a complementarity. Any two variables that do not commute cannot be measured simultaneously—the more precisely one is known, the less precisely the other can be known. Heisenberg wrote:

It can be expressed in its simplest form as follows: One can never know with perfect accuracy both of those two important factors which determine the movement of one of the smallest particles—its position and its velocity. It is impossible to determine accurately *both* the position and the direction and speed of a particle *at the same instant.*

In his celebrated 1927 paper, "Über den anschaulichen Inhalt der quantentheoretischen Kinematik und Mechanik" ("On the Perceptual Content of Quantum Theoretical Kinematics and Mechanics"), Heisenberg established this expression as the minimum amount of unavoidable momentum disturbance caused by any position measurement, but he did not give a precise definition for the uncertainties Δx and Δp. Instead, he gave some plausible estimates in each case separately. In his Chicago lecture he refined his principle:

$$\Delta x \Delta p \gtrsim h \qquad \qquad \textbf{(1)}$$

Kennard in 1927 first proved the modern inequality:

$$\sigma_x \sigma_p \geq \frac{\hbar}{2} \tag{2}$$

where $\hbar = h/2\pi$, and σ_x, σ_p are the standard deviations of position and momentum. Heisenberg only proved relation (2) for the special case of Gaussian states.

Terminology and Translation

Throughout the main body of his original 1927 paper, written in German, Heisenberg used the word, "Ungenauigkeit" ("indeterminacy"), to describe the basic theoretical principle. Only in the endnote did he switch to the word, "Unsicherheit" ("uncertainty"). When the English-language version of Heisenberg's textbook, *The Physical Principles of the Quantum Theory*, was published in 1930, however, the translation "uncertainty" was used, and it became the more commonly used term in the English language thereafter.

Heisenberg's Microscope

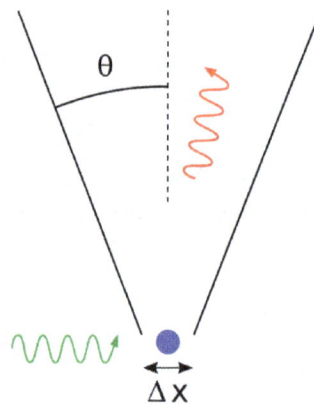

Heisenberg's gamma-ray microscope for locating an electron (shown in blue). The incoming gamma ray (shown in green) is scattered by the electron up into the microscope's aperture angle θ. The scattered gamma-ray is shown in red. Classical optics shows that the electron position can be resolved only up to an uncertainty Δx that depends on θ and the wavelength λ of the incoming light.

The principle is quite counter-intuitive, so the early students of quantum theory had to be reassured that naive measurements to violate it were bound always to be unworkable. One way in which Heisenberg originally illustrated the intrinsic impossibility of violating the uncertainty principle is by using an imaginary microscope as a measuring device.

He imagines an experimenter trying to measure the position and momentum of an electron by shooting a photon at it.

Problem 1 – If the photon has a short wavelength, and therefore, a large momentum, the position can be measured accurately. But the photon scatters in a random direction, transferring a large and uncertain amount of momentum to the electron. If the photon has a long wavelength and low momentum, the collision does not disturb the electron's momentum very much, but the scattering will reveal its position only vaguely.

Problem 2 – If a large aperture is used for the microscope, the electron's location can be well resolved; but by the principle of conservation of momentum, the transverse momentum of the incoming photon affects the electrons beamline momentum and hence, the new momentum of the electron resolves poorly. If a small aperture is used, the accuracy of both resolutions is the other way around.

The combination of these trade-offs imply that no matter what photon wavelength and aperture size are used, the product of the uncertainty in measured position and measured momentum is greater than or equal to a lower limit, which is (up to a small numerical factor) equal to Planck's constant. Heisenberg did not care to formulate the uncertainty principle as an exact limit (which is elaborated below), and preferred to use it instead, as a heuristic quantitative statement, correct up to small numerical factors, which makes the radically new noncommutativity of quantum mechanics inevitable.

Critical Reactions

The Copenhagen interpretation of quantum mechanics and Heisenberg's Uncertainty Principle were, in fact, seen as twin targets by detractors who believed in an underlying determinism and realism. According to the Copenhagen interpretation of quantum mechanics, there is no fundamental reality that the quantum state describes, just a prescription for calculating experimental results. There is no way to say what the state of a system fundamentally is, only what the result of observations might be.

Albert Einstein believed that randomness is a reflection of our ignorance of some fundamental property of reality, while Niels Bohr believed that the probability distributions are fundamental and irreducible, and depend on which measurements we choose to perform. Einstein and Bohr debated the uncertainty principle for many years. Some experiments within the first decade of the twenty-first century have cast doubt on observer effect aspects of the uncertainty principle.

Einstein's Slit

The first of Einstein's thought experiments challenging the uncertainty principle went as follows:

> Consider a particle passing through a slit of width d. The slit introduces an uncertainty in momentum of approximately h/d because the particle passes through the wall. But let us determine the momentum of the particle by measuring the recoil of the wall. In doing so, we find the momentum of the particle to arbitrary accuracy by conservation of momentum.

Bohr's response was that the wall is quantum mechanical as well, and that to measure the recoil to accuracy Δp, the momentum of the wall must be known to this accuracy before the particle passes through. This introduces an uncertainty in the position of the wall and therefore the position of the slit equal to $h/\Delta p$, and if the wall's momentum is known precisely enough to measure the recoil, the slit's position is uncertain enough to disallow a position measurement.

A similar analysis with particles diffracting through multiple slits is given by Richard Feynman.

Einstein's Box

Bohr was present when Einstein proposed the thought experiment which has become known as

Einstein's box. Einstein argued that "Heisenberg's uncertainty equation implied that the uncertainty in time was related to the uncertainty in energy, the product of the two being related to Planck's constant." Consider, he said, an ideal box, lined with mirrors so that it can contain light indefinitely. The box could be weighed before a clockwork mechanism opened an ideal shutter at a chosen instant to allow one single photon to escape. "We now know, explained Einstein, precisely the time at which the photon left the box." "Now, weigh the box again. The change of mass tells the energy of the emitted light. In this manner, said Einstein, one could measure the energy emitted and the time it was released with any desired precision, in contradiction to the uncertainty principle."

Bohr spent a sleepless night considering this argument, and eventually realized that it was flawed. He pointed out that if the box were to be weighed, say by a spring and a pointer on a scale, "since the box must move vertically with a change in its weight, there will be uncertainty in its vertical velocity and therefore an uncertainty in its height above the table. ... Furthermore, the uncertainty about the elevation above the earth's surface will result in an uncertainty in the rate of the clock," because of Einstein's own theory of gravity's effect on time. "Through this chain of uncertainties, Bohr showed that Einstein's light box experiment could not simultaneously measure exactly both the energy of the photon and the time of its escape."

EPR Paradox for Entangled Particles

Bohr was compelled to modify his understanding of the uncertainty principle after another thought experiment by Einstein. In 1935, Einstein, Podolsky and Rosen published an analysis of widely separated entangled particles. Measuring one particle, Einstein realized, would alter the probability distribution of the other, yet here the other particle could not possibly be disturbed. This example led Bohr to revise his understanding of the principle, concluding that the uncertainty was not caused by a direct interaction.

But Einstein came to much more far-reaching conclusions from the same thought experiment. He believed the "natural basic assumption" that a complete description of reality, would have to predict the results of experiments from "locally changing deterministic quantities", and therefore, would have to include more information than the maximum possible allowed by the uncertainty principle.

In 1964, John Bell showed that this assumption can be falsified, since it would imply a certain inequality between the probabilities of different experiments. Experimental results confirm the predictions of quantum mechanics, ruling out Einstein's basic assumption that led him to the suggestion of his *hidden variables*. Ironically this fact is one of the best pieces of evidence supporting Karl Popper's philosophy of invalidation of a theory by falsification-experiments. That is to say, here Einstein's "basic assumption" became falsified by experiments based on Bell's inequalities.

While it is possible to assume that quantum mechanical predictions are due to nonlocal, hidden variables, and in fact David Bohm invented such a formulation, this resolution is not satisfactory to the vast majority of physicists. The question of whether a random outcome is predetermined by a nonlocal theory can be philosophical, and it can be potentially intractable. If the hidden variables are not constrained, they could just be a list of random digits that are used to produce the measurement outcomes. To make it sensible, the assumption of nonlocal hidden variables is sometimes augmented by a second assumption—that the size of the observable universe puts a limit on the computations that these variables can do. A nonlocal theory of this sort predicts that a quantum computer would encounter fundamental obstacles when attempting to factor numbers of approximately 10,000 digits or more; a potentially achievable task in quantum mechanics.

Popper's Criticism

Karl Popper approached the problem of indeterminacy as a logician and metaphysical realist. He disagreed with the application of the uncertainty relations to individual particles rather than to ensembles of identically prepared particles, referring to them as "statistical scatter relations". In this statistical interpretation, a *particular* measurement may be made to arbitrary precision without invalidating the quantum theory. This directly contrasts with the Copenhagen interpretation of quantum mechanics, which is non-deterministic but lacks local hidden variables.

In 1934, Popper published *Zur Kritik der Ungenauigkeitsrelationen* (*Critique of the Uncertainty Relations*) in *Naturwissenschaften*, and in the same year *Logik der Forschung* (translated and updated by the author as *The Logic of Scientific Discovery* in 1959), outlining his arguments for the statistical interpretation. In 1982, he further developed his theory in *Quantum theory and the schism in Physics*, writing:

[Heisenberg's] formulae are, beyond all doubt, derivable *statistical formulae* of the quantum theory. But they have been *habitually misinterpreted* by those quantum theorists who said that these formulae can be interpreted as determining some upper limit to the *precision of our measurements*.[original emphasis]

Popper proposed an experiment to falsify the uncertainty relations, although he later withdrew his initial version after discussions with Weizsäcker, Heisenberg, and Einstein; this experiment may have influenced the formulation of the EPR experiment.

Many-worlds Uncertainty

The many-worlds interpretation originally outlined by Hugh Everett III in 1957 is partly meant to reconcile the differences between Einstein's and Bohr's views by replacing Bohr's wave function collapse with an ensemble of deterministic and independent universes whose *distribution* is governed by wave functions and the Schrödinger equation. Thus, uncertainty in the many-worlds interpretation follows from each observer within any universe having no knowledge of what goes on in the other universes.

Free Will

Some scientists including Arthur Compton and Martin Heisenberg have suggested that the uncertainty principle, or at least the general probabilistic nature of quantum mechanics, could be evidence for the two-stage model of free will. One critique, however, is that apart from the basic role of quantum mechanics as a foundation for chemistry, nontrivial biological mechanisms requiring quantum mechanics are unlikely, due to the rapid decoherence time of quantum systems at room temperature. The standard view, however, is that this decoherence is overcome by both screening and decoherence-free subspaces found in biological cells.

Wave Function

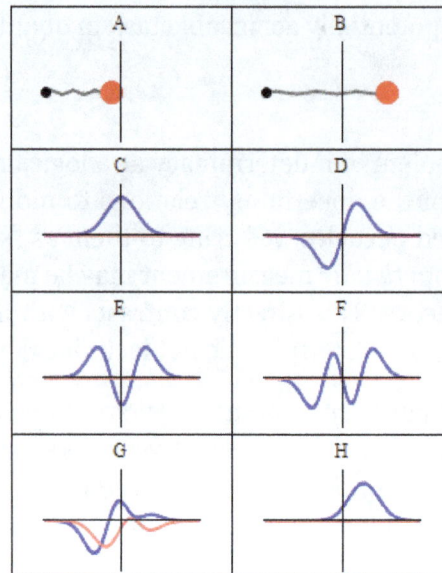

Comparison of classical and quantum harmonic oscillator conceptions for a single spinless particle. The two processes differ greatly. The classical process (A–B) is represented as the motion of a particle along a trajectory. The quantum process (C–H) has no such trajectory. Rather, it is represented as a wave. Panels (C–F) show four different standing wave solutions of the Schrödinger equation. Panels (G–H) further show two different wave functions that are solutions of the Schrödinger equation but not standing waves.

A wave function in quantum mechanics is a description of the quantum state of a system. The wave function is a complex-valued probability amplitude, and the probabilities for the possible results of measurements made on the system can be derived from it. The most common symbols for a wave function are the Greek letters ψ or Ψ (lower-case and capital psi).

The wave function is a function of the degrees of freedom corresponding to some maximal set of commuting observables. Once such a representation is chosen, the wave function can be derived from the quantum state.

For a given system, the choice of which commuting degrees of freedom to use is not unique, and correspondingly the domain of the wave function is not unique. For instance it may be taken to be a function of all the position coordinates of the particles over *position space*, or the momenta of all the particles over *momentum space*, the two are related by a Fourier transform. Some particles, like electrons and photons, have nonzero spin, and the wave function for such particles includes spin as an intrinsic, discrete degree of freedom. Other discrete variables can also be included, such as isospin. When a system has internal degrees of freedom, the wave function at each point in the continuous degrees of freedom (e.g. a point in space) assigns a complex number for *each* possible value of the discrete degrees of freedom (e.g. z-component of spin). These values are often displayed in a column matrix (e.g. a 2×1 column vector for a non-relativistic electron with spin $\frac{1}{2}$).

According to the superposition principle of quantum mechanics, wave functions can be added together and multiplied by complex numbers to form new wave functions and form a Hilbert space. The inner product between two wave functions is a measure of the overlap between the corresponding physical states and is used in the foundational probabilistic interpretation of quantum mechanics, the Born rule, relating transition probabilities to inner products. The Schrödinger equation determines how wave functions evolve over time. A wave function behaves qualitatively like other waves, such as water waves or waves on a string, because the Schrödinger equation is mathematically a type of wave equation. This explains the name "wave function", and gives rise to wave–particle duality. However, the wave function in quantum mechanics describes a kind of physical phenomenon, still open to different interpretations, which fundamentally differs from that of classic mechanical waves.

In Born's statistical interpretation in non-relativistic quantum mechanics, the squared modulus of the wave function, $|\psi|^2$, is a real number interpreted as the probability density of measuring a particle's being detected at a given place, or having a given momentum, at a given time, and possibly having definite values for discrete degrees of freedom. The integral of this quantity, over all the system's degrees of freedom, must be 1 in accordance with the probability interpretation, this general requirement a wave function must satisfy is called the *normalization condition*. Since the wave function is complex valued, only its relative phase and relative magnitude can be measured. Its value does not in isolation tell anything about the magnitudes or directions of measurable observables; one has to apply quantum operators, whose eigenvalues correspond to sets of possible results of measurements, to the wave function ψ and calculate the statistical distributions for measurable quantities.

Historical Background

In 1905 Einstein postulated the proportionality between the frequency of a photon and its energy, $E = hf$, and in 1916 the corresponding relation between photon momentum and wavelength, $\lambda = h/p$. In 1923, De Broglie was the first to suggest that the relation $\lambda = h/p$, now called the De Broglie relation, holds for *massive* particles, the chief clue being Lorentz invariance, and this can be viewed as the starting point for the modern development of quantum mechanics. The equations represent wave–particle duality for both massless and massive particles.

In the 1920s and 1930s, quantum mechanics was developed using calculus and linear algebra. Those who used the techniques of calculus included Louis de Broglie, Erwin Schrödinger, and others, developing "wave mechanics". Those who applied the methods of linear algebra included Werner Heisenberg, Max Born, and others, developing "matrix mechanics". Schrödinger subsequently showed that the two approaches were equivalent.

In 1926, Schrödinger published the famous wave equation now named after him, indeed the Schrödinger equation, based on classical Conservation of energy using quantum operators and the de Broglie relations such that the solutions of the equation are the wave functions for the quantum system. However, no one was clear on how to *interpret it*. At first, Schrödinger and others thought that wave functions represent particles that are spread out with most of the particle being where the wave function is large. This was shown to be incompatible with how elastic scattering of a wave packet representing a particle off a target appears; it spreads out in all directions. While a scattered particle may scatter in any direction, it does not break up and take off in all directions. In 1926, Born provided the perspective of probability amplitude. This relates calculations of quantum mechanics directly to probabilistic experimental observations. It is accepted as part of the Copenhagen interpretation of quantum mechanics. There are many other interpretations of quantum mechanics. In 1927, Hartree and Fock made the first step in an attempt to solve the N-body wave function, and developed the *self-consistency cycle*: an iterative algorithm to approximate the solution. Now it is also known as the Hartree–Fock method. The Slater determinant and permanent (of a matrix) was part of the method, provided by John C. Slater.

Schrödinger did encounter an equation for the wave function that satisfied relativistic energy conservation *before* he published the non-relativistic one, but discarded it as it predicted negative probabilities and negative energies. In 1927, Klein, Gordon and Fock also found it, but incorporated the electromagnetic interaction and proved that it was Lorentz invariant. De Broglie also arrived at the same equation in 1928. This relativistic wave equation is now most commonly known as the Klein–Gordon equation.

In 1927, Pauli phenomenologically found a non-relativistic equation to describe spin-1/2 particles in electromagnetic fields, now called the Pauli equation. Pauli found the wave function was not described by a single complex function of space and time, but needed two complex numbers, which respectively correspond to the spin +1/2 and −1/2 states of the fermion. Soon after in 1928, Dirac found an equation from the first successful unification of special relativity and quantum mechanics applied to the electron, now called the Dirac equation. In this, the wave function is a *spinor* represented by four complex-valued components: two for the electron and two for the electron's antiparticle, the positron. In the non-relativistic limit, the Dirac wave function resembles the Pauli wave function for the electron. Later, other relativistic wave equations were found.

Wave Functions and Wave Equations in Modern Theories

All these wave equations are of enduring importance. The Schrödinger equation and the Pauli equation are under many circumstances excellent approximations of the relativistic variants. They are considerably easier to solve in practical problems than the relativistic counterparts.

The Klein-Gordon equation and the Dirac equation, while being relativistic, do not represent full reconciliation of quantum mechanics and special relativity. The branch of quantum mechanics where these equations are studied the same way as the Schrödinger equation, often called relativistic quantum mechanics, while very successful, has its limitations and conceptual problems.

Relativity makes it inevitable that the number of particles in a system is not constant. For full reconciliation, quantum field theory is needed. In this theory, the wave equations and the wave functions have their place, but in a somewhat different guise. The main objects of interest are not the wave functions, but rather operators, so called *field operators* (or just fields where "operator" is understood) on the Hilbert space of states (to be described next section). It turns out that the original relativistic wave equations and their solutions are still needed to build the Hilbert space. Moreover, the *free fields operators*, i.e. when interactions are assumed not to exist, turn out to (formally) satisfy the same equation as do the fields (wave functions) in many cases.

Thus the Klein-Gordon equation (spin 0) and the Dirac equation (spin $\frac{1}{2}$) in this guise remain in the theory. Higher spin analogues include the Proca equation (spin 1), Rarita–Schwinger equation (spin $\frac{3}{2}$), and, more generally, the Bargmann–Wigner equations. For *massless* free fields two examples are the free field Maxwell equation (spin 1) and the free field Einstein equation (spin 2) for the field operators. All of them are essentially a direct consequence of the requirement of Lorentz invariance. Their solutions must transform under Lorentz transformation in a prescribed way, i.e. under a particular representation of the Lorentz group and that together with few other reasonable demands, e.g. the *cluster decomposition principle*, with implications for causality is enough to fix the equations.

It should be emphasized that this applies to free field equations; interactions are not included. If a Lagrangian density (including interactions) is available, then the Lagrangian formalism will yield an equation of motion at the classical level. This equation may be very complex and not amenable to solution. Any solution would refer to a *fixed* number of particles and would not account for the term "interaction" as referred to in these theories, which involves the creation and annihilation of particles and not external potentials as in ordinary "first quantized" quantum theory.

In string theory, the situation remains analogous. For instance, a wave function in momentum space has the role of Fourier expansion coefficient in a general state of a particle (string) with momentum that is not sharply defined.

Definition (One Spinless Particle in 1d)

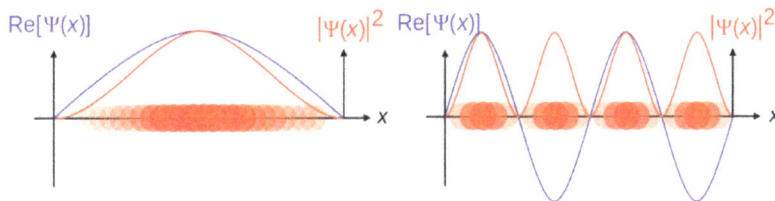

Standing waves for a particle in a box, examples of stationary states.

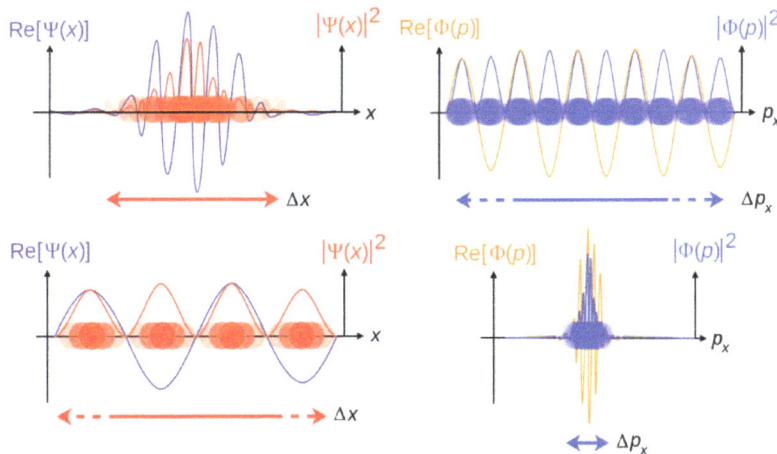

Travelling waves of a free particle.

The real parts of position wave function $\Psi(x)$ and momentum wave function $\Phi(p)$, and corresponding probability densities $|\Psi(x)|^2$ and $|\Phi(p)|^2$, for one spin-0 particle in one x or p dimension. The colour opacity of the particles corresponds to the probability density (*not* the wave function) of finding the particle at position x or momentum p.

For now, consider the simple case of a non-relativistic single particle, without spin, in one spatial dimension. More general cases are discussed below.

Position-space Wave Functions

The state of such a particle is completely described by its wave function,

$$\Psi(x,t),$$

where x is position and t is time. This is a complex-valued function of two real variables x and t.

For one spinless particle in 1d, if the wave function is interpreted as a probability amplitude, the square modulus of the wave function, the positive real number

$$\left|\Psi(x,t)\right|^2 = \Psi(x,t)^*\,\Psi(x,t) = \rho(x,t),$$

is interpreted as the probability density that the particle is at x. The asterisk indicates the complex conjugate. If the particle's position is measured, its location cannot be determined from the wave function, but is described by a probability distribution. The probability that its position x will be in the interval $a \le x \le b$ is the integral of the density over this interval:

$$P_{a \le x \le b}(t) = \int dx\, |\,\Psi(x,t)\,|$$

where t is the time at which the particle was measured. This leads to the normalization condition:

$$\int_{-\infty}^{\infty} dx\, |\,\Psi(x,t)\,|^2 = 1,$$

because if the particle is measured, there is 100% probability that it will be *somewhere*.

For a given system, the set of all possible normalizable wave functions (at any given time) forms an abstract mathematical vector space, meaning that it is possible to add together different wave functions, and multiply wave functions by complex numbers. Technically, because of the normalization condition, wave functions form a projective space rather than an ordinary vector space. This vector space is infinite-dimensional, because there is no finite set of functions which can be added together in various combinations to create every possible function. Also, it is a Hilbert space, because the inner product of two wave functions Ψ_1 and Ψ_2 can be defined as the complex number (at time t)

$$(\Psi_1, \Psi_2) = \int_{-\infty}^{\infty} dx\, \Psi_1^*(x,t)\Psi_2(x,t).$$

More details are given below. Although the inner product of two wave functions is a complex number, the inner product of a wave function Ψ with itself,

$$(\Psi, \Psi) = \|\,\Psi\,\|^2,$$

is *always* a positive real number. The number $\|\Psi\|$ (not $\|\Psi\|^2$) is called the norm of the wave function Ψ, and is not the same as the modulus $|\Psi|$.

If $(\Psi, \Psi) = 1$, then Ψ is normalized. If Ψ is not normalized, then dividing by its norm gives the normalized function $\Psi / ||\Psi||$. Two wave functions Ψ_1 and Ψ_2 are orthogonal if $(\Psi_1, \Psi_2) = 0$. If they are normalized *and* orthogonal, they are orthonormal. Orthogonality (hence also orthonormality) of wave functions is not a necessary condition wave functions must satisfy, but is instructive to consider since this guarantees linear independence of the functions. In a linear combination of orthogonal wave functions Ψ_n we have,

$$\Psi = \sum_n a_n \Psi_n, \quad a_n = \frac{(\Psi_n, \Psi)}{(\Psi_n, \Psi_n)}$$

If the wave functions Ψ_n were nonorthogonal, the coefficients would be less simple to obtain.

In the Copenhagen interpretation, the modulus squared of the inner product (a complex number) gives a real number

$$\left|(\Psi_1, \Psi_2)\right|^2 = P\left(\Psi_2 \rightarrow \Psi_1\right),$$

which, assuming both wave functions are normalized, is interpreted as the probability of the wave function Ψ_2 "collapsing" to the new wave function Ψ_1 upon measurement of an observable, whose eigenvalues are the possible results of the measurement, with Ψ_1 being an eigenvector of the resulting eigenvalue. This is the Born rule, and is one of the fundamental postulates of quantum mechanics.

At a particular instant of time, all values of the wave function $\Psi(x, t)$ are components of a vector. There are uncountably infinitely many of them and integration is used in place of summation. In Bra–ket notation, this vector is written

$$|\Psi(t)\rangle = \int dx \Psi(x,t)\,|x\rangle$$

and is referred to as a "quantum state vector", or simply "quantum state".There are several advantages to understanding wave functions as representing elements of an abstract vector space:

- All the powerful tools of linear algebra can be used to manipulate and understand wave functions. For example:

    - Linear algebra explains how a vector space can be given a basis, and then any vector in the vector space can be expressed in this basis. This explains the relationship between a wave function in position space and a wave function in momentum space, and suggests that there are other possibilities too.

    - Bra–ket notation can be used to manipulate wave functions.

- The idea that quantum states are vectors in an abstract vector space is completely general in all aspects of quantum mechanics and quantum field theory, whereas the idea that quantum states are complex-valued "wave" functions of space is only true in certain situations.

The time parameter is often suppressed, and will be in the following. The x coordinate is a continuous index. The $|x\rangle$ are the basis vectors, which are orthonormal so their inner product is a delta function;

$$\langle x' \mid x \rangle = \delta(x' - x)$$

thus

$$\langle x' \mid \Psi \rangle = \int dx \Psi(x) \langle x' \mid x \rangle = \Psi(x')$$

and

$$\mid \Psi \rangle = \int dx \mid x \rangle \langle x \mid \Psi \rangle = \left(\int dx \mid x \rangle \langle x \mid \right) \mid \Psi \rangle$$

which illuminates the identity operator

$$I = \int dx \mid x \rangle \langle x \mid.$$

Finding the identity operator in a basis allows the abstract state to be expressed explicitly in a basis, and more (the inner product between two state vectors, and other operators for observables, can be expressed in the basis).

Momentum-space Wave Functions

The particle also has a wave function in momentum space:

$$\Phi(p, t)$$

where p is the momentum in one dimension, which can be any value from $-\infty$ to $+\infty$, and t is time.

Analogous to the position case, the inner product of two wave functions $\Phi_1(p, t)$ and $\Phi_2(p, t)$ can be defined as:

$$(\Phi_1, \Phi_2) = \int_{-\infty}^{\infty} dp \Phi_1^*(p,t) \Phi_2(p,t).$$

One particular solution to the time-independent Schrödinger equation is

$$\Psi_p(x) = e^{ipx/\hbar},$$

a plane wave, which can be used in the description of a particle with momentum exactly p, since it is an eigenfunction of the momentum operator. These functions are not normalizable to unity (they aren't square-integrable), so they are not really elements of physical Hilbert space. The set

$$\{\Psi_p(x,t), -\infty \le p \le \infty\}$$

forms what is called the momentum basis. This "basis" is not a basis in the usual mathematical sense. For one thing, since the functions aren't normalizable, they are instead normalized to a delta function,

$$(\Psi_p, \Psi_{p'}) = \delta(p - p').$$

For another thing, though they are linearly independent, there are too many of them (they form an uncountable set) for a basis for physical Hilbert space. They can still be used to express all functions in it using Fourier transforms as described next.

Relations between Position and Momentum Representations

The x and p representations are

$$|\Psi\rangle = I\,|\,\Psi\rangle \ = \int |x\rangle\langle x\,|\,\Psi\rangle dx = \int \Psi(x)\,|\,x\rangle dx, |\,\Psi\rangle = I\,|\,\Psi\rangle \ = \int |p\rangle\langle p\,|\,\Psi\rangle dp = \int \Phi(p)\,|\,p\rangle dp.$$

Now take the projection of the state Ψ onto eigenfunctions of momentum using the last expression in the two equations,

$$\int \Psi(x)\langle p\,|\,x\rangle dx = \int \Phi(p')\langle p\,|\,p'\rangle dp' = \int \Phi(p')\delta(p-p')dp' = \Phi(p).$$

Then utilizing the known expression for suitably normalized eigenstates of momentum in the position representation solutions of the free Schrödinger equation

$$\langle x\,|\,p\rangle = p(x) = \frac{1}{\sqrt{2\pi\hbar}}e^{\frac{i}{\hbar}px} \Rightarrow \langle p\,|\,x\rangle = \frac{1}{\sqrt{2\pi\hbar}}e^{-\frac{i}{\hbar}px},$$

one obtains

$$\Phi(p) = \frac{1}{\sqrt{2\pi\hbar}}\int \Psi(x)e^{-\frac{i}{\hbar}px}dx.$$

Likewise, using eigenfunctions of position,

$$\Psi(x) = \frac{1}{\sqrt{2\pi\hbar}}\int \Phi(p)e^{\frac{i}{\hbar}px}dp.$$

The position-space and momentum-space wave functions are thus found to be Fourier transforms of each other. The two wave functions contain the same information, and either one alone is sufficient to calculate any property of the particle. As representatives of elements of abstract physical Hilbert space, whose elements are the possible states of the system under consideration, they represent the same state vector, hence *identical physical states*, but they are not generally equal when viewed as square-integrable functions.

In practice, the position-space wave function is used much more often than the momentum-space wave function. The potential entering the relevant equation (Schrödinger, Dirac, etc.) determines in which basis the description is easiest. For the harmonic oscillator, x and p enter symmetrically, so there it doesn't matter which description one uses. The same equation (modulo constants) results. From this follows, with a little bit of afterthought, a factoid: The solutions to the wave equation of the harmonic oscillator are eigenfunctions of the Fourier transform in L^2.[nb 2]

Definitions (Other Cases)

Following are the general forms of the wave function for systems in higher dimensions and more particles, as well as including other degrees of freedom than position coordinates or momentum components.

One-particle States in 3d Position Space

The position-space wave function of a single particle without spin in three spatial dimensions is similar to the case of one spatial dimension above:

$$\Psi(\mathbf{r},t)$$

where r is the position vector in three-dimensional space, and t is time. As always $\Psi(\mathbf{r},t)$ is a complex-valued function of real variables. As a single vector in Dirac notation

$$|\Psi(t)\rangle = \int d^3\mathbf{r}\,\Psi(\mathbf{r},t)\,|\mathbf{r}\rangle$$

All the previous remarks on inner products, momentum space wave functions, Fourier transforms, and so on extend to higher dimensions.

For a particle with spin, ignoring the position degrees of freedom, the wave function is a function of spin only (time is a parameter);

$$\xi(s_z,t)$$

where s_z is the spin projection quantum number along the z axis. (The z axis is an arbitrary choice; other axes can be used instead if the wave function is transformed appropriately.) The s_z parameter, unlike r and t, is a *discrete variable*. For example, for a spin-1/2 particle, s_z can only be +1/2 or −1/2, and not any other value. (In general, for spin s, s_z can be s, s − 1, ... , −s + 1, −s). Inserting each quantum number gives a complex valued function of space and time, there are 2s + 1 of them. These can be arranged into a column vector.

$$\xi = \begin{bmatrix} \xi(s,t) \\ \xi(s-1,t) \\ \vdots \\ \xi(-(s-1),t) \\ \xi(-s,t) \end{bmatrix} = \xi(s,t)\begin{bmatrix} 1 \\ 0 \\ \vdots \\ 0 \\ 0 \end{bmatrix} + \xi(s-1,t)\begin{bmatrix} 0 \\ 1 \\ \vdots \\ 0 \\ 0 \end{bmatrix} + \cdots + \xi(-(s-1),t)\begin{bmatrix} 0 \\ 0 \\ \vdots \\ 1 \\ 0 \end{bmatrix} + \xi(-s,t)\begin{bmatrix} 0 \\ 0 \\ \vdots \\ 0 \\ 1 \end{bmatrix}$$

In bra ket notation, these easily arrange into the components of a vector

$$|\xi(t)\rangle = \sum_{s_z=-s}^{s} \xi(s_z,t)\,|s_z\rangle$$

The entire vector ξ is a solution of the Schrödinger equation (with a suitable Hamiltonian), which unfolds to a coupled system of 2s + 1 ordinary differential equations with solutions $\xi(s, t)$, $\xi(s − 1, t)$, ..., $\xi(−s, t)$. The term "spin function" instead of "wave function" is used by some authors. This contrasts the solutions to position space wave functions, the position coordinates being continuous degrees of freedom, because then the Schrödinger equation does take the form of a wave equation.

More generally, for a particle in 3d with any spin, the wave function can be written in "position–spin space" as:

$$\Psi(\mathbf{r}, s_z, t)$$

and these can also be arranged into a column vector

$$\Psi(\mathbf{r}, t) = \begin{bmatrix} \Psi(\mathbf{r}, s, t) \\ \Psi(\mathbf{r}, s-1, t) \\ \vdots \\ \Psi(\mathbf{r}, -(s-1), t) \\ \Psi(\mathbf{r}, -s, t) \end{bmatrix}$$

in which the spin dependence is placed in indexing the entries, and the wave function is a complex vector-valued function of space and time only.

All values of the wave function, not only for discrete but continuous variables also, collect into a single vector

$$|\Psi(t)\rangle = \sum_{s_z} \int d^3\mathbf{r} \Psi(\mathbf{r}, s_z, t) |\mathbf{r}, s_z\rangle$$

For a single particle, the tensor product $\otimes$ of its position state vector $|\psi\rangle$ and spin state vector $|\xi\rangle$ gives the composite position-spin state vector

$$|\psi(t)\rangle \otimes |\xi(t)\rangle = \sum_{s_z} \int d^3\mathbf{r} \psi(\mathbf{r}, t) \xi(s_z, t) |\mathbf{r}\rangle \otimes |s_z\rangle$$

with the identifications

$$|\psi(t)\rangle = |\psi(t)\rangle \otimes |\xi(t)\rangle$$
$$\Psi(\mathbf{r}, s_z, t) = \psi(\mathbf{r}, t) \xi(s_z, t)$$
$$|\mathbf{r}, s_z\rangle = |\mathbf{r}\rangle \otimes |s_z\rangle$$

The tensor product factorization is only possible if the orbital and spin angular momenta of the particle are separable in the Hamiltonian operator underlying the system's dynamics (in other words, the Hamiltonian can be split into the sum of orbital and spin terms). The time dependence can be placed in either factor, and time evolution of each can be studied separately. The factorization is not possible for those interactions where an external field or any space-dependent quantity couples to the spin; examples include a particle in a magnetic field, and spin-orbit coupling.

The preceding discussion is not limited to spin as a discrete variable, the total angular momentum J may also be used. Other discrete degrees of freedom, like isospin, can expressed similarly to the case of spin above.

Many Particle States in 3d Position Space

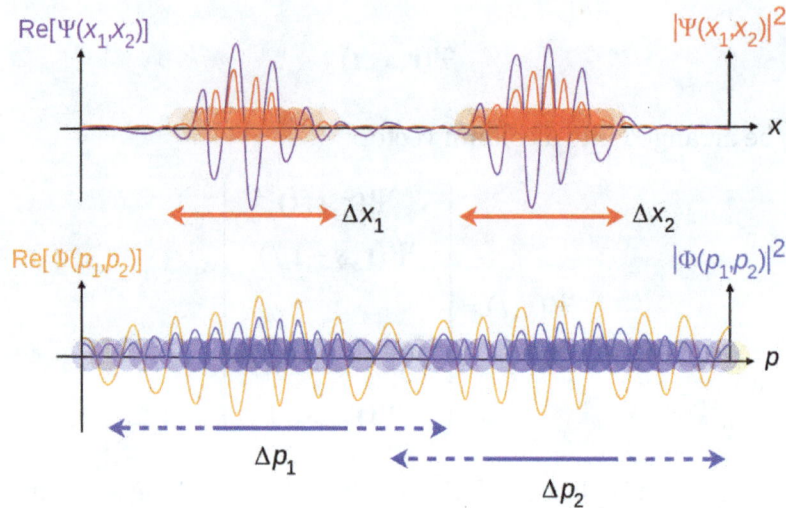

Traveling waves of two free particles, with two of three dimensions suppressed. Top is position space wave function, bottom is momentum space wave function, with corresponding probability densities.

If there are many particles, in general there is only one wave function, not a separate wave function for each particle. The fact that *one* wave function describes *many* particles is what makes quantum entanglement and the EPR paradox possible. The position-space wave function for N particles is written:

$$\Psi(\mathbf{r}_1, \mathbf{r}_2 \cdots \mathbf{r}_N, t)$$

where $\mathbf{r}_i$ is the position of the ith particle in three-dimensional space, and t is time. Altogether, this is a complex-valued function of $3N + 1$ real variables.

In quantum mechanics there is a fundamental distinction between *identical particles* and *distinguishable* particles. For example, any two electrons are identical and fundamentally indistinguishable from each other; the laws of physics make it impossible to "stamp an identification number" on a certain electron to keep track of it. This translates to a requirement on the wave function for a system of identical particles:

$$\Psi\left(\ldots \mathbf{r}_a, \ldots, \mathbf{r}_b, \ldots\right) = \pm \Psi\left(\ldots \mathbf{r}_b, \ldots, \mathbf{r}_a, \ldots\right)$$

where the + sign occurs if the particles are *all bosons* and − sign if they are *all fermions*. In other words, the wave function is either totally symmetric in the positions of bosons, or totally antisymmetric in the positions of fermions. The physical interchange of particles corresponds to mathematically switching arguments in the wave function. The antisymmetry feature of fermionic wave functions leads to the Pauli principle. Generally, bosonic and fermionic symmetry requirements are the manifestation of particle statistics and are present in other quantum state formalisms.

For *N distinguishable* particles (no two being identical, i.e. no two having the same set of quantum numbers), there is no requirement for the wave function to be either symmetric or antisymmetric.

For a collection of particles, some identical with coordinates $\mathbf{r}_1$, $\mathbf{r}_2$, ... and others distinguishable $\mathbf{x}_1$,

x_2, ... (not identical with each other, and not identical to the aforementioned identical particles), the wave function is symmetric or antisymmetric in the identical particle coordinates r_i only:

$$\psi\left(\ldots r_a,\ldots,r_b,\ldots,x_1,x,\ldots\right)=\pm\Psi\left(\ldots r_b,\ldots,r_a,\ldots,x_1,x_2,\ldots\right)$$

Again, there is no symmetry requirement for the distinguishable particle coordinates x_i.

The wave function for N particles each with spin is the complex-valued function

$$\Psi(r_1,r_2\cdots r_N,s_{z1},s_{z2}\cdots s_{zN},t)$$

For identical particles, symmetry requirements apply to both position and spin arguments of the wave function so it has the overall correct symmetry.

The formulae for the inner products are integrals over all coordinates or momenta and sums over all spin quantum numbers. For the general case of N particles with spin in 3d,

$$(\Psi_1,\Psi_2)=\sum_{s_{zN}}\cdots\sum_{s_{z2}}\sum_{s_{z1}}\int_{all\,space}d^3r_1\int_{all\,space}d^3r_2\cdots\int_{all\,space}d^3r_N\Psi_1^*\left(r_1\cdots r_N,s_{z1}\cdots s_{zN},t\right)\Psi_2\left(r_1\cdots r_N,s_{z1}\cdots s_{zN},t\right)$$

this is altogether N three-dimensional volume integrals and N sums over the spins. The differential volume elements d^3r_i are also written "dV_i" or "$dx_i\,dy_i\,dz_i$".

The multidimensional Fourier transforms of the position or position–spin space wave functions yields momentum or momentum–spin space wave functions.

Probability Interpretation

For the general case of N particles with spin in 3d, if Ψ is interpreted as a probability amplitude, the probability density is

$$\rho\left(r_1\cdots r_N,s_{z1}\cdots s_{zN},t\right)=\left|\Psi\left(r_1\cdots r_N,s_{z1}\cdots s_{zN},t\right)\right|^2$$

and the probability that particle 1 is in region R_1 with spin $s_{z1}=m_1$ and particle 2 is in region R_2 with spin $s_{z2}=m_2$ etc. at time t is the integral of the probability density over these regions and evaluated at these spin numbers:

$$P_{r_1\in R_1,s_{z1}=m_1,\ldots,r_N\in R_N,s_{zN}=m_N}(t)=\int_{R_1}d^3r_1\int_{R_2}d^3r_2\cdots\int_{R_N}d^3r_N\left|\Psi\left(r_1\cdots r_N,m_1\cdots m_N,t\right)\right|^2$$

Time Dependence

For systems in time-independent potentials, the wave function can always be written as a function of the degrees of freedom multiplied by a time-dependent phase factor, the form of which is given by the Schrödinger equation. For N particles, considering their positions only and suppressing other degrees of freedom,

$$\Psi(r_1,r_2,\ldots,r_N,t)=e^{-iEt/\hbar}\psi(r_1,r_2,\ldots,r_N),$$

where E is the energy eigenvalue of the system corresponding to the eigenstate Ψ. Wave functions of this form are called stationary states.

The time dependence of the quantum state and the operators can be placed according to unitary transformations on the operators and states. For any quantum state $|\Psi\rangle$ and operator O, in the Schrödinger picture $|\Psi(t)\rangle$ changes with time according to the Schrödinger equation while O is constant. In the Heisenberg picture it is the other way round, $|\Psi\rangle$ is constant while $O(t)$ evolves with time according to the Heisenberg equation of motion. The Dirac (or interaction) picture is intermediate, time dependence is places in both operators and states which evolve according to equations of motion. It is useful primarily in computing S-matrix elements.

Non-relativistic Examples

The following are solutions to the Schrödinger equation for one nonrelativistic spinless particle.

Particle in a Box

A simple model is the particle in a box, a particle is restricted to a 1D region between $x = 0$ and $x = L$ subject to a potential

$$V(x) = \begin{cases} 0 & |x| \leq L \\ \infty & |x| > L \end{cases}$$

has the normalized wave function

$$\Psi(x,t) = \frac{1}{\sqrt{L}} e^{i(kx - \omega t)}, |x| \leq L$$

$$\Psi(x,t) = 0, |x| > L$$

Finite Potential Barrier

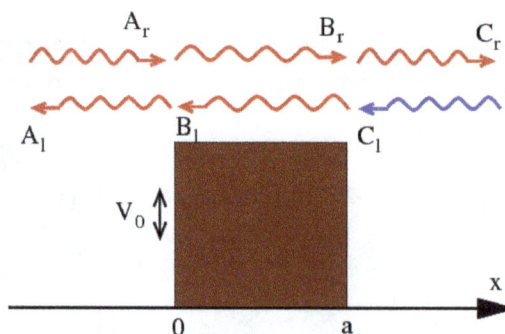

Scattering at a finite potential barrier of height V_o. The amplitudes and direction of left and right moving waves are indicated. In red, those waves used for the derivation of the reflection and transmission amplitude. $E > V_o$ for this illustration.

One of most prominent features of the wave mechanics is a possibility for a particle to reach a location with a prohibitive (in classical mechanics) force potential. A common model is the "potential barrier", the one-dimensional case has the potential

$$V(x) = \begin{cases} V_0 & |x| < a \\ 0 & |x| \geq L \end{cases}$$

and the steady-state solutions to the wave equation have the form (for some constants k, κ)

$$\Psi(x) = \begin{cases} A_r e^{ikx} + A_l e^{-ikx} & x < -a, \\ B_r e^{\kappa x} + B_l e^{-\kappa x} & |x| \leq a, \\ C_r e^{ikx} + C_l e^{-ikx} & x > a. \end{cases}$$

The standard interpretation of this is as a stream of particles being fired at the step from the left (the direction of negative x): setting $A_r = 1$ corresponds to firing particles singly; the terms containing A_r and C_r signify motion to the right, while A_l and C_l – to the left. Under this beam interpretation, put $C_l = 0$ since no particles are coming from the right. By applying the continuity of wave functions and their derivatives at the boundaries, it is hence possible to determine the constants above.

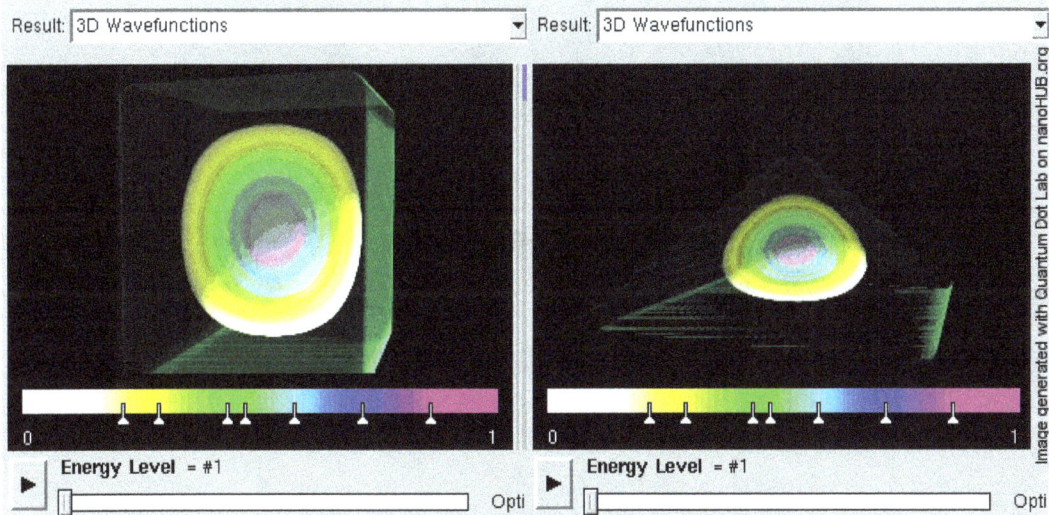

3D confined electron wave functions in a quantum dot. Here, rectangular and triangular-shaped quantum dots are shown. Energy states in rectangular dots are more *s-type* and *p-type*. However, in a triangular dot the wave functions are mixed due to confinement symmetry.

In a semiconductor crystallite whose radius is smaller than the size of its exciton Bohr radius, the excitons are squeezed, leading to quantum confinement. The energy levels can then be modeled using the particle in a box model in which the energy of different states is dependent on the length of the box.

Quantum Harmonic Oscillator

The wave functions for the quantum harmonic oscillator can be expressed in terms of Hermite polynomials H_n, they are

$$\Psi_n(x) = \sqrt{\frac{1}{2 \quad !}} \cdot \left(\frac{m\omega}{\hbar}\right)^{1/4} \cdot e^{\frac{m \quad x}{\quad}} \cdot H_n\left(\sqrt{\frac{m\omega}{\hbar}}x\right)$$

where $n = 0,1,2,....$

Hydrogen Atom

The wave functions of an electron in a Hydrogen atom are expressed in terms of spherical harmonics and generalized Laguerre polynomials.

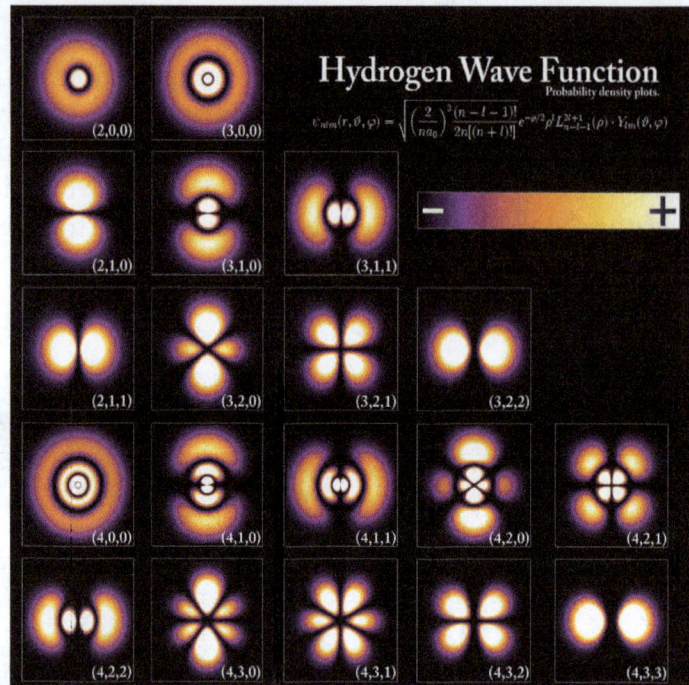

The electron probability density for the first few hydrogen atom electron orbitals shown as cross-sections. These orbitals form an orthonormal basis for the wave function of the electron. Different orbitals are depicted with different scale.

It is convenient to use spherical coordinates, and the wavefunction can be separated into functions of each coordinate,

$$\Psi(r,\theta,\phi) = R(r)Y_\ell^m(\theta,\phi)$$

where R are radial functions and $Ym\ell(\theta, \varphi)$ are spherical harmonics of degree ℓ and order m. This is the only atom for which the Schrödinger equation has been solved for exactly. Multi-electron atoms require approximative methods. The family of solutions are:

$$\Psi_{n\ell m}(r,\theta,\phi) = \sqrt{\left(\frac{2}{na_0}\right)^3 \frac{(n-\ell-1)!}{2n[(n+\ell)!]}}e^{-r/na_0}\left(\frac{2r}{na_0}\right)^\ell L_{n-\ell-1}^{2\ell+1}\left(\frac{2r}{na_0}\right) \cdot Y_\ell^m(\theta,\phi)$$

where $a_0 = 4\pi\varepsilon_0\hbar^2/m_e e^2$ is the Bohr radius, $L2\ell + 1n - \ell - 1$ are the generalized Laguerre polynomials of degree $n - \ell - 1$, $n = 1, 2, \ldots$ is the principal quantum number, $\ell = 1, 2, \ldots n - 1$ the azimuthal quantum number, $m = -\ell, -\ell + 1, \ldots, \ell - 1, \ell$ the magnetic quantum number. Hydrogen-like atoms have very similar solutions.

This solution does not take into account the spin of the electron.

In the figure of the hydrogen orbitals, the 19 sub-images are images of wave functions in position space (their norm squared). The wave functions each represent the abstract state characterized by the triple of quantum numbers (n, l, m), in the lower right of each image. These are the principal quantum number, the orbital angular momentum quantum number and the magnetic quantum number. Together with one spin-projection quantum number of the electron, this is a complete set of observables.

The figure can serve to illustrate some further properties of the function spaces of wave functions.

- In this case, the wave functions are square integrable. One can initially take the function space as the space of square integrable functions, usually denoted L^2.

- The displayed functions are solutions to the Schrödinger equation. Obviously, not every function in L^2 satisfies the Schrödinger equation for the hydrogen atom. The function space is thus a subspace of L^2.

- The displayed functions form part of a basis for the function space. To each triple (n, l, m), there corresponds a basis wave function. If spin is taken into account, there are two basis functions for each triple. The function space thus has a countable basis.

- The basis functions are mutually orthonormal.

Wave Functions and Function Spaces

The concept of function spaces enters naturally in the discussion about wave functions. A function space is a set of functions, usually with some defining requirements on the functions (in the present case that they are square integrable), sometimes with an algebraic structure on the set (in the present case a vector space structure with an inner product), together with a topology on the set. The latter will sparsely be used here, it is only needed to obtain a precise definition of what it means for a subset of a function space to be closed. It will be concluded below that the function space of wave functions is a Hilbert space. This observation is the foundation of the predominant mathematical formulation of quantum mechanics.

Vector Space Structure

A wave function is an element of a function space partly characterized by the following concrete and abstract descriptions.

- The Schrödinger equation is linear. This means that the solutions to it, wave functions, can be added and multiplied by scalars to form a new solution. The set of solutions to the Schrödinger equation is a vector space.

- The superposition principle of quantum mechanics. If Ψ and Φ are two states in the abstract space of states of a quantum mechanical system, and a and b are any two complex numbers, then $a\Psi + b\Phi$ is a valid state as well. (Whether the null vector counts as a valid state ("no system present") is a matter of definition. The null vector does *not* at any rate describe the vacuum state in quantum field theory.) The set of allowable states is a vector space.

This similarity is of course not accidental. There are also a distinctions between the spaces to keep in mind.

Representations

Basic states are characterized by a set of quantum numbers. This is a set of eigenvalues of a maximal set of commuting observables. Physical observables are represented by linear operators, also called observables, on the vectors space. Maximality means that there can be added to the set no further algebraically independent observables that commute with the ones already present. A choice of such a set may be called a choice of representation.

- It is a postulate of quantum mechanics that a physically observable quantity of a system, such as position, momentum, or spin, is represented by a linear Hermitian operator on the state space. The possible outcomes of measurement of the quantity are the eigenvalues of the operator. At a deeper level, most observables, perhaps all, arise as generators of symmetries.

- The physical interpretation is that such a set represents what can – in theory – be simultaneously be measured with arbitrary precision. The Heisenberg uncertainty relation prohibits simultaneous exact measurements of two non-commuting observables.

- The set is non-unique. It may for a one-particle system, for example, be position and spin z-projection, (x, S_z), or it may be momentum and spin y-projection, (p, S_y). In this case, the operator corresponding to position (a multiplication operator in the position representation) and the operator corresponding to momentum (a differential operator in the position the position representation) do not commute.

- Once a representation is chosen, there is still arbitrariness. It remains to choose a coordinate system. This may, for example, correspond to a choice of x, y- and z-axis, or a choice of curvilinear coordinates as exemplified by the spherical coordinates used for the Hydrogen atomic wave functions. This final choice also fixes a basis in abstract Hilbert space. The basic states are labeled by the quantum numbers corresponding to the maximal set of commuting observables and an appropriate coordinate system.

The abstract states are "abstract" only in that an arbitrary choice necessary for a particular *explicit* description of it is not given. This is the same as saying that no choice of maximal set of commuting observables has been given. This is analogous to a vector space without a specified basis. Wave functions corresponding to a state are accordingly not unique. This non-uniqueness reflects the non-uniqueness in the choice of a maximal set of commuting observables. For one spin particle in one dimension, to a particular state there corresponds two wave functions, $\Psi(x, S_z)$ and $\Psi(p, S_y)$, both describing the *same* state.

- For each choice of maximal commuting sets of observables for the abstract state space, there is a corresponding representation that is associated to a function space of wave functions.

- Between all these different function spaces and the abstract state space, there are one-to-one correspondences (here disregarding normalization and unobservable phase factors), the common denominator here being a particular abstract state. The relationship between the momentum and position space wave functions, for instance, describing the same state is the Fourier transform.

Each choice of representation should be thought of as specifying a unique function space in which wave functions corresponding to that choice of representation lives. This distinction is best kept, even if one could argue that two such function spaces are mathematically equal, e.g. being the set of square integrable functions. One can then think of the function spaces as two distinct copies of that set.

Inner Product

There is additional algebraic structure on the vector spaces of wave functions and the abstract state space.

- Physically, different wave functions are interpreted to overlap to some degree. A system in a state Ψ that does *not* overlap with a state Φ cannot be found to be in the state Φ upon measurement. But if Φ_1, Φ_2, ... overlap Ψ to *some* degree, there is a chance that measurement of a system described by Ψ will be found un states Φ_1, Φ_2, ... Also selection rules are observed apply. These are usually formulated in the preservation of some quantum numbers. This means that certain processes allowable from some perspectives (e.g. energy and momentum conservation) do not occur because the initial and final *total* wave functions don't overlap.

- Mathematically, it turns out that solutions to the Schrödinger equation for particular potentials are orthogonal in some manner, this is usually described by an integral

$$\int \Psi_m^* \Psi_n \, wdV = \delta_{nm},$$

where m, n are (sets of) indices (quantum numbers) labeling different solutions, the strictly positive function w is called a weight function, and δ_{mn} is the Kronecker delta. The integration is taken over all of the relevant space.

This motivates the introduction of an inner product on the vector space of abstract quantum states, compatible with the mathematical observations of above when passing to a representation. It is denoted (Ψ, Φ), or in the Bra–ket notation $\langle \Psi | \Phi \rangle$. It yields a complex number. With the inner product, the function space is an inner product space. The explicit appearance of the inner product (usually an integral or a sum of integrals) depends on the choice of representation, but the complex number (Ψ, Φ) does not. Much of the physical interpretation of quantum mechanics stems from the Born rule. It states that the probability p of finding upon measurement the state Φ given the system is in the state Ψ is

$$p = |(\Phi, \Psi)|^2,$$

where Φ and Ψ are assumed normalized. Consider a scattering experiment. In quantum field theory, if Φ_{out} describes a state in the "distant future" (an "out state") after interactions between scattering particles have ceased, and Ψ_{in} an "in state" in the "distant past", then the quantities (Φ_{out}, Ψ_{in}), with Φ_{out} and Ψ_{in} varying over a complete set of in states and out states respectively, is called the S-matrix or scattering matrix. Knowledge of it is, effectively, having *solved* the theory at hand, at least as far as predictions go. Measurable quantities such as decay rates and scattering cross sections are calculable from the S-matric.

Hilbert Space

The above observations encapsulate the essence of the function spaces of which wave functions are elements. However the description is not yet complete. There is a further technical requirement on the function space, that of completeness, that allows one to take limits of sequences in the function space, and be ensured that, if the limit exists, it is an element of the function space. A complete inner product space is called a Hilbert space. The property of completeness is crucial in advanced treatments and applications of quantum mechanics. For instance, the existence of projection operators or orthogonal projections relies on the completeness of the space. These projection operators, in turn, are essential for the statement and proof of many useful theorems, e.g. the spectral theorem. It is not very important in the in introductory quantum mechanics, and technical details and links may be found in footnotes like the one that follows. The space L^2 is a Hilbert space, with inner product presented later. The function space of the example of the figure is a subspace of L^2. A subspace of a Hilbert space is a Hilbert space if it is closed.

In summary, the set of all possible normalizable wave functions for a system with a particular choice of basis, together with the null vector, constitute a Hilbert space.

Not all functions of interest are elements of some Hilbert space, say L^2. The most glaring example is the set of functions $e^{2\pi i p \cdot x/h}$. These are plane wave solutions of the Schrödinger equation for a free particle, but are not normalizable, hence not in L^2. But they are nonetheless fundamental for the description. One can, using them, express functions that *are* normalizable using wave packets. They are, in a sense, a basis (but not a Hilbert space basis, nor a Hamel basis) in which wave functions of interest can be expressed. There is also the artifact "normalization to a delta function" that is frequently employed for notational convenience. The delta functions themselves aren't square integrable either.

The above description of the function space containing the wave functions is mostly mathematically motivated. The function spaces are, due to completeness, very *large* in a certain sense. Not all functions are realistic descriptions of any physical system. For instance, in the function space L^2 one can find the function that takes on the value 0 for all rational numbers and $-i$ for the irrationals in the interval [0, 1]. This *is* square integrable,[nb 8] but can hardly represent a physical state.

Common Hilbert Spaces

While the space of solutions as a whole is a Hilbert space there are many other Hilbert spaces that commonly occur as ingredients.

- Square integrable complex valued functions on the interval $[0, 2\pi]$. The set $\{e^{int}/2\pi, n \in Z\}$ is a Hilbert space basis, i.e. a maximal orthonormal set.

- The Fourier transform takes functions in the above space to elements of $l^2(Z)$, the space of *square summable* functions $Z \rightarrow C$. The latter space is a Hilbert space and the Fourier transform is an isomorphism of Hilbert spaces.[nb 9] Its basis is $\{e_i, i \in Z\}$ with $e_i(j) = \delta_{ij}$, $i, j \in Z$.

- The most basic example of spanning polynomials is in the space of square integrable functions on the interval $[-1, 1]$ for which the Legendre polynomials is a Hilbert space basis (complete orthonormal set).

- The square integrable functions on the unit sphere S^2 is a Hilbert space. The basis functions in this case are the spherical harmonics. The Legendre polynomials are ingredients in the spherical harmonics. Most problems with rotational symmetry will have "the same" (known) solution with respect to that symmetry, so the original problem is reduced to a problem of lower dimensionality.

- The associated Laguerre polynomials appear in the hydrogenic wave function problem after factoring out the spherical harmonics. These span the Hilbert space of square integrable functions on the semi-infinite interval $[0, \infty)$.

More generally, one may consider a unified treatment of all second order polynomial solutions to the Sturm–Liouville equations in the setting of Hilbert space. These include the Legendre and Laguerre polynomials as well as Chebyshev polynomials, Jacobi polynomials and Hermite polynomials. All of these actually appear in physical problems, the latter ones in the harmonic oscillator, and what is otherwise a bewildering maze of properties of special functions becomes an organized body of facts.

There occurs also finite-dimensional Hilbert spaces. The space C^n is a Hilbert space of dimension n. The inner product is the standard inner product on these spaces. In it, the "spin part" of a single particle wave function resides.

- In the non-relativistic description of an electron one has $n = 2$ and the total wave function is a solution of the Pauli equation.

- In the corresponding relativistic treatment, $n = 4$ and the wave function solves the Dirac equation.

With more particles, the situations is more complicated. One has to employ tensor products and use representation theory of the symmetry groups involved (the rotation group and the Lorentz group respectively) to extract from the tensor product the spaces in which the (total) spin wave functions reside. (Further problems arise in the relativistic case unless the particles are free. See the Bethe–Salpeter equation.) Corresponding remarks apply to the concept of isospin, for which the symmetry group is SU(2). The models of the nuclear forces of the sixties (still useful today, see nuclear force) used the symmetry group SU(3). In this case as well, the part of the wave functions corresponding to the inner symmetries reside in some C^n or subspaces of tensor products of such spaces.

- In quantum field theory the underlying Hilbert space is Fock space. It is built from free single-particle states, i.e. wave functions when a representation is chosen, and can accommodate any finite, not necessarily constant in time, number of particles. The interesting (or rather the *tractable*) dynamics lies not in the wave functions but in the field operators that are operators acting on Fock space. Thus the Heisenberg picture is the most common choice (constant states, time varying operators).

Due to the infinite-dimensional nature of the system, the appropriate mathematical tools are objects of study in functional analysis.

Simplified Description

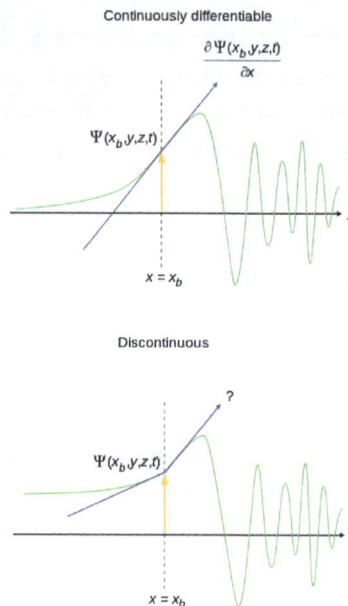

Continuously differentiable

$$\frac{\partial \Psi(x_b, y, z, t)}{\partial x}$$

$\Psi(x_b, y, z, t)$

x

$x = x_b$

Discontinuous

?

$\Psi(x_b, y, z, t)$

x

$x = x_b$

Continuity of the wave function and its first spatial derivative (in the x direction, y and z coordinates not shown), at some time t.

Not all introductory textbooks take the long route and introduce the full Hilbert space machinery, but the focus is on the non-relativistic Schrödinger equation in position representation for certain standard potentials. The following constraints on the wave function are sometimes explicitly formulated for the calculations and physical interpretation to make sense:

- The wave function must be square integrable. This is motivated by the Copenhagen interpretation of the wave function as a probability amplitude.

- It must be everywhere continuous a nd everywhere continuously differentiable. This is motivated by the appearance of the Schrödinger equation for most physically reasonable potentials.

It is possible to relax these conditions somewhat for special purposes. If these requirements are not met, it is not possible to interpret the wave function as a probability amplitude.

This does not alter the structure of the Hilbert space that these particular wave functions inhabit, but it should be pointed out that the subspace of the square-integrable functions L^2, which is a Hilbert space, satisfying the second requirement *is not closed* in L^2, hence not a Hilbert space in itself.[nb 11] The functions that does not meet the requirements are still needed for both technical and practical reasons.

More on Wave Functions and Abstract State Space

As has been demonstrated, the set of all possible wave functions in some representation for a system constitute an in general infinite-dimensional Hilbert space. Due to the multiple possible choices of representation basis, these Hilbert spaces are not unique. One therefore talks about an abstract Hilbert space, state space, where the choice of representation and basis is left undetermined. Specifically, each state is represented as an abstract vector in state space. A quantum state $|\Psi\rangle$ in any representation is generally expressed as a vector

$$|\psi\rangle = \sum_\alpha \int d^m\omega\, \psi(\alpha,\omega,t)\,|\alpha,\omega\rangle$$

where $\alpha = (\alpha_1, \alpha_2, ..., \alpha_n)$ are (dimensionless) discrete quantum numbers, and $\omega = (\omega_1, \omega_2, ..., \omega_m)$ are continuous variables (not necessarily dimensionless). All of them index the components of the vector, and $|\alpha, \omega\rangle$ are the basis vectors in this representation. All α are in an n-dimensional set $A = A_1 \times A_2 \times ... A_n$ where each A_i is the set of allowed values for α_i, likewise all ω are in an m-dimensional "volume" $\Omega \subseteq R^m$ where $\Omega = \Omega_1 \times \Omega_2 \times ... \Omega_m$ and each $\Omega_i \subseteq R$ is the set of allowed values for ω_i, a subset of the real numbers R. For generality n and m are not necessarily equal.

For example, for a single particle in 3d with spin s, neglecting other degrees of freedom, using Cartesian coordinates, we could take $\alpha = (s_z)$ for the spin quantum number of the particle along the z direction, and $\omega = (x, y, z)$ for the particle's position coordinates. Here $A = \{-s, -s + 1, ..., s - 1, s\}$ is the set of allowed spin quantum numbers and $\Omega = R^3$ is the set of all possible particle positions throughout 3d position space. An alternative choice is $\alpha = (s_y)$ for the spin quantum number along the y direction and $\omega = (p_x, p_y, p_z)$ for the particle's momentum components. In this case A and Ω are the same.

Then, a component $\Psi(\alpha, \omega, t)$ of the vector $|\Psi\rangle$ is referred to as the "wave function" of the system.

When interpreted as a probability amplitude (non-relativistic systems with constant number of particles), the probability density of finding the system at α, ω is

$$\rho = |\psi(\alpha,\omega,t)|^2$$

The probability of finding system with α in some or all possible discrete-variable configurations, $D \subseteq A$, and ω in some or all possible continuous-variable configurations, $C \subseteq \Omega$, is the sum and integral over the density,

$$P = \sum_{\alpha \in D} \int_C \rho\, d^m\omega$$

where $d^m\omega = d\omega_1 d\omega_2 ... d\omega_m$ is a "differential volume element" in the continuous degrees of freedom. Since the sum of all probabilities must be 1, the normalization condition

$$1 = \sum_{\alpha \in A\dot{U}} \int \rho \, d^m \omega$$

must hold at all times during the evolution of the system.

The normalisation condition requires $\rho \, d^m\omega$ to be dimensionless, by dimensional analysis Ψ must have the same units as $(\omega_1 \omega_2 ... \omega_m)^{-1/2}$.

Ontology

Whether the wave function really exists, and what it represents, are major questions in the interpretation of quantum mechanics. Many famous physicists of a previous generation puzzled over this problem, such as Schrödinger, Einstein and Bohr. Some advocate formulations or variants of the Copenhagen interpretation (e.g. Bohr, Wigner and von Neumann) while others, such as Wheeler or Jaynes, take the more classical approach and regard the wave function as representing information in the mind of the observer, i.e. a measure of our knowledge of reality. Some, including Schrödinger, Bohm and Everett and others, argued that the wave function must have an objective, physical existence. Einstein thought that a complete description of physical reality should refer directly to physical space and time, as distinct from the wave function, which refers to an abstract mathematical space.

Old Quantum Theory

The old quantum theory is a collection of results from the years 1900–1925 which predate modern quantum mechanics. The theory was never complete or self-consistent, but was a set of heuristic prescriptions which are now understood to be the first quantum corrections to classical mechanics. The Bohr model was the focus of study, and Arnold Sommerfeld made a crucial contribution by quantizing the z-component of the angular momentum, which in the old quantum era was called *space quantization* (Richtungsquantelung). This allowed the orbits of the electron to be ellipses instead of circles, and introduced the concept of quantum degeneracy. The theory would have correctly explained the Zeeman effect, except for the issue of electron spin.

The main tool was Bohr–Sommerfeld quantization, a procedure for selecting out certain discrete set of states of a classical integrable motion as allowed states. These are like the allowed orbits of the Bohr model of the atom; the system can only be in one of these states and not in any states in between.

Basic Principles

The basic idea of the old quantum theory is that the motion in an atomic system is quantized, or discrete. The system obeys classical mechanics except that not every motion is allowed, only those motions which obey the *old quantum condition*:

$$\oint_{H(p,q)=E} p_i \, dq_i = n_i h$$

where the p_i are the momenta of the system and the q_i are the corresponding coordinates. The quantum numbers n_i are *integers* and the integral is taken over one period of the motion at constant energy (as described by the Hamiltonian). The integral is an area in phase space, which is a quantity called the action and is quantized in units of Planck's constant. For this reason, Planck's constant was often called the *quantum of action*.

In order for the old quantum condition to make sense, the classical motion must be separable, meaning that there are separate coordinates in terms of which the motion is periodic. The periods of the different motions do not have to be the same, they can even be incommensurate, but there must be a set of coordinates where the motion decomposes in a multi-periodic way.

The motivation for the old quantum condition was the correspondence principle, complemented by the physical observation that the quantities which are quantized must be adiabatic invariants. Given Planck's quantization rule for the harmonic oscillator, either condition determines the correct classical quantity to quantize in a general system up to an additive constant.

Examples

Thermal Properties of the Harmonic Oscillator

The simplest system in the old quantum theory is the harmonic oscillator, whose Hamiltonian is:

$$H = \frac{p^2}{2m} + \frac{m\omega^2 q^2}{2}.$$

The old quantum theory yields a recipe for the quantization of the energy levels of the harmonic oscillator, which, when combined with the Boltzmann probability distribution of thermodynamics, yields the correct expression for the stored energy and specific heat of a quantum oscillator both at low and at ordinary temperatures. Applied as a model for the specific heat of solids, this resolved a discrepancy in pre-quantum thermodynamics that had troubled 19th-century scientists. Let us now describe this.

The level sets of H are the orbits, and the quantum condition is that the area enclosed by an orbit in phase space is an integer. It follows that the energy is quantized according to the Planck rule:

$$E = n\hbar\omega,$$

a result which was known well before, and used to formulate the old quantum condition. This result differs by $\frac{1}{2}\hbar\omega$ from the results found with the help of quantum mechanics. This constant is neglected in the derivation of the *old quantum theory*, and its value can not be determined using it.

The thermal properties of a quantized oscillator may be found by averaging the energy in each of the discrete states assuming that they are occupied with a Boltzmann weight:

$$U = \frac{\sum_n \hbar\omega n e^{-\beta n\hbar\omega}}{\sum_n e^{-\beta n\hbar\omega}} = \frac{\hbar\omega e^{-\beta\hbar\omega}}{1 - e^{-\beta\hbar\omega}}, \quad \text{where } \beta = \frac{1}{kT},$$

kT is Boltzmann constant times the absolute temperature, which is the temperature as measured in more natural units of energy. The quantity β is more fundamental in thermodynamics than the temperature, because it is the thermodynamic potential associated to the energy.

From this expression, it is easy to see that for large values of β, for very low temperatures, the average energy U in the Harmonic oscillator approaches zero very quickly, exponentially fast. The reason is that kT is the typical energy of random motion at temperature T, and when this is smaller than $\hbar\omega$, there is not enough energy to give the oscillator even one quantum of energy. So the oscillator stays in its ground state, storing next to no energy at all.

This means that at very cold temperatures, the change in energy with respect to beta, or equivalently the change in energy with respect to temperature, is also exponentially small. The change in energy with respect to temperature is the specific heat, so the specific heat is exponentially small at low temperatures, going to zero like

$$\exp(-\hbar\omega / kT)$$

At small values of β, at high temperatures, the average energy U is equal to $1/\beta = kT$. This reproduces the equipartition theorem of classical thermodynamics: every harmonic oscillator at temperature *T* has energy *kT* on average. This means that the specific heat of an oscillator is constant in classical mechanics and equal to *k*. For a collection of atoms connected by springs, a reasonable model of a solid, the total specific heat is equal to the total number of oscillators times *k*. There are overall three oscillators for each atom, corresponding to the three possible directions of independent oscillations in three dimensions. So the specific heat of a classical solid is always 3k per atom, or in chemistry units, 3R per mole of atoms.

Monatomic solids at room temperatures have approximately the same specific heat of 3k per atom, but at low temperatures they don't. The specific heat is smaller at colder temperatures, and it goes to zero at absolute zero. This is true for all material systems, and this observation is called the third law of thermodynamics. Classical mechanics cannot explain the third law, because in classical mechanics the specific heat is independent of the temperature.

This contradiction between classical mechanics and the specific heat of cold materials was noted by James Clerk Maxwell in the 19th century, and remained a deep puzzle for those who advocated an atomic theory of matter. Einstein resolved this problem in 1906 by proposing that atomic motion is quantized. This was the first application of quantum theory to mechanical systems. A short while later, Peter Debye gave a quantitative theory of solid specific heats in terms of quantized oscillators with various frequencies.

One-dimensional Potential: U=0

One-dimensional problems are easy to solve. At any energy E, the value of the momentum p is found from the conservation equation:

$$\sqrt{2m(E - V(q))} = p$$

which is integrated over all values of q between the classical *turning points*, the places where the momentum vanishes. The integral is easiest for a *particle in a box* of length L, where the quantum condition is:

$$2\int_0^L p\,dq = nh$$

which gives the allowed momenta:

$$p = \frac{nh}{2L}$$

and the energy levels

$$E_n = \frac{p^2}{2m} = \frac{n^2 h^2}{8mL^2}$$

One-dimensional Potential: U=Fx

Another easy case to solve with the old quantum theory is a linear potential on the positive halfline, the constant confining force F binding a particle to an impenetrable wall. This case is much more difficult in the full quantum mechanical treatment, and unlike the other examples, the semiclassical answer here is not exact but approximate, becoming more accurate at large quantum numbers.

$$2\int_0^{\frac{E}{F}} \sqrt{2m(E - Fx)}\,dx = nh$$

so that the quantum condition is

$$\frac{4}{3}\sqrt{2m}\,\frac{E^{3/2}}{F} = nh$$

which determines the energy levels,

$$E_n = \left(\frac{3nhF}{4\sqrt{2m}}\right)^{2/3}$$

In the specific case F=mg, the particle is confined by the gravitational potential of the earth and the "wall" here is the surface of the earth.

One-dimensional Potential: U=(1/2)kx^2

This case is also easy to solve, and the semiclassical answer here agrees with the quantum one to within the ground-state energy. Its quantization-condition integral is

$$2\int_{-\sqrt{\frac{2E}{k}}}^{\sqrt{\frac{2E}{k}}} \sqrt{2m\left(E - \frac{1}{2}kx^2\right)}\,dx = nh$$

with solution

$$E = n\frac{h}{2\pi}\sqrt{\frac{k}{m}} = n\hbar\omega$$

for oscillation angular frequency ω, as before.

Rotator

Another simple system is the rotator. A rotator consists of a mass M at the end of a massless rigid rod of length R and in two dimensions has the Lagrangian:

$$L = \frac{MR^2}{2}\dot{\theta}^2$$

which determines that the angular momentum J conjugate to θ, the polar angle, $J = MR^2\dot{\theta}$. The old quantum condition requires that J multiplied by the period of θ is an integer multiple of Planck's constant:

$$2\pi J = nh$$

the angular momentum to be an integer multiple of $\hbar$. In the Bohr model, this restriction imposed on circular orbits was enough to determine the energy levels.

In three dimensions, a rigid rotator can be described by two angles — θ and ϕ, where θ is the inclination relative to an arbitrarily chosen z-axis while ϕ is the rotator angle in the projection to the x–y plane. The kinetic energy is again the only contribution to the Lagrangian:

$$L = \frac{MR^2}{2}\dot{\theta}^2 + \frac{MR^2}{2}(\sin(\theta)\dot{\phi})^2$$

And the conjugate momenta are $p_\theta = \dot{\theta}$ and $p_\phi = \sin(\theta)^2\dot{\phi}$. The equation of motion for ϕ is trivial: p_ϕ is a constant:

$$p_\phi = l_\phi$$

which is the z-component of the angular momentum. The quantum condition demands that the integral of the constant l_ϕ as ϕ varies from 0 to 2π is an integer multiple of h:

$$l_\phi = m\hbar$$

And m is called the magnetic quantum number, because the z component of the angular momentum is the magnetic moment of the rotator along the z direction in the case where the particle at the end of the rotator is charged.

Since the three-dimensional rotator is rotating about an axis, the total angular momentum should be restricted in the same way as the two-dimensional rotator. The two quantum conditions restrict the total angular momentum and the z-component of the angular momentum to be the integers l, m. This condition is reproduced in modern quantum mechanics, but in the era of the old quantum theory it led to a paradox: how can the orientation of the angular momentum relative to the arbitrarily chosen z-axis be quantized? This seems to pick out a direction in space.

This phenomenon, the quantization of angular momentum about an axis, was given the name *space quantization*, because it seemed incompatible with rotational invariance. In modern quantum mechanics, the angular momentum is quantized the same way, but the discrete states of definite angular momentum in any one orientation are quantum superpositions of the states in other orientations, so that the process of quantization does not pick out a preferred axis. For this reason, the name "space quantization" fell out of favor, and the same phenomenon is now called the quantization of angular momentum.

Hydrogen Atom

The angular part of the hydrogen atom is just the rotator, and gives the quantum numbers l and m. The only remaining variable is the radial coordinate, which executes a periodic one-dimensional potential motion, which can be solved.

For a fixed value of the total angular momentum L, the Hamiltonian for a classical Kepler problem is (the unit of mass and unit of energy redefined to absorb two constants):

$$ H = \frac{p^2}{2} + \frac{l^2}{2r^2} - \frac{1}{r}. $$

Fixing the energy to be (a negative) constant and solving for the radial momentum p, the quantum condition integral is:

$$ 2 \oint \sqrt{2E - \frac{l^2}{r^2} + \frac{2}{r}} \, dr = kh $$

which can be solved with the method of residues, and gives a new quantum number k which determines the energy in combination with l. The energy is:

$$ E = -\frac{1}{2(k+l)^2} $$

and it only depends on the sum of k and l, which is the *principal quantum number n*. Since k is positive, the allowed values of l for any given n are no bigger than n. The energies reproduce those in the Bohr model, except with the correct quantum mechanical multiplicities, with some ambiguity at the extreme values.

The semiclassical hydrogen atom is called the Sommerfeld model, and its orbits are ellipses of various sizes at discrete inclinations. The Sommerfeld model predicted that the magnetic moment of an atom measured along an axis will only take on discrete values, a result which seems to contradict rotational invariance but which was confirmed by the Stern–Gerlach experiment. This Bohr–Sommerfeld theory is a significant step in the development of quantum mechanics. It also describes the possibility of atomic energy levels being split by a magnetic field (called the Zeeman effect).

Relativistic Orbit

Arnold Sommerfeld derived the relativistic solution of atomic energy levels. We will start this derivation with the relativistic equation for energy in the electric potential

$$W = m_0 c^2 \left(\frac{1}{\sqrt{1 - \frac{v^2}{c^2}}} - 1 \right) - k\frac{Ze^2}{r}$$

After substitution $u = \frac{1}{r}$ we get

$$\frac{1}{\sqrt{1 - \frac{v^2}{c^2}}} = 1 + \frac{W}{m_0 c^2} + k\frac{Ze^2}{m_0 c^2}u$$

For momentum $p_r = m\dot{r}$, $p_\varphi = mr^2\dot{\varphi}$ and their ratio $\dfrac{p_r}{p_\varphi} = -\dfrac{du}{d\varphi}$ the equation of motion is

$$\frac{d^2u}{d\varphi^2} = -\left(1 - k^2\frac{Z^2 e^4}{c^2 p_\varphi^2}\right)u + \frac{m_0 kZe^2}{p_\varphi^2}\left(1 + \frac{W}{m_0 c^2}\right) = -\omega_0^2 u + K$$

with solution

$$u = \frac{1}{r} = K + A\cos\omega_0\varphi$$

The angular shift of periapsis per revolution is given by

$$\varphi_s = 2\pi\left(\frac{1}{\omega_0} - 1\right) \approx 4\pi^3 k^2 \frac{Z^2 e^4}{c^2 n_\varphi^2 h^2}$$

With the quantum conditions

$$\oint p_\varphi\, d\varphi = 2\pi p_\varphi = n_\varphi h$$

and

$$\oint p_r\, dr = p_\varphi \oint \left(\frac{1}{r}\frac{dr}{d\varphi}\right)^2 d\varphi = n_r h$$

we will obtain energies

$$\frac{W}{m_0 c^2} = \left(1 + \frac{\alpha^2 Z^2}{(n_r + \sqrt{n_\varphi^2 - \alpha^2 Z^2})^2}\right)^{-1/2} - 1$$

where α is the fine-structure constant. This solution (using substitutions for quantum numbers) is equivalent to the solution of the Dirac equation. Nevertheless, both solutions fail to predict the Lamb shifts.

De Broglie Waves

In 1905, Einstein noted that the entropy of the quantized electromagnetic field oscillators in a box is, for short wavelength, equal to the entropy of a gas of point particles in the same box. The number of point particles is equal to the number of quanta. Einstein concluded that the quanta could be treated as if they were localizable objects, particles of light, and named them photons.

Einstein's theoretical argument was based on thermodynamics, on counting the number of states, and so was not completely convincing. Nevertheless, he concluded that light had attributes of both waves and particles, more precisely that an electromagnetic standing wave with frequency ω with the quantized energy:

$$E = n\hbar\omega$$

should be thought of as consisting of n photons each with an energy $\hbar\omega$. Einstein could not describe how the photons were related to the wave.

The photons have momentum as well as energy, and the momentum had to be $\hbar k$ where k is the wavenumber of the electromagnetic wave. This is required by relativity, because the momentum and energy form a four-vector, as do the frequency and wave-number.

In 1924, as a PhD candidate, Louis de Broglie proposed a new interpretation of the quantum condition. He suggested that all matter, electrons as well as photons, are described by waves obeying the relations.

$$p = \hbar k$$

or, expressed in terms of wavelength λ instead,

$$p = \frac{h}{\lambda}$$

He then noted that the quantum condition:

$$\int p \, dx = \hbar \int k \, dx = 2\pi\hbar n$$

counts the change in phase for the wave as it travels along the classical orbit, and requires that it be an integer multiple of 2π. Expressed in wavelengths, the number of wavelengths along a classical orbit must be an integer. This is the condition for constructive interference, and it explained the reason for quantized orbits—the matter waves make standing waves only at discrete frequencies, at discrete energies.

For example, for a particle confined in a box, a standing wave must fit an integer number of wavelengths between twice the distance between the walls. The condition becomes:

$$n\lambda = 2L$$

so that the quantized momenta are:

$$p = \frac{nh}{2L}$$

reproducing the old quantum energy levels.

This development was given a more mathematical form by Einstein, who noted that the phase function for the waves: $\theta(J, x)$ in a mechanical system should be identified with the solution to the Hamilton–Jacobi equation, an equation which even Hamilton considered to be the short-wavelength limit of wave mechanics.

These ideas led to the development of the Schrödinger equation.

Kramers Transition Matrix

The old quantum theory was formulated only for special mechanical systems which could be separated into action angle variables which were periodic. It did not deal with the emission and absorption of radiation. Nevertheless, Hendrik Kramers was able to find heuristics for describing how emission and absorption should be calculated.

Kramers suggested that the orbits of a quantum system should be Fourier analyzed, decomposed into harmonics at multiples of the orbit frequency:

$$X_n(t) = \sum_{k=-\infty}^{\infty} e^{ik\omega t} X_{n;k}$$

The index n describes the quantum numbers of the orbit, it would be n–l–m in the Sommerfeld model. The frequency ω is the angular frequency of the orbit ω while k is an index for the Fourier mode. Bohr had suggested that the k-th harmonic of the classical motion correspond to the transition from level n to level n–k.

Kramers proposed that the transition between states were analogous to classical emission of radiation, which happens at frequencies at multiples of the orbit frequencies. The rate of emission of radiation is proportional to $|X_k|^2$, as it would be in classical mechanics. The description was approximate, since the Fourier components did not have frequencies that exactly match the energy spacings between levels.

This idea led to the development of matrix mechanics.

Limitations of the Old Quantum Theory

The old quantum theory had some limitations:

- The old quantum theory provides no means to calculate the intensities of the spectral lines.

- It fails to explain the anomalous Zeeman effect (that is, where the spin of the electron cannot be neglected).

- It cannot quantize "chaotic" systems, i.e. dynamical systems in which trajectories are neither closed nor periodic and whose analytical form does not exist. This presents a problem for systems as simple as a 2-electron atom which is classically chaotic analogously to the famous gravitational three-body problem.

However it can be used to describe atoms with more than one electron (e.g. Helium) and the Zeeman effect. It was later proposed that the old quantum theory is in fact the semi-classical approximation to the canonical quantum mechanics but its limitations are still under investigation.

History

The old quantum theory was sparked by the 1900 work of Max Planck on the emission and absorption of light, and began in earnest after the work of Albert Einstein on the specific heats of solids. Einstein, followed by Debye, applied quantum principles to the motion of atoms, explaining the specific heat anomaly.

In 1913, Niels Bohr identified the correspondence principle and used it to formulate a model of the hydrogen atom which explained the line spectrum. In the next few years Arnold Sommerfeld extended the quantum rule to arbitrary integrable systems making use of the principle of adiabatic invariance of the quantum numbers introduced by Lorentz and Einstein. Sommerfeld's model was much closer to the modern quantum mechanical picture than Bohr's.

Throughout the 1910s and well into the 1920s, many problems were attacked using the old quantum theory with mixed results. Molecular rotation and vibration spectra were understood and the electron's spin was discovered, leading to the confusion of half-integer quantum numbers. Max Planck introduced the zero point energy and Arnold Sommerfeld semiclassically quantized the relativistic hydrogen atom. Hendrik Kramers explained the Stark effect. Bose and Einstein gave the correct quantum statistics for photons.

Kramers gave a prescription for calculating transition probabilities between quantum states in terms of Fourier components of the motion, ideas which were extended in collaboration with Werner Heisenberg to a semiclassical matrix-like description of atomic transition probabilities. Heisenberg went on to reformulate all of quantum theory in terms of a version of these transition matrices, creating matrix mechanics.

In 1924, Louis de Broglie introduced the wave theory of matter, which was extended to a semiclassical equation for matter waves by Albert Einstein a short time later. In 1926 Erwin Schrödinger found a completely quantum mechanical wave-equation, which reproduced all the successes of the old quantum theory without ambiguities and inconsistencies. Schrödinger's wave mechanics developed separately from matrix mechanics until Schrödinger and others proved that the two methods predicted the same experimental consequences. Paul Dirac later proved in 1926 that both methods can be obtained from a more general method called transformation theory.

Double-slit Experiment

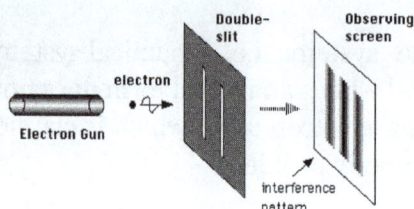

Photons or particles of matter (like an electron) produce a wave pattern when two slits are used

The modern double-slit experiment is a demonstration that light and matter can display characteristics of both classically defined waves and particles; moreover, it displays the fundamentally probabilistic nature of quantum mechanical phenomena. A simpler form of the double-slit experiment was performed originally by Thomas Young in 1801 (well before quantum mechanics). He believed it demonstrated that the wave theory of light was correct and his experiment is sometimes referred to as *Young's experiment* or *Young's slits*. The experiment belongs to a general class of "double path" experiments, in which a wave is split into two separate waves that later combine into a single wave. Changes in the path lengths of both waves result in a phase shift, creating an interference pattern. Another version is the Mach–Zehnder interferometer, which splits the beam with a mirror.

In the basic version of this experiment, a coherent light source, such as a laser beam, illuminates a plate pierced by two parallel slits, and the light passing through the slits is observed on a screen behind the plate. The wave nature of light causes the light waves passing through the two slits to interfere, producing bright and dark bands on the screen—a result that would not be expected if light consisted of classical particles. However, the light is always found to be absorbed at the screen at discrete points, as individual particles (not waves), the interference pattern appearing via the varying density of these particle hits on the screen. Furthermore, versions of the experiment that include detectors at the slits find that each detected photon passes through one slit (as would a classical particle), and not through both slits (as would a wave). However, such experiments demonstrate that particles do *not* form the interference pattern if one detects which slit they pass through. These results demonstrate the principle of wave–particle duality.

Other atomic-scale entities such as electrons are found to exhibit the same behavior when fired towards a double slit. Additionally, the detection of individual discrete impacts is observed to be inherently probabilistic, which is inexplicable using classical mechanics.

The experiment can be done with entities much larger than electrons and photons, although it becomes more difficult as size increases. The largest entities for which the double-slit experiment has been performed were molecules that each comprised 810 atoms (whose total mass was over 10,000 atomic mass units).

Overview

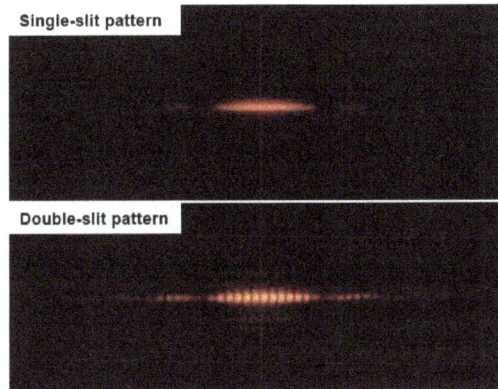

Same double-slit assembly (0.7 mm between slits); in top image, one slit is closed. In the single-slit image, a diffraction pattern (the faint spots on either side of the main band) forms due to the nonzero width of the slit. A diffraction pattern is also seen in the double-slit image, but at twice the intensity and with the addition of many smaller interference fringes.

If light consisted strictly of ordinary or classical particles, and these particles were fired in a straight line through a slit and allowed to strike a screen on the other side, we would expect to see a pattern corresponding to the size and shape of the slit. However, when this "single-slit experiment" is actually performed, the pattern on the screen is a diffraction pattern in which the light is spread out. The smaller the slit, the greater the angle of spread. The top portion of the image shows the central portion of the pattern formed when a red laser illuminates a slit and, if one looks carefully, two faint side bands. More bands can be seen with a more highly refined apparatus. Diffraction explains the pattern as being the result of the interference of light waves from the slit.

Simulation of a particle wave function: double slit experiment. The white blur represents the particle. The whiter the pixel, the greater the probability of finding a particle in that place if measured.

If one illuminates *two* parallel slits, the light from the two slits again interferes. Here the interference is a more pronounced pattern with a series of light and dark bands. The width of the bands is a property of the frequency of the illuminating light. When Thomas Young (1773–1829) first demonstrated this phenomenon, it indicated that light consists of waves, as the distribution of brightness can be explained by the alternately additive and subtractive interference of wavefronts. Young's experiment, performed in the early 1800s, played a vital part in the acceptance of the wave theory of light, vanquishing the corpuscular theory of light proposed by Isaac Newton, which had been the accepted model of light propagation in the 17th and 18th centuries. However, the later discovery of the photoelectric effect demonstrated that under different circumstances, light can behave as if it is composed of discrete particles. These seemingly contradictory discoveries made it necessary to go beyond classical physics and take the quantum nature of light into account.

The double-slit experiment (and its variations) has become a classic thought experiment, for its clarity in expressing the central puzzles of quantum mechanics. Because it demonstrates the fundamental limitation of the ability of the observer to predict experimental results, Richard Feynman called it "a phenomenon which is impossible [...] to explain in any classical way, and which has in it the heart of quantum mechanics. In reality, it contains the *only* mystery [of quantum mechanics].” Feynman was fond of saying that all of quantum mechanics can be gleaned from carefully thinking through the implications of this single experiment. Richard Feynman also proposed (as a thought experiment) that if detectors were placed before each slit, the interference pattern would disappear.

The Englert–Greenberger duality relation provides a detailed treatment of the mathematics of double-slit interference in the context of quantum mechanics.

A low-intensity double-slit experiment was first performed by G. I. Taylor in 1909, by reducing the level of incident light until photon emission/absorption events were mostly nonoverlapping. A double-slit experiment was not performed with anything other than light until 1961, when Claus Jönsson of the University of Tübingen performed it with electron beams. In 1974, the Italian physicists Pier Giorgio Merli, Gian Franco Missiroli, and Giulio Pozzi repeated the experiment using single electrons and biprism (instead of slits), showing that each electron interferes with itself as predicted by quantum theory. In 2002, the single-electron version of the experiment was voted “the most beautiful experiment” by readers of *Physics World*.

Variations of the Experiment

Interference of Individual Particles

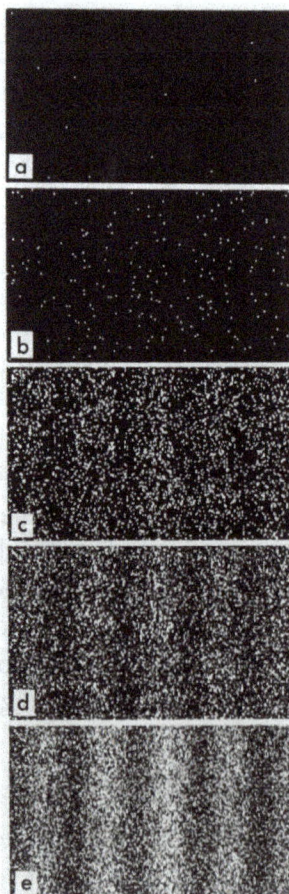

Electron buildup over time

An important version of this experiment involves single particles (or waves—for consistency, they are called particles here). Sending particles through a double-slit apparatus one at a time results in single particles appearing on the screen, as expected. Remarkably, however, an interference pattern emerges when these particles are allowed to build up one by one. This demonstrates the wave-particle duality, which states that all matter exhibits both wave and particle properties: the particle is measured as a single pulse at a single position, while the wave describes the probability of absorbing the particle at a specific place of the detector. This phenomenon has been shown to occur with photons, electrons, atoms and even some molecules, including buckyballs. So experiments with electrons add confirmatory evidence to the view that electrons, protons, neutrons, and even larger entities that are ordinarily called particles nevertheless have their own wave nature and even their own specific frequencies.

The probability of detection is the square of the amplitude of the wave and can be calculated with classical waves. The particles do not arrive at the screen in a predictable order, so knowing where all the previous particles appeared on the screen and in what order tells nothing about where a future particle will be detected. If there is a cancellation of waves at some point, that does not mean that a particle disappears; it will appear somewhere else. Ever since the origination of quantum mechanics, some theorists have searched for ways to incorporate additional determinants or "hidden variables" that, were they to become known, would account for the location of each individual impact with the target.

More complicated systems that involve two or more particles in superposition are not amenable to the above explanation.

"Which-way" Experiments and the Principle of Complementarity

A well-known thought experiment predicts that if particle detectors are positioned at the slits, showing through which slit a photon goes, the interference pattern will disappear. This *which-way* experiment illustrates the complementarity principle that photons can behave as either particles or waves, but cannot be observed as both at the same time. Despite the importance of this *gedanken* in the history of quantum mechanics, technically feasible realizations of this experiment were not proposed until the 1970s. (Naive implementations of the textbook *gedanken* are not possible because photons cannot be detected without absorbing the photon.) Currently, multiple experiments have been performed illustrating various aspects of complementarity.

An experiment performed in 1987 produced results that demonstrated that information could be obtained regarding which path a particle had taken without destroying the interference altogether. This showed the effect of measurements that disturbed the particles in transit to a lesser degree and thereby influenced the interference pattern only to a comparable extent. In other words, if one does not insist that the method used to determine which slit each photon passes through be completely reliable, one can still detect a (degraded) interference pattern.

Delayed Choice and Quantum Eraser Variations

Wheeler's delayed choice experiments demonstrate that extracting "which path" information *after* a particle passes through the slits can seem to retroactively alter its previous behavior at the slits.

Quantum eraser experiments demonstrate that wave behavior can be restored by erasing or otherwise making permanently unavailable the "which path" information.

A simple do-it-at-home demonstration of the quantum eraser phenomenon was given in an article in *Scientific American*. If one sets polarizers before each slit with their axes orthogonal to each other, the interference pattern will be eliminated. The polarizers can be considered as introducing which-path information to each beam. Introducing a third polarizer in front of the detector with an axis of 45° relative to the other polarizers "erases" this information, allowing the interference pattern to reappear. This can also be accounted for by considering the light to be a classical wave, and also when using circular polarizers and single photons. Implementations of the polarizers using entangled photon pairs have no classical explanation.

Weak Measurement

In a highly publicized experiment in 2012, researchers claimed to have identified the path each particle had taken without any adverse effects at all on the interference pattern generated by the particles. In order to do this, they used a setup such that particles coming to the screen were not from a point-like source, but from a source with two intensity maxima. However, commentators such as Motl and Svensson have pointed out that there is in fact no conflict between the weak measurements performed in this variant of the double-slit experiment and the Heisenberg uncertainty principle. Weak measurement followed by post-selection did not allow simultaneous position and momentum measurements for each individual particle, but rather allowed measurement of the average trajectory of the particles that arrived at different positions. In other words, the experimenters were creating a statistical map of the full trajectory landscape.

Other Variations

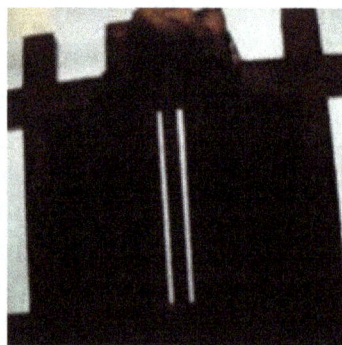

A laboratory double-slit assembly; distance between top posts approximately 2.5 cm (one inch).

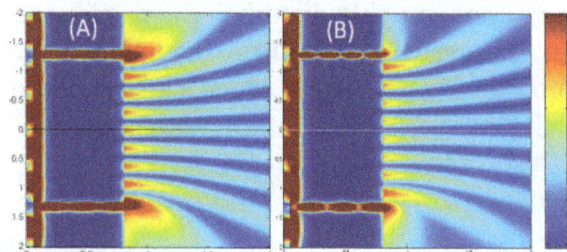

Near-field intensity distribution patterns for plasmonic slits with equal widths (A) and non-equal widths (B).

In 1967, Pfleegor and Mandel demonstrated two-source interference using two separate lasers as light sources.

It was shown experimentally in 1972 that in a double-slit system where only one slit was open at any time, interference was nonetheless observed provided the path difference was such that the detected photon could have come from either slit. The experimental conditions were such that the photon density in the system was much less than unity.

In 1999, the double-slit experiment was successfully performed with buckyball molecules (each of which comprises 60 carbon atoms). A buckyball is large enough (diameter about 0.7 nm, nearly half a million times larger than a proton) to be seen under an electron microscope.

In 2005, E. R. Eliel presented an experimental and theoretical study of the optical transmission of a thin metal screen perforated by two subwavelength slits, separated by many optical wavelengths. The total intensity of the far-field double-slit pattern is shown to be reduced or enhanced as a function of the wavelength of the incident light beam.

In 2012, researchers at the University of Nebraska–Lincoln performed the double-slit experiment with electrons as described by Richard Feynman, using new instruments that allowed control of the transmission of the two slits and the monitoring of single-electron detection events. Electrons were fired by an electron gun and passed through one or two slits of 62 nm wide × 4 μm tall.

In 2013, the double-slit experiment was successfully performed with molecules that each comprised 810 atoms (whose total mass was over 10,000 atomic mass units).

Hydrodynamic Pilot Wave Analogs

Hydrodynamic analogs have been developed that can recreate various aspects of quantum mechanical systems, including single-particle interference through a double-slit. A silicone oil droplet, bouncing along the surface of a liquid, self-propels via resonant interactions with its own wave field. The droplet gently sloshes the liquid with every bounce. At the same time, ripples from past bounces affect its course. The droplet's interaction with its own ripples, which form what is known as a pilot wave, causes it to exhibit behaviors previously thought to be peculiar to elementary particles — including behaviors customarily taken as evidence that elementary particles are spread through space like waves, without any specific location, until they are measured.

Behaviors mimicked via this hydrodynamic pilot-wave system include quantum single particle diffraction, tunneling, quantized orbits, orbital level splitting, spin, and multimodal statistics. It is also possible to infer uncertainty relations and exclusion principles. Videos are available illustrating various features of this system.

However, more complicated systems that involve two or more particles in superposition are not amenable to such a simple, classically intuitive explanation. Accordingly, no hydrodynamic analog of entanglement has been developed. Nevertheless, optical analogs are possible.

Classical Wave-Optics Formulation

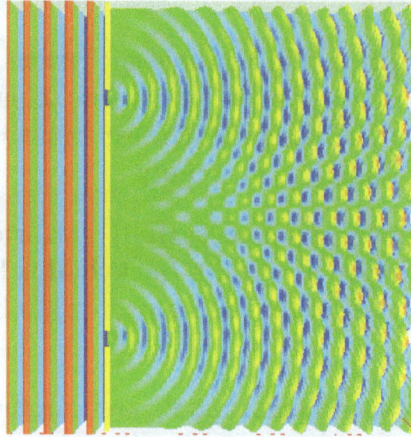

Two-slit diffraction pattern by a plane wave

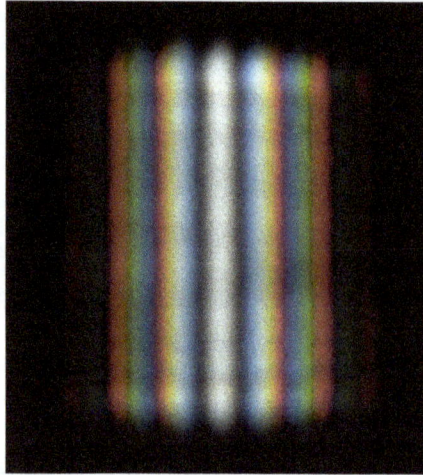

Photo of the double-slit interference of the sunlight.

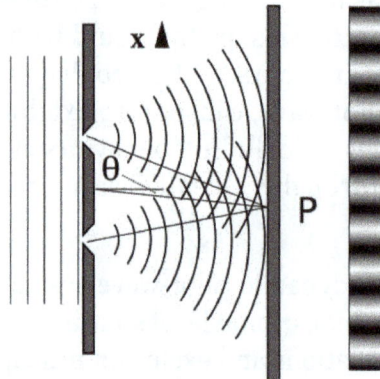

Two slits are illuminated by a plane wave.

Much of the behaviour of light can be modelled using classical wave theory. The Huygens–Fresnel principle is one such model; it states that each point on a wavefront generates a secondary wavelet, and that the disturbance at any subsequent point can be found by summing the contributions of the individual wavelets at that point. This summation needs to take into account the phase as well as the amplitude of the individual wavelets. It should be noted that only the intensity of a light field can be measured—this is proportional to the square of the amplitude.

In the double-slit experiment, the two slits are illuminated by a single laser beam. If the width of the slits is small enough (less than the wavelength of the laser light), the slits diffract the light into cylindrical waves. These two cylindrical wavefronts are superimposed, and the amplitude, and therefore the intensity, at any point in the combined wavefronts depends on both the magnitude and the phase of the two wavefronts. The difference in phase between the two waves is determined by the difference in the distance travelled by the two waves.

If the viewing distance is large compared with the separation of the slits (the far field), the phase difference can be found using the geometry shown in the figure below right. The path difference between two waves travelling at an angle θ is given by:

$$d \sin \theta \approx d\theta$$

Where d is the distance between the two slits. When the two waves are in phase, i.e. the path difference is equal to an integral number of wavelengths, the summed amplitude, and therefore the summed intensity is maximum, and when they are in anti-phase, i.e. the path difference is equal to half a wavelength, one and a half wavelengths, etc., then the two waves cancel and the summed intensity is zero. This effect is known as interference. The interference fringe maxima occur at angles

$$d\theta_n = n\lambda, n = 0, 1, 2, \ldots$$

where λ is the wavelength of the light. The angular spacing of the fringes, θ_f, is given by

$$\theta_f \approx \lambda / d$$

The spacing of the fringes at a distance z from the slits is given by

$$w = z\theta_f = z\lambda / d$$

For example, if two slits are separated by 0.5 mm (d), and are illuminated with a 0.6µm wavelength laser (λ), then at a distance of 1m (z), the spacing of the fringes will be 1.2 mm.

If the width of the slits b is greater than the wavelength, the Fraunhofer diffraction equation gives the intensity of the diffracted light as:

$$I(\theta) \quad \infty \cos^2 \left[\frac{\pi d \sin \theta}{\lambda} \right] \operatorname{sinc}^2 \left[\frac{\pi b \sin \theta}{\lambda} \right]$$

Where the sinc function is defined as $\operatorname{sinc}(x) = \sin(x)/(x)$ for $x \neq 0$, and $\operatorname{sinc}(0) = 1$.

This is illustrated in the figure above, where the first pattern is the diffraction pattern of a single slit, given by the sinc function in this equation, and the second figure shows the combined intensity of the light diffracted from the two slits, where the cos function represent the fine structure, and the coarser structure represents diffraction by the individual slits as described by the sinc function.

Similar calculations for the near field can be done using the Fresnel diffraction equation. As the plane of observation gets closer to the plane in which the slits are located, the diffraction patterns associated with each slit decrease in size, so that the area in which interference occurs is reduced, and may vanish altogether when there is no overlap in the two diffracted patterns.

Interpretations of the Experiment

Like the Schrödinger's cat thought experiment, the double-slit experiment is often used to highlight the differences and similarities between the various interpretations of quantum mechanics.

Copenhagen Interpretation

The Copenhagen interpretation, put forth by some of the pioneers in the field of quantum mechanics, asserts that it is undesirable to posit anything that goes beyond the mathematical formulae and the kinds of physical apparatus and reactions that enable us to gain some knowledge of what goes on at the atomic scale. One of the mathematical constructs that enables experimenters to predict very accurately certain experimental results is sometimes called a probability wave. In its mathematical form it is analogous to the description of a physical wave, but its "crests" and "troughs" indicate levels of probability for the occurrence of certain phenomena (e.g., a spark of light at a certain point on a detector screen) that can be observed in the macro world of ordinary human experience.

The probability "wave" can be said to "pass through space" because the probability values that one can compute from its mathematical representation are dependent on time. One cannot speak of the location of any particle such as a photon between the time it is emitted and the time it is detected simply because in order to say that something is located somewhere at a certain time one has to detect it. The requirement for the eventual appearance of an interference pattern is that particles be emitted, and that there be a screen with at least two distinct paths for the particle to take from the emitter to the detection screen. Experiments observe nothing whatsoever between the time of emission of the particle and its arrival at the detection screen. If a ray tracing is next made as if a light wave (as understood in classical physics) is wide enough to take both paths, then that ray tracing will accurately predict the appearance of maxima and minima on the detector screen when many particles pass through the apparatus and gradually "paint" the expected interference pattern.

Path-integral Formulation

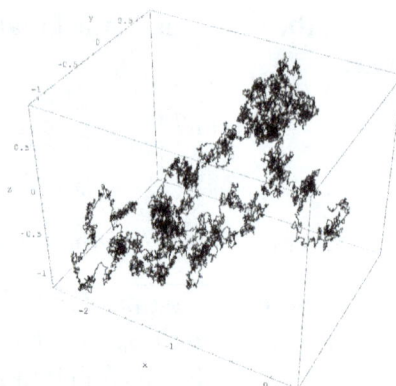

One of an infinite number of equally likely paths used in the Feynman path integral.

The Copenhagen interpretation is similar to the path integral formulation of quantum mechanics provided by Feynman. The path integral formulation replaces the classical notion of a single, unique trajectory for a system, with a sum over all possible trajectories. The trajectories are added together by using functional integration.

Each path is considered equally likely, and thus contributes the same amount. However, the phase of this contribution at any given point along the path is determined by the action along the path:

$$A_{\text{path}}(x, y, z, t) = e^{iS(x,y,z,t)}$$

All these contributions are then added together, and the magnitude of the final result is squared, to get the probability distribution for the position of a particle:

$$p(x, y, z, t) \propto \left| \int_{\text{all paths}} e^{iS(x,y,z,t)} \right|^2$$

As is always the case when calculating probability, the results must then be normalized by imposing:

$$\iiint_{\text{all space}} p(x, y, z, t) \mathrm{d}V = 1$$

To summarize, the probability distribution of the outcome is the normalized square of the norm of the superposition, over all paths from the point of origin to the final point, of waves propagating proportionally to the action along each path. The differences in the cumulative action along the different paths (and thus the relative phases of the contributions) produces the interference pattern observed by the double-slit experiment. Feynman stressed that his formulation is merely a mathematical description, not an attempt to describe a real process that we can measure.

Relational Interpretation

According to the relational interpretation of quantum mechanics, first proposed by Carlo Rovelli, observations such as those in the double-slit experiment result specifically from the interaction between the observer (measuring device) and the object being observed (physically interacted with), not any absolute property possessed by the object. In the case of an electron, if it is initially "observed" at a particular slit, then the observer–particle (photon–electron) interaction includes information about the electron's position. This partially constrains the particle's eventual location at the screen. If it is "observed" (measured with a photon) not at a particular slit but rather at the screen, then there is no "which path" information as part of the interaction, so the electron's "observed" position on the screen is determined strictly by its probability function. This makes the resulting pattern on the screen the same as if each individual electron had passed through both slits. It has also been suggested that space and distance themselves are relational, and that an electron can appear to be in "two places at once"—for example, at both slits—because its spatial relations to particular points on the screen remain identical from both slit locations.

Many-worlds Interpretation

Physicist David Deutsch argues in his book *The Fabric of Reality* that the double-slit experiment is evidence for the many-worlds interpretation.

Canonical Quantization

In physics, canonical quantization is a procedure for quantizing a classical theory, while attempting to preserve the formal structure, such as symmetries, of the classical theory, to the greatest extent possible.

Historically, this was not quite Werner Heisenberg's route to obtaining quantum mechanics, but Paul Dirac introduced it in his 1926 doctoral thesis, the "method of classical analogy" for quantization, and detailed it in his classic text. The word *canonical* arises from the Hamiltonian approach to classical mechanics, in which a system's dynamics is generated via canonical Poisson brackets, a structure which is *only partially preserved* in canonical quantization.

This method was further used in the context of quantum field theory by Paul Dirac, in his construction of quantum electrodynamics. In the field theory context, it is also called second quantization, in contrast to the semi-classical first quantization for single particles.

History

When it was first developed, quantum physics dealt only with the quantization of the motion of particles, leaving the electromagnetic field classical, hence the name quantum mechanics.

Later the electromagnetic field was also quantized, and even the particles themselves became represented through quantized fields, resulting in the development of quantum electrodynamics (QED) and quantum field theory in general. Thus, by convention, the original form of particle quantum mechanics is denoted first quantization, while quantum field theory is formulated in the language of second quantization.

First Quantization

Single Particle Systems

The following exposition is based on Dirac's treatise on quantum mechanics. In the classical mechanics of a particle, there are dynamic variables which are called coordinates (x) and momenta (p). These specify the *state* of a classical system. The canonical structure (also known as the symplectic structure) of classical mechanics consists of Poisson brackets between these variables, such as $\{x,p\} = 1$. All transformations of variables which preserve these brackets are allowed as canonical transformations in classical mechanics. Motion itself is such a canonical transformation.

By contrast, in quantum mechanics, all significant features of a particle are contained in a state $|\psi\rangle$, called a quantum state. Observables are represented by operators acting on a Hilbert space of such quantum states.

The (eigen)value of an operator acting on one of its eigenstates represents the value of a measurement on the particle thus represented. For example, the energy is read off by the Hamiltonian operator $\hat{H}$ acting on a state $|\psi_n\rangle$, yielding

$$\hat{H}|\psi_n\rangle = E_n|\psi_n\rangle,$$

where E_n is the characteristic energy associated to this $|\psi_n\rangle$ eigenstate.

Any state could be represented as a linear combination of eigenstates of energy; for example,

$$|\psi\rangle = \sum_{n=0}^{\infty} a_n|\psi_n\rangle,$$

where a_n are constant coefficients.

As in classical mechanics, all dynamical operators can be represented by functions of the position and momentum ones, $\hat{X}$ and $\hat{P}$, respectively. The connection between this representation and the more usual wavefunction representation is given by the eigenstate of the position operator $\hat{X}$ representing a particle at position x, which is denoted by an element $|x\rangle$ in the Hilbert space, and which satisfies $\hat{X}|x\rangle = x|x\rangle$. Then, $\psi(x) = \langle x|\psi\rangle$.

Likewise, the eigenstates $|p\rangle$ of the momentum operator $\hat{P}$ specify the momentum representation: $\psi(p) = \langle p|\psi\rangle$..

The central relation between these operators is a quantum analog of the above Poisson bracket of classical mechanics, the canonical commutation relation,

$$[\hat{X}, \hat{P}] = \hat{X}\hat{P} - \hat{P}\hat{X} = i\hbar.$$

This relation encodes (and formally leads to) the uncertainty principle, in the form $\Delta x\,\Delta p \geq \hbar/2$. This algebraic structure may be thus considered as the quantum analog of the *canonical structure* of classical mechanics.

Many-Particle Systems

When turning to N-particle systems, i.e., systems containing N identical particles (particles characterized by the same quantum numbers such as mass, charge and spin), it is necessary to extend the single-particle state function $\psi(\mathbf{r})$ to the N-particle state function $\psi(\mathbf{r}_1, \mathbf{r}_2, ..., \mathbf{r}_N)$.. A fundamental difference between classical and quantum mechanics concerns the concept of indistinguishability of identical particles. Only two species of particles are thus possible in quantum physics, the so-called bosons and fermions which obey the rules:

$$\psi(\mathbf{r}_1, ..., \mathbf{r}_j, ..., \mathbf{r}_k, ..., \mathbf{r}_N) = +\psi(\mathbf{r}_1, ..., \mathbf{r}_k, ..., \mathbf{r}_j, ..., \mathbf{r}_N) \text{ (bosons)},$$

$$\psi(\mathbf{r}_1, ..., \mathbf{r}_j, ..., \mathbf{r}_k, ..., \mathbf{r}_N) = -\psi(\mathbf{r}_1, ..., \mathbf{r}_k, ..., \mathbf{r}_j, ..., \mathbf{r}_N) \text{ (fermions)}.$$

Where we have interchanged two coordinates $(\mathbf{r}_j, \mathbf{r}_k)$ of the state function. The usual wave function is obtained using the Slater determinant and the identical particles theory. Using this basis, it is possible to solve various many-particle problems.

Issues and Limitations

Dirac's book details his popular rule of supplanting Poisson brackets by commutators:

$$\{A, B\} \mapsto \tfrac{1}{i\hbar}[\hat{A}, \hat{B}] .$$

This rule is not as simple or well-defined as it appears. It is ambiguous when products of classical observables are involved which correspond to noncommuting products of the analog operators, and fails in polynomials of sufficiently high order.

For example, the reader is encouraged to check the following pair of equalities introduced by Groenewold, assuming only the commutation relation $[x\hat{\,}, p\hat{\,}] = i\hbar$:

$$\{x^3, p^3\} + \tfrac{1}{12}\{\{p^2, x^3\}, \{x^2, p^3\}\} = 0$$

$$\tfrac{1}{i\hbar}[\hat{x}^3, \hat{p}^3] + \tfrac{1}{12i\hbar}\left[\tfrac{1}{i\hbar}[\hat{p}^2, \hat{x}^3], \tfrac{1}{i\hbar}[\hat{x}^2, \hat{p}^3]\right] = -3\hbar^2 .$$

The right-hand-side "anomaly" term $-3\hbar^2$ is not predicted by application of the above naive quantization rule. In order to make this procedure more rigorous, one might hope to take an axiomatic approach to the problem.

If Q represents the quantization map that acts on functions f in classical phase space, then the following properties are usually considered desirable:

1. $Q_x\psi = x\psi$ and $Q_p\psi = -i\hbar\partial_x\psi$ (elementary position/momentum operators)

2. $f \mapsto Q_f$ is a linear map

3. $[Q_f, Q_g] = i\hbar Q_{\{f,g\}}$ (Poisson bracket)

4. $Q_{g\circ f} = g(Q_f)$ (von Neumann rule).

However, not only are these four properties mutually inconsistent, *any three* of them are also inconsistent! As it turns out, the only pairs of these properties that lead to self-consistent, nontrivial solutions are 2 & 3, and possibly 1 & 3 or 1 & 4. Accepting properties 1 & 2, along with a weaker condition that 3 be true only asymptotically in the limit $\hbar \to 0$, leads to deformation quantization, and some extraneous information must be provided, as in the standard theories utilized in most of physics. Accepting properties 1 & 2 & 3 but restricting the space of quantizable observables to exclude terms such as the cubic ones in the above example amounts to geometric quantization.

Second Quantization: Field Theory

Quantum mechanics was successful at describing non-relativistic systems with fixed numbers of particles, but a new framework was needed to describe systems in which particles can be created or destroyed, for example, the electromagnetic field, considered as a collection of photons. It was eventually realized that special relativity was inconsistent with single-particle quantum mechanics, so that all particles are now described relativistically by quantum fields.

When the canonical quantization procedure is applied to a field, such as the electromagnetic field, the classical field variables become *quantum operators*. Thus, the normal modes comprising the amplitude of the field become quantized, and the quanta are identified with individual particles or excitations. For example, the quanta of the electromagnetic field are identified with photons. Unlike first quantization, conventional second quantization is completely unambiguous, in effect a functor.

Historically, quantizing the classical theory of a single particle gave rise to a wavefunction. The classical equations of motion of a field are typically identical in form to the (quantum) equations for the wave-function of *one of its quanta*. For example, the Klein–Gordon equation is the classical equation of motion for a free scalar field, but also the quantum equation for a scalar particle wave-function. This meant that quantizing a field *appeared* to be similar to quantizing a theory that was already quantized, leading to the fanciful term second quantization in the early literature, which is still used to describe field quantization, even though the modern interpretation detailed is different.

One drawback to canonical quantization for a relativistic field is that by relying on the Hamiltonian to determine time dependence, relativistic invariance is no longer manifest. Thus it is necessary to check that relativistic invariance is not lost. Alternatively, the Feynman integral approach is available for quantizing relativistic fields, and is manifestly invariant. For non-relativistic field theories, such as those used in condensed matter physics, Lorentz invariance is not an issue.

Field Operators

Quantum mechanically, the variables of a field (such as the field's amplitude at a given point) are represented by operators on a Hilbert space. In general, all observables are constructed as operators on the Hilbert space, and the time-evolution of the operators is governed by the Hamiltonian, which must be a positive operator. A state annihilated by the Hamiltonian must be identified as the vacuum state, which is the basis for building all other states. In a non-interacting (free) field theory, the vacuum is normally identified as a state containing zero particles. In a theory with interacting particles, identifying the vacuum is more subtle, due to vacuum polarization, which implies that the physical vacuum in quantum field theory is never really empty. For further elaboration, see the articles on the quantum mechanical vacuum and the vacuum of quantum chromodynamics.

Real Scalar Field

A scalar field theory provides a good example of the canonical quantization procedure. Classically, a scalar field is a collection of an infinity of oscillator normal modes. It suffices to consider a 1+1-dimensional space-time $R \times S_1$, in which the spatial direction is compactified to a circle of circumference 2π, rendering the momenta discrete. The classical Lagrangian density is then

$$\mathcal{L}(\phi) = \frac{1}{2}(\partial_t\phi)^2 - \frac{1}{2}(\partial_x\phi)^2 - \frac{1}{2}m^2\phi^2 - V(\phi),$$

where φ is classical field, $V(\varphi)$ is a potential term, often taken to be a polynomial or monomial of degree 3 or higher. The action functional is

$$S(\phi) = \int \mathcal{L}(\phi)dxdt = \int L(\phi, \partial_t\phi)dt.$$

The canonical momentum obtained via the Legendre transform using the action L is $\pi = \partial_t\phi,$, and the classical Hamiltonian is found to be

$$H(\phi, \pi) = \int dx \left[\frac{1}{2}\pi^2 + \frac{1}{2}(\partial_x\phi)^2 + \frac{1}{2}m^2\phi^2 + V(\phi)\right].$$

Canonical quantization treats the variables $\phi(x)$ and $\pi(x)$ as operators with canonical commutation relations at time $t = 0$, given by

$$[\phi(x), \phi(y)] = 0, \quad [\pi(x), \pi(y)] = 0, \quad [\phi(x), \pi(y)] = i\hbar\delta(x - y).$$

Operators constructed from ϕ and π can then formally be defined at other times via the time-evolution generated by the Hamiltonian:

$$\mathcal{O}(t) = e^{itH}\mathcal{O}e^{-itH}.$$

However, since φ and π do not commute, this expression is ambiguous at the quantum level. The problem is to construct a representation of the relevant operators $\mathcal{O}$ on a Hilbert space $\mathcal{H}$ and to construct a positive operator H as a quantum operator on this Hilbert space in such a way that it gives this evolution for the operators $\mathcal{O}$ as given by the preceding equation, and to show that $\mathcal{H}$ contains a vacuum state $|0\rangle$ on which H has zero eigenvalue. In practice, this construction is a difficult problem for interacting field theories, and has been solved completely only in a few simple cases via the methods of constructive quantum field theory. Many of these issues can be sidestepped using the Feynman integral as described for a particular $V(\varphi)$ in the article on scalar field theory.

In the case of a free field, with $V(\varphi) = 0$, the quantization procedure is relatively straightforward. It is convenient to Fourier transform the fields, so that

$$\phi_k = \int \phi(x)e^{-ikx}dx, \quad \pi_k = \int \pi(x)e^{-ikx}dx.$$

The reality of the fields implies that

$$\phi_{-k} = \phi_k^\dagger, \quad \pi_{-k} = \pi_k^\dagger.$$

The classical Hamiltonian may be expanded in Fourier modes as

$$H = \frac{1}{2}\sum_{k=-\infty}^{\infty}\left[\pi_k\pi_k^\dagger + \omega_k^2\phi_k\phi_k^\dagger\right],$$

where $\omega_k = \sqrt{k^2 + m^2}$.

This Hamiltonian is thus recognizable as an infinite sum of classical normal mode oscillator excitations φ_k, each one of which is quantized in the standard manner, so the free quantum Hamiltonian looks identical. It is the φ_ks that have become operators obeying the standard commutation relations, $[\varphi_k, \pi_k^\dagger] = [\varphi_k^\dagger, \pi_k] = i\hbar$, with all others vanishing. The collective Hilbert space of all these oscillators is thus constructed using creation and annihilation operators constructed from these modes,

$$a_k = \frac{1}{\sqrt{2\hbar\omega_k}}\left(\omega_k \phi_k + i\pi_k\right), \ \ a_k^\dagger = \frac{1}{\sqrt{2\hbar\omega_k}}\left(\omega_k \phi_k^\dagger - i\pi_k^\dagger\right),$$

for which $[a_k, a_k^\dagger] = 1$ for all k, with all other commutators vanishing.

The vacuum $|0\rangle$ is taken to be annihilated by all of the a_k, and $\mathcal{H}$ is the Hilbert space constructed by applying any combination of the infinite collection of creation operators $a_k^\dagger$ to $|0\rangle$. This Hilbert space is called Fock space. For each k, this construction is identical to a quantum harmonic oscillator. The quantum field is an infinite array of quantum oscillators. The quantum Hamiltonian then amounts to

$$H = \sum_{k=-\infty}^{\infty} \hbar\omega_k a_k^\dagger a_k = \sum_{k=-\infty}^{\infty} \hbar\omega_k N_k,$$

where N_k may be interpreted as the *number operator* giving the number of particles in a state with momentum k.

This Hamiltonian differs from the previous expression by the subtraction of the zero-point energy $\hbar\omega_k/2$ of each harmonic oscillator. This satisfies the condition that H must annihilate the vacuum, without affecting the time-evolution of operators via the above exponentiation operation. This subtraction of the zero-point energy may be considered to be a resolution of the quantum operator ordering ambiguity, since it is equivalent to requiring that *all creation operators appear to the left of annihilation operators* in the expansion of the Hamiltonian. This procedure is known as Wick ordering or normal ordering.

Other Fields

All other fields can be quantized by a generalization of this procedure. Vector or tensor fields simply have more components, and independent creation and destruction operators must be introduced for each independent component. If a field has any internal symmetry, then creation and destruction operators must be introduced for each component of the field related to this symmetry as well. If there is a gauge symmetry, then the number of independent components of the field must be carefully analyzed to avoid over-counting equivalent configurations, and gauge-fixing may be applied if needed.

It turns out that commutation relations are useful only for quantizing *bosons*, for which the occupancy number of any state is unlimited. To quantize *fermions*, which satisfy the Pauli exclusion principle, anti-commutators are needed. These are defined by {A,B} = AB+BA.

When quantizing fermions, the fields are expanded in creation and annihilation operators, $\theta_k^{\dagger}$, θ_k, which satisfy

$$\{\theta_k, \theta_l^{\dagger}\} = \delta_{kl}, \quad \{\theta_k, \theta_l\} = 0, \quad \{\theta_k^{\dagger}, \theta_l^{\dagger}\} = 0.$$

The states are constructed on a vacuum |0> annihilated by the θ_k, and the Fock space is built by applying all products of creation operators $\theta_k^{\dagger}$ to |0>. Pauli's exclusion principle is satisfied, because $(\theta_k^{\dagger})^2 |0\rangle = 0$, by virtue of the anti-commutation relations.

Condensates

The construction of the scalar field states above assumed that the potential was minimized at $\varphi = 0$, so that the vacuum minimizing the Hamiltonian satisfies $\langle\ \varphi\ \rangle = 0$, indicating that the vacuum expectation value (VEV) of the field is zero. In cases involving spontaneous symmetry breaking, it is possible to have a non-zero VEV, because the potential is minimized for a value $\varphi = v$. This occurs for example, if $V(\varphi) = g\varphi^4$ and $m^2 < 0$, for which the minimum energy is found at $v = \pm m/\sqrt{g}$. The value of v in one of these vacua may be considered as *condensate* of the field φ. Canonical quantization then can be carried out for the *shifted field* $\varphi(x,t) - v$, and particle states with respect to the shifted vacuum are defined by quantizing the shifted field. This construction is utilized in the Higgs mechanism in the standard model of particle physics.

Mathematical Quantization

The classical theory is described using a spacelike foliation of spacetime with the state at each slice being described by an element of a symplectic manifold with the time evolution given by the symplectomorphism generated by a Hamiltonian function over the symplectic manifold. The *quantum algebra* of "operators" is an $\hbar$-deformation of the algebra of smooth functions over the symplectic space such that the leading term in the Taylor expansion over $\hbar$ of the commutator $[A, B]$ expressed in the phase space formulation is $i\hbar\{A, B\}$. (Here, the curly braces denote the Poisson bracket. The subleading terms are all encoded in the Moyal bracket, the suitable quantum deformation of the Poisson bracket.) In general, for the quantities (observables) involved, and providing the arguments of such brackets, $\hbar$-deformations are highly nonunique—quantization is an "art", and is specified by the physical context. (Two *different* quantum systems may represent two different, inequivalent, deformations of the same classical limit, $\hbar \to 0$.)

Now, one looks for unitary representations of this quantum algebra. With respect to such a unitary representation, a symplectomorphism in the classical theory would now deform to a (metaplectic) unitary transformation. In particular, the time evolution symplectomorphism generated by the classical Hamiltonian deforms to a unitary transformation generated by the corresponding quantum Hamiltonian.

A further generalization is to consider a Poisson manifold instead of a symplectic space for the classical theory and perform an $\hbar$-deformation of the corresponding Poisson algebra or even Poisson supermanifolds.

References

- Gerlach, W.; Stern, O. (1922). "Der experimentelle Nachweis der Richtungsquantelung im Magnetfeld". Zeitschrift für Physik. 9: 349–352. Bibcode:1922ZPhy....9..349G. doi:10.1007/BF01326983.

- Gerlach, W.; Stern, O. (1922). "Das magnetische Moment des Silberatoms". Zeitschrift für Physik. 9: 353–355. Bibcode:1922ZPhy....9..353G. doi:10.1007/BF01326984.

- Gerlach, W.; Stern, O. (1922). "Der experimentelle Nachweis des magnetischen Moments des Silberatoms". Zeitschrift für Physik. 8: 110–111. Bibcode:1922ZPhy....9..349G. doi:10.1007/BF01329580.

- George H. Rutherford and Rainer Grobe (1997). "Comment on "Stern-Gerlach Effect for Electron Beams"". Phys.Rev.Lett. 81 (4772). Bibcode:1998PhRvL..81.4772R. doi:10.1103/PhysRevLett.81.4772.

- Stern, O. (1921). "Ein Weg zur experimentellen Pruefung der Richtungsquantelung im Magnetfeld". Zeitschrift für Physik. 7: 249–253. Bibcode:1921ZPhy....7..249S. doi:10.1007/BF01332793.

- Weinert, F. (1995). "Wrong theory—right experiment: The significance of the Stern–Gerlach experiments". Studies in History and Philosophy of Modern Physics. 26B: 75–86. doi:10.1016/1355-2198(95)00002-B.

- Phipps, T.E.; Taylor, J.B. (1927). "The Magnetic Moment of the Hydrogen Atom". Physical Review. 29 (2): 309–320. Bibcode:1927PhRv...29..309P. doi:10.1103/PhysRev.29.309.

- PAM Dirac (1939). "A new notation for quantum mechanics". Mathematical Proceedings of the Cambridge Philosophical Society. 35 (3): 416–418. Bibcode:1939PCPS...35..416D. doi:10.1017/S0305004100021162.

- H. Grassmann (1862). Extension Theory. History of Mathematics Sources. American Mathematical Society, London Mathematical Society, 2000 translation by Lloyd C. Kannenberg.

- Cajori, Florian (1929). A History Of Mathematical Notations Volume II. Open Court Publishing. p. 134. ISBN 978-0-486-67766-8.

- Kennard, E. H. (1927), "Zur Quantenmechanik einfacher Bewegungstypen", Zeitschrift für Physik (in German), 44 (4–5): 326, Bibcode:1927ZPhy...44..326K, doi:10.1007/BF01391200.

Schrödinger Equation: An Overview

Schrödinger equation explains how the quantum state of a quantum system changes, and how it changes with time. A Schrödinger field is a quantum field which goes as per the Schrödinger equation. The section on Schrödinger equation offers an insightful focus, keeping in mind the complex subject matter.

Schrödinger Equation

Schrödinger equation as part of a monument in front of Warsaw University's Centre of New Technologies

In quantum mechanics, the Schrödinger equation is a partial differential equation that describes how the quantum state of a quantum system changes with time. It was formulated in late 1925, and published in 1926, by the Austrian physicist Erwin Schrödinger.

In classical mechanics, Newton's second law (F = ma) is used to make a mathematical prediction as to what path a given system will take following a set of known initial conditions. In quantum mechanics, the analogue of Newton's law is Schrödinger's equation for a quantum system (usually atoms, molecules, and subatomic particles whether free, bound, or localised). It is not a simple algebraic equation, but in general a linear partial differential equation, describing the time-evolution of the system's wave function (also called a "state function").

The concept of a wavefunction is a fundamental postulate of quantum mechanics. Although Schrödinger's equation is often presented as a separate postulate, some authors[Chapter 3] show that some properties resulting from Schrödinger's equation may be deduced just from symmetry principles alone, for example the commutation relations. Generally, "derivations" of the Schrödinger equation demonstrate its mathematical plausibility for describing wave–particle duality, but to date there is no universally accepted derivation of Schrödinger's equation from appropriate axioms.

In the Copenhagen interpretation of quantum mechanics, the wave function is the most complete description that can be given of a physical system. Solutions to Schrödinger's equation describe not only molecular, atomic, and subatomic systems, but also macroscopic systems, possibly even the whole universe. The Schrödinger equation is consistent with classical mechanics but not with special relativity. Making quantum mechanics consistent with special relativity is one of the great achievements of Quantum Field Theory and requires the introduction of the creation and annihilation operators so that the number of particles can no longer be considered constant.

The Schrödinger equation is not the only way to make predictions in quantum mechanics—other formulations can be used, such as Werner Heisenberg's matrix mechanics, and Richard Feynman's path integral formulation.

Equation

Time-dependent Equation

The form of the Schrödinger equation depends on the physical situation. The most general form is the time-dependent Schrödinger equation, which gives a description of a system evolving with time:

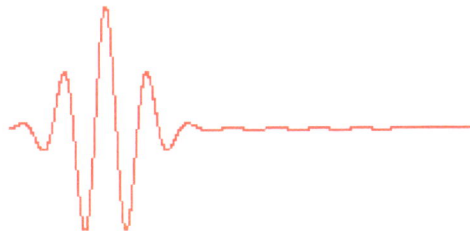

A wave function that satisfies the non-relativistic Schrödinger equation with $V = 0$. In other words, this corresponds to a particle traveling freely through empty space. The real part of the wave function is plotted here.

Time-dependent Schrödinger equation
(general)

$$i\hbar\frac{\partial}{\partial t}\Psi(\mathbf{r},t) = \hat{H}\Psi(\mathbf{r},t)$$

where i is the imaginary unit, $\hbar$ is the Planck constant divided by 2π, the symbol $\partial/\partial t$ indicates a partial derivative with respect to time t, Ψ (the Greek letter psi) is the wave function of the quantum system, r and t are the position vector and time respectively, and $\hat{H}$ is the Hamiltonian operator (which characterizes the total energy of any given wave function and takes different forms depending on the situation).

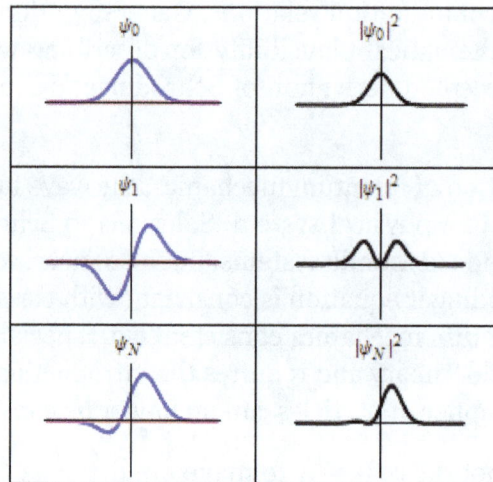

Each of these three rows is a wave function which satisfies the time-dependent Schrödinger equation for a harmonic oscillator. Left: The real part (blue) and imaginary part (red) of the wave function. Right: The probability distribution of finding the particle with this wave function at a given position. The top two rows are examples of stationary states, which correspond to standing waves. The bottom row is an example of a state which is *not* a stationary state. The right column illustrates why stationary states are called "stationary".

The most famous example is the non-relativistic Schrödinger equation for a single particle moving in an electric field (but not a magnetic field):

> **Time-dependent Schrödinger equation**
> *(single non-relativistic particle)*
>
> $$i\hbar\frac{\partial}{\partial t}\Psi(\mathbf{r},t) = \left[\frac{-\hbar^2}{2\mu}\nabla^2 + V(\mathbf{r},t)\right]\Psi(\mathbf{r},t)$$

where μ is the particle's "reduced mass", V is its potential energy, ∇^2 is the Laplacian (a differential operator), and Ψ is the wave function (more precisely, in this context, it is called the "position-space wave function"). In plain language, it means "total energy equals kinetic energy plus potential energy", but the terms take unfamiliar forms for reasons explained below.

Given the particular differential operators involved, this is a linear partial differential equation. It is also a diffusion equation, but unlike the heat equation, this one is also a wave equation given the imaginary unit present in the transient term.

The term *"Schrödinger equation"* can refer to both the general equation (first box above), or the specific nonrelativistic version (second box above and variations thereof). The general equation is indeed quite general, used throughout quantum mechanics, for everything from the Dirac equation to quantum field theory, by plugging in various complicated expressions for the Hamiltonian. The specific nonrelativistic version is a simplified approximation to reality, which is quite accurate in many situations, but very inaccurate in others.

To apply the Schrödinger equation, the Hamiltonian operator is set up for the system, accounting for the kinetic and potential energy of the particles constituting the system, then inserted into the

Schrödinger equation. The resulting partial differential equation is solved for the wave function, which contains information about the system.

Time-independent Equation

The time-dependent Schrödinger equation described above predicts that wave functions can form standing waves, called stationary states (also called "orbitals", as in atomic orbitals or molecular orbitals). These states are important in their own right, and if the stationary states are classified and understood, then it becomes easier to solve the time-dependent Schrödinger equation for *any* state. Stationary states can also be described by a simpler form of the Schrödinger equation, the *time-independent Schrödinger equation*. (This is only used when the Hamiltonian itself is not dependent on time explicitly. However, even in this case the total wave function still has a time dependency.)

> **Time-independent Schrödinger equation**
> (*general*)
> $$\hat{H}\Psi = E\Psi$$

In words, the equation states:

When the Hamiltonian operator acts on a certain wave function Ψ, and the result is proportional to the same wave function Ψ, then Ψ is a stationary state, and the proportionality constant, E, is the energy of the state Ψ.

The time-independent Schrödinger equation is discussed further below. In linear algebra terminology, this equation is an eigenvalue equation.

As before, the most famous manifestation is the non-relativistic Schrödinger equation for a single particle moving in an electric field (but not a magnetic field):

> **Time-independent Schrödinger equation** (*single non-relativistic particle*)
> $$\left[\frac{-\hbar^2}{2\mu} \nabla^2 + V(\mathbf{r}) \right] \Psi(\mathbf{r}) = E\Psi(\mathbf{r})$$

with definitions as above.

Implications

The Schrödinger equation and its solutions introduced a breakthrough in thinking about physics. Schrödinger's equation was the first of its type, and solutions led to consequences that were very unusual and unexpected for the time.

Total, Kinetic, and Potential Energy

The *overall* form of the equation is *not* unusual or unexpected, as it uses the principle of the conservation of energy. The terms of the nonrelativistic Schrödinger equation can be interpreted as total energy of the system, equal to the system kinetic energy plus the system potential energy. In this respect, it is just the same as in classical physics.

Quantization

The Schrödinger equation predicts that if certain properties of a system are measured, the result may be *quantized*, meaning that only specific discrete values can occur. One example is *energy quantization*: the energy of an electron in an atom is always one of the quantized energy levels, a fact discovered via atomic spectroscopy. (Energy quantization is discussed below.) Another example is quantization of angular momentum. This was an *assumption* in the earlier Bohr model of the atom, but it is a *prediction* of the Schrödinger equation.

Another result of the Schrödinger equation is that not every measurement gives a quantized result in quantum mechanics. For example, position, momentum, time, and (in some situations) energy can have any value across a continuous range.

Measurement and Uncertainty

In classical mechanics, a particle has, at every moment, an exact position and an exact momentum. These values change deterministically as the particle moves according to Newton's laws. Under the Copenhagen interpretation of quantum mechanics, particles do not have exactly determined properties, and when they are measured, the result is randomly drawn from a probability distribution. The Schrödinger equation predicts what the probability distributions are, but fundamentally cannot predict the exact result of each measurement.

The Heisenberg uncertainty principle is the statement of the inherent measurement uncertainty in quantum mechanics. It states that the more precisely a particle's position is known, the less precisely its momentum is known, and vice versa.

The Schrödinger equation describes the (deterministic) evolution of the wave function of a particle. However, even if the wave function is known exactly, the result of a specific measurement on the wave function is uncertain.

Quantum Tunneling

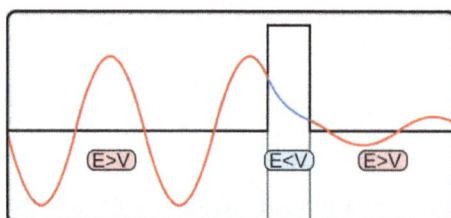

Quantum tunneling through a barrier. A particle coming from the left does not have enough energy to climb the barrier. However, it can sometimes "tunnel" to the other side.

In classical physics, when a ball is rolled slowly up a large hill, it will come to a stop and roll back, because it doesn't have enough energy to get over the top of the hill to the other side. However, the Schrödinger equation predicts that there is a small probability that the ball will get to the other side of the hill, even if it has too little energy to reach the top. This is called quantum tunneling. It is related to the distribution of energy: although the ball's assumed position seems to be on one side of the hill, there is a chance of finding it on the other side.

Particles as Waves

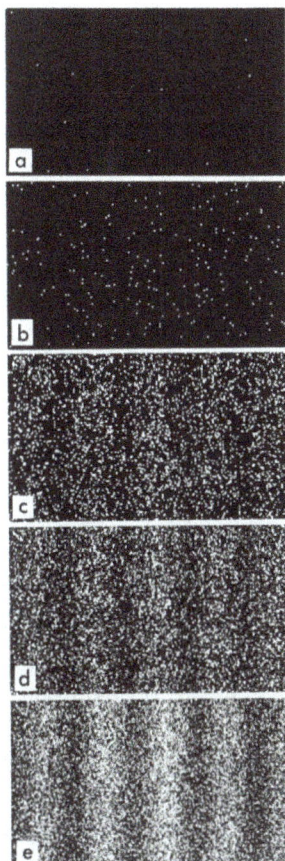

A double slit experiment showing the accumulation of electrons on a screen as time passes.

The nonrelativistic Schrödinger equation is a type of partial differential equation called a wave equation. Therefore, it is often said particles can exhibit behavior usually attributed to waves. In some modern interpretations this description is reversed – the quantum state, i.e. wave, is the only genuine physical reality, and under the appropriate conditions it can show features of particle-like behavior. However, Ballentine shows that such an interpretation has problems. Ballentine points out that whilst it is arguable to associate a physical wave with a single particle, there is still only *one* Schrödinger wave equation for many particles. He points out:

> "If a physical wave field were associated with a particle, or if a particle were identified with a wave packet, then corresponding to N interacting particles there should be N interacting waves in ordinary three-dimensional space. But according to (4.6) that is not the case; instead there is one "wave" function in an abstract 3N-dimensional configuration space. The misinterpretation of psi as a physical wave in ordinary space is possible only because the most common applications of quantum mechanics are to one-particle states, for which configuration space and ordinary space are isomorphic."

Two-slit diffraction is a famous example of the strange behaviors that waves regularly display, that are not intuitively associated with particles. The overlapping waves from the two slits cancel each other out in some locations, and reinforce each other in other locations, causing a complex pattern to emerge. Intuitively, one would not expect this pattern from firing a single particle at the slits, because the particle should pass through one slit or the other, not a complex overlap of both.

However, since the Schrödinger equation is a wave equation, a single particle fired through a double-slit *does* show this same pattern (figure on right). Note: The experiment must be repeated many times for the complex pattern to emerge. Although this is counterintuitive, the prediction is correct; in particular, electron diffraction and neutron diffraction are well understood and widely used in science and engineering.

Related to diffraction, particles also display superposition and interference.

The superposition property allows the particle to be in a quantum superposition of two or more quantum states at the same time. However, it is noted that a "quantum state" in QM means the *probability* that a system will be, for example at a position x, not that the system will actually be at position x. It does not imply that the particle itself may be in two classical states at once. Indeed, QM is generally unable to assign values for properties prior to measurement at all.

Multiverse

In Dublin in 1952 Erwin Schrödinger gave a lecture in which at one point he jocularly warned his audience that what he was about to say might "seem lunatic". It was that, when his Nobel equations seem to be describing several different histories, they are "not alternatives but all really happen simultaneously". This is the earliest known reference to the multiverse.

Interpretation of the Wave Function

The Schrödinger equation provides a way to calculate the wave function of a system and how it changes dynamically in time. However, the Schrödinger equation does not directly say *what*, exactly, the wave function is. Interpretations of quantum mechanics address questions such as what the relation is between the wave function, the underlying reality, and the results of experimental measurements.

An important aspect is the relationship between the Schrödinger equation and wavefunction collapse. In the oldest Copenhagen interpretation, particles follow the Schrödinger equation *except* during wavefunction collapse, during which they behave entirely differently. The advent of quantum decoherence theory allowed alternative approaches (such as the Everett many-worlds interpretation and consistent histories), wherein the Schrödinger equation is *always* satisfied, and wavefunction collapse should be explained as a consequence of the Schrödinger equation.

Historical Background and Development

Following Max Planck's quantization of light, Albert Einstein interpreted Planck's quanta to be photons, particles of light, and proposed that the energy of a photon is proportional to its frequency, one of the first signs of wave–particle duality. Since energy and momentum are related in the same way as frequency and wavenumber in special relativity, it followed that the momentum p of a photon is inversely proportional to its wavelength λ, or proportional to its wavenumber k.

$$p = \frac{h}{\lambda} = \hbar k$$

Erwin Schrödinger

where h is Planck's constant. Louis de Broglie hypothesized that this is true for all particles, even particles which have mass such as electrons. He showed that, assuming that the matter waves propagate along with their particle counterparts, electrons form standing waves, meaning that only certain discrete rotational frequencies about the nucleus of an atom are allowed. These quantized orbits correspond to discrete energy levels, and de Broglie reproduced the Bohr model formula for the energy levels. The Bohr model was based on the assumed quantization of angular momentum L according to:

$$L = n\frac{h}{2\pi} = n\hbar.$$

According to de Broglie the electron is described by a wave and a whole number of wavelengths must fit along the circumference of the electron's orbit:

$$n\lambda = 2\pi r.$$

This approach essentially confined the electron wave in one dimension, along a circular orbit of radius r.

In 1921, prior to de Broglie, Arthur C. Lunn at the University of Chicago had used the same argument based on the completion of the relativistic energy–momentum 4-vector to derive what we now call the de Broglie relation. Unlike de Broglie, Lunn went on to formulate the differential equation now known as the Schrödinger equation, and solve for its energy eigenvalues for the hydrogen atom. Unfortunately the paper was rejected by the Physical Review, as recounted by Kamen.

Following up on de Broglie's ideas, physicist Peter Debye made an offhand comment that if particles behaved as waves, they should satisfy some sort of wave equation. Inspired by Debye's remark, Schrödinger decided to find a proper 3-dimensional wave equation for the electron. He was guided by William R. Hamilton's analogy between mechanics and optics, encoded in the observation that the zero-wavelength limit of optics resembles a mechanical system—the trajectories of light rays become sharp tracks that obey Fermat's principle, an analog of the principle of least action. A modern version of his reasoning is reproduced below. The equation he found is:

$$i\hbar\frac{\partial}{\partial t}\Psi(\mathbf{r},t) = -\frac{\hbar^2}{2m}\nabla^2\Psi(\mathbf{r},t) + V(\mathbf{r})\Psi(\mathbf{r},t).$$

However, by that time, Arnold Sommerfeld had refined the Bohr model with relativistic corrections. Schrödinger used the relativistic energy momentum relation to find what is now known as the Klein–Gordon equation in a Coulomb potential (in natural units):

$$\left(E + \frac{e^2}{r}\right)^2 \psi(x) = -\nabla^2\psi(x) + m^2\psi(x).$$

He found the standing waves of this relativistic equation, but the relativistic corrections disagreed with Sommerfeld's formula. Discouraged, he put away his calculations and secluded himself in an isolated mountain cabin in December 1925.

While at the cabin, Schrödinger decided that his earlier non-relativistic calculations were novel enough to publish, and decided to leave off the problem of relativistic corrections for the future. Despite the difficulties in solving the differential equation for hydrogen (he had sought help from his friend the mathematician Hermann Weyl[3]) Schrödinger showed that his non-relativistic version of the wave equation produced the correct spectral energies of hydrogen in a paper published in 1926. In the equation, Schrödinger computed the hydrogen spectral series by treating a hydrogen atom's electron as a wave $\Psi(x, t)$, moving in a potential well V, created by the proton. This computation accurately reproduced the energy levels of the Bohr model. In a paper, Schrödinger himself explained this equation as follows:

> The already ... mentioned psi-function.... is now the means for predicting probability of measurement results. In it is embodied the momentarily attained sum of theoretically based future expectation, somewhat as laid down in a catalog.
>
> — Erwin Schrödinger

This 1926 paper was enthusiastically endorsed by Einstein, who saw the matter-waves as an intuitive depiction of nature, as opposed to Heisenberg's matrix mechanics, which he considered overly formal.

The Schrödinger equation details the behavior of Ψ but says nothing of its *nature*. Schrödinger tried to interpret it as a charge density in his fourth paper, but he was unsuccessful. In 1926, just a few days after Schrödinger's fourth and final paper was published, Max Born successfully interpreted

Ψ as the probability amplitude, whose absolute square is equal to probability density. Schrödinger, though, always opposed a statistical or probabilistic approach, with its associated discontinuities—much like Einstein, who believed that quantum mechanics was a statistical approximation to an underlying deterministic theory—and never reconciled with the Copenhagen interpretation.

Louis de Broglie in his later years proposed a real valued wave function connected to the complex wave function by a proportionality constant and developed the De Broglie–Bohm theory.

The Wave Equation for Particles

The Schrödinger equation is a diffusion equation, the *solutions* are functions which describe wave-like motions. Wave equations in physics can normally be derived from other physical laws – the wave equation for mechanical vibrations on strings and in matter can be derived from Newton's laws – where the wave function represents the displacement of matter, and electromagnetic waves from Maxwell's equations, where the wave functions are electric and magnetic fields. The basis for Schrödinger's equation, on the other hand, is the energy of the system and a separate postulate of quantum mechanics: the wave function is a description of the system. The Schrödinger equation is therefore a new concept in itself; as Feynman put it:

> Where did we get that (equation) from? Nowhere. It is not possible to derive it from anything you know. It came out of the mind of Schrödinger.

> — Richard Feynman

The foundation of the equation is structured to be a linear differential equation based on classical energy conservation, and consistent with the De Broglie relations. The solution is the wave function ψ, which contains all the information that can be known about the system. In the Copenhagen interpretation, the modulus of ψ is related to the probability the particles are in some spatial configuration at some instant of time. Solving the equation for ψ can be used to predict how the particles will behave under the influence of the specified potential and with each other.

The Schrödinger equation was developed principally from the De Broglie hypothesis, a wave equation that would describe particles, and can be constructed as shown informally in the following sections.

Consistency with Energy Conservation

The total energy E of a particle is the sum of kinetic energy T and potential energy V, this sum is also the frequent expression for the Hamiltonian H in classical mechanics:

$$E = T + V = H$$

Explicitly, for a particle in one dimension with position x, mass m and momentum p, and potential energy V which generally varies with position and time t:

$$E = \frac{p^2}{2m} + V(x,t) = H.$$

For three dimensions, the position vector r and momentum vector p must be used:

$$E = \frac{\mathbf{p} \cdot \mathbf{p}}{2m} + V(\mathbf{r},t) = H$$

This formalism can be extended to any fixed number of particles: the total energy of the system is then the total kinetic energies of the particles, plus the total potential energy, again the Hamiltonian. However, there can be interactions between the particles (an N-body problem), so the potential energy V can change as the spatial configuration of particles changes, and possibly with time. The potential energy, in general, is *not* the sum of the separate potential energies for each particle, it is a function of all the spatial positions of the particles. Explicitly:

$$E = \sum_{n=1}^{N} \frac{\mathbf{P}_n \cdot \mathbf{P}_n}{2m_n} + V(\mathbf{r}_1, \mathbf{r}_2 \cdots \mathbf{r}_N, t) = H$$

Linearity

The simplest wavefunction is a plane wave of the form:

$$\Psi(\mathbf{r},t) = A e^{i(\mathbf{k} \cdot \mathbf{r} - \omega t)}$$

where the A is the amplitude, k the wavevector, and ω the angular frequency, of the plane wave. In general, physical situations are not purely described by plane waves, so for generality the superposition principle is required; any wave can be made by superposition of sinusoidal plane waves. So if the equation is linear, a linear combination of plane waves is also an allowed solution. Hence a necessary and separate requirement is that the Schrödinger equation is a linear differential equation.

For discrete k the sum is a superposition of plane waves:

$$\Psi(\mathbf{r},t) = \sum_{n=1}^{\infty} A_n e^{i(\mathbf{k}_n \cdot \mathbf{r} - \omega_n t)}$$

for some real amplitude coefficients A_n, and for continuous k the sum becomes an integral, the Fourier transform of a momentum space wavefunction:

$$\Psi(\mathbf{r},t) = \frac{1}{(\sqrt{2\pi})^3} \int \Phi(\mathbf{k}) e^{i(\mathbf{k} \cdot \mathbf{r} - \omega t)} d^3\mathbf{k}$$

where $d^3k = dk_x dk_y dk_z$ is the differential volume element in k-space, and the integrals are taken over all k-space. The momentum wavefunction $\Phi(k)$ arises in the integrand since the position and momentum space wavefunctions are Fourier transforms of each other.

Consistency with the De Broglie Relations

Einstein's light quanta hypothesis (1905) states that the energy E of a photon is proportional to the frequency v (or angular frequency, $\omega = 2\pi v$) of the corresponding quantum wavepacket of light:

$$E = hv = \hbar\omega$$

$$-\frac{\hbar^2}{2m}\frac{\partial^2\Psi}{\partial x^2}+V\Psi=E\Psi$$

	$E-V(x)<0$	$E-V(x)>0$
$\Psi>0$	$\big[E-V(x)\big]\Psi<0$ $\dfrac{\partial^2\Psi}{\partial x^2}>0 \quad T<0$	$\big[E-V(x)\big]\Psi>0$ $\dfrac{\partial^2\Psi}{\partial x^2}<0 \quad T>0$
$\Psi<0$	$\big[E-V(x)\big]\Psi>0$ $\dfrac{\partial^2\Psi}{\partial x^2}<0 \quad T>0$	$\big[E-V(x)\big]\Psi<0$ $\dfrac{\partial^2\Psi}{\partial x^2}>0 \quad T<0$

Diagrammatic summary of the quantities related to the wavefunction, as used in De broglie's hypothesis and development of the Schrödinger equation.

Likewise De Broglie's hypothesis (1924) states that any particle can be associated with a wave, and that the momentum p of the particle is inversely proportional to the wavelength λ of such a wave (or proportional to the wavenumber, $k=2\pi/\lambda$), in one dimension, by:

$$p=\frac{h}{\lambda}=\hbar k \,,$$

while in three dimensions, wavelength λ is related to the magnitude of the wavevector k:

$$\mathbf{p}=\hbar\mathbf{k}, \quad |\mathbf{k}|=\frac{2\pi}{\lambda}.$$

The Planck–Einstein and de Broglie relations illuminate the deep connections between energy with time, and space with momentum, and express wave–particle duality. In practice, natural units comprising $\hbar=1$ are used, as the De Broglie *equations* reduce to *identities*: allowing momentum, wavenumber, energy and frequency to be used interchangeably, to prevent duplication of quantities, and reduce the number of dimensions of related quantities. For familiarity SI units are still used in this article.

Schrödinger's insight, late in 1925, was to express the phase of a plane wave as a complex phase factor using these relations:

$$\varnothing=Ae^{i(\mathbf{k}\cdot\mathbf{r}-\omega t)}=Ae^{i(\mathbf{p}\cdot\mathbf{r}-Et)/\hbar}$$

and to realize that the first order partial derivatives were:

with respect to space:

$$\psi=Ae^{i(\mathbf{k}\cdot\mathbf{r}-\omega t)}=Ae^{i(\mathbf{p}\cdot\mathbf{r}-Et)/\hbar}$$

with respect to time:

$$\frac{\partial \Psi}{\partial t} = -\frac{iE}{\hbar} A e^{i(\mathbf{p}\cdot\mathbf{r}-Et)/\hbar} = -\frac{iE}{\hbar}\Psi$$

Another postulate of quantum mechanics is that all observables are represented by linear Hermitian operators which act on the wavefunction, and the eigenvalues of the operator are the values the observable takes. The previous derivatives are consistent with the energy operator, corresponding to the time derivative,

$$\hat{E}\Psi = i\hbar \frac{\partial}{\partial t}\Psi = E\Psi$$

where E are the energy eigenvalues, and the momentum operator, corresponding to the spatial derivatives (the gradient ∇),

$$\hat{\mathbf{p}}\Psi = -i\hbar\nabla\Psi = \mathbf{p}\Psi$$

where p is a vector of the momentum eigenvalues. In the above, the "hats" (ˆ) indicate these observables are operators, not simply ordinary numbers or vectors. The energy and momentum operators are *differential operators*, while the potential energy function V is just a multiplicative factor.

Substituting the energy and momentum operators into the classical energy conservation equation obtains the operator:

$$E = \frac{\mathbf{p}\cdot\mathbf{p}}{2m} + V \quad \rightarrow \quad \hat{E} = \frac{\hat{\mathbf{p}}\cdot\hat{\mathbf{p}}}{2m} + V$$

so in terms of derivatives with respect to time and space, acting this operator on the wavefunction Ψ immediately led Schrödinger to his equation:

$$i\hbar\frac{\partial \Psi}{\partial t} = -\frac{\hbar^2}{2m}\nabla^2\Psi + V\Psi$$

Wave–particle duality can be assessed from these equations as follows. The kinetic energy T is related to the square of momentum p. As the particle's momentum increases, the kinetic energy increases more rapidly, but since the wavenumber $|k|$ increases the wavelength λ decreases. In terms of ordinary scalar and vector quantities (not operators):

$$\mathbf{p}\cdot\mathbf{p} \propto \mathbf{k}\cdot\mathbf{k} \propto T \propto \frac{1}{\lambda^2}$$

The kinetic energy is also proportional to the second spatial derivatives, so it is also proportional to the magnitude of the *curvature* of the wave, in terms of operators:

$$\hat{T}\Psi = \frac{-\hbar^2}{2m}\nabla\cdot\nabla\Psi \propto \nabla^2\Psi.$$

As the curvature increases, the amplitude of the wave alternates between positive and negative more rapidly, and also shortens the wavelength. So the inverse relation between momentum and wavelength is consistent with the energy the particle has, and so the energy of the particle has a connection to a wave, all in the same mathematical formulation.

Wave and Particle Motion

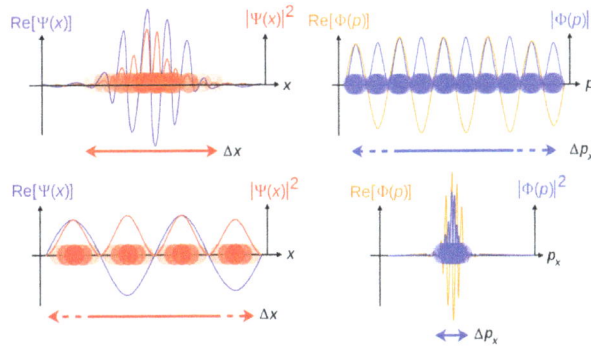

Increasing levels of wavepacket localization, meaning the particle has a more localized position.

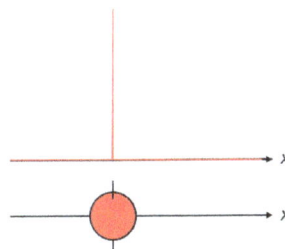

In the limit ℏ → 0, the particle's position and momentum become known exactly. This is equivalent to the classical particle.

Schrödinger required that a wave packet solution near position r with wavevector near k will move along the trajectory determined by classical mechanics for times short enough for the spread in k (and hence in velocity) not to substantially increase the spread in r. Since, for a given spread in k, the spread in velocity is proportional to Planck's constant ℏ, it is sometimes said that in the limit as ℏ approaches zero, the equations of classical mechanics are restored from quantum mechanics. Great care is required in how that limit is taken, and in what cases.

The limiting short-wavelength is equivalent to ℏ tending to zero because this is limiting case of increasing the wave packet localization to the definite position of the particle. Using the Heisenberg uncertainty principle for position and momentum, the products of uncertainty in position and momentum become zero as ℏ → 0:

$$\sigma(x)\sigma(p_x) \geqslant \frac{\hbar}{2} \quad \rightarrow \quad \sigma(x)\sigma(p_x) \geqslant 0$$

where σ denotes the (root mean square) measurement uncertainty in x and p_x (and similarly for the y and z directions) which implies the position and momentum can only be known to arbitrary precision in this limit.

The Schrödinger equation in its general form

$$ i\hbar \frac{\partial}{\partial t} \Psi(\mathbf{r},t) = \hat{H} \Psi(\mathbf{r},t) $$

is closely related to the Hamilton–Jacobi equation (HJE)

$$ \frac{\partial}{\partial t} S(q_i,t) = H\left(q_i, \frac{\partial S}{\partial q_i}, t \right) $$

where S is action and H is the Hamiltonian function (not operator). Here the generalized coordinates q_i for i = 1, 2, 3 (used in the context of the HJE) can be set to the position in Cartesian coordinates as $\mathbf{r}$ = (q_1, q_2, q_3) = (x, y, z).

Substituting

$$ \Psi = \sqrt{\rho(\mathbf{r},t)} e^{iS(\mathbf{r},t)/\hbar} $$

where ρ is the probability density, into the Schrödinger equation and then taking the limit $\hbar \to 0$ in the resulting equation, yields the Hamilton–Jacobi equation.

The implications are:

- The motion of a particle, described by a (short-wavelength) wave packet solution to the Schrödinger equation, is also described by the Hamilton–Jacobi equation of motion.

- The Schrödinger equation includes the wavefunction, so its wave packet solution implies the position of a (quantum) particle is fuzzily spread out in wave fronts. On the contrary, the Hamilton–Jacobi equation applies to a (classical) particle of definite position and momentum, instead the position and momentum at all times (the trajectory) are deterministic and can be simultaneously known.

Non-relativistic Quantum Mechanics

The quantum mechanics of particles without accounting for the effects of special relativity, for example particles propagating at speeds much less than light, is known as non-relativistic quantum mechanics. Following are several forms of Schrödinger's equation in this context for different situations: time independence and dependence, one and three spatial dimensions, and one and N particles.

In actuality, the particles constituting the system do not have the numerical labels used in theory. The language of mathematics forces us to label the positions of particles one way or another, otherwise there would be confusion between symbols representing which variables are for which particle.

Time Independent

If the Hamiltonian is not an explicit function of time, the equation is separable into a product of spatial and temporal parts. In general, the wavefunction takes the form:

$$\Psi(\text{space coords}, t) = \psi(\text{space coords})\tau(t).$$

where $\psi(\text{space coords})$ is a function of all the spatial coordinate(s) of the particle(s) constituting the system only, and $\tau(t)$ is a function of time only.

Substituting for ψ into the Schrödinger equation for the relevant number of particles in the relevant number of dimensions, solving by separation of variables implies the general solution of the time-dependent equation has the form:

$$\Psi(\text{space coords}, t) = \psi(\text{space coords})e^{-iEt/\hbar}.$$

Since the time dependent phase factor is always the same, only the spatial part needs to be solved for in time independent problems. Additionally, the energy operator $\hat{E} = i\hbar\partial/\partial t$ can always be replaced by the energy eigenvalue E, thus the time independent Schrödinger equation is an eigenvalue equation for the Hamiltonian operator:

$$\hat{H}\psi = E\psi$$

This is true for any number of particles in any number of dimensions (in a time independent potential). This case describes the standing wave solutions of the time-dependent equation, which are the states with definite energy (instead of a probability distribution of different energies). In physics, these standing waves are called "stationary states" or "energy eigenstates"; in chemistry they are called "atomic orbitals" or "molecular orbitals". Superpositions of energy eigenstates change their properties according to the relative phases between the energy levels.

The energy eigenvalues from this equation form a discrete spectrum of values, so mathematically energy must be quantized. More specifically, the energy eigenstates form a basis – any wavefunction may be written as a sum over the discrete energy states or an integral over continuous energy states, or more generally as an integral over a measure. This is the spectral theorem in mathematics, and in a finite state space it is just a statement of the completeness of the eigenvectors of a Hermitian matrix.

One-dimensional Examples

For a particle in one dimension, the Hamiltonian is:

$$\hat{H} = \frac{\hat{p}^2}{2m} + V(x), \quad \hat{p} = -i\hbar\frac{d}{dx}$$

and substituting this into the general Schrödinger equation gives:

$$-\frac{\hbar^2}{2m}\frac{d^2}{dx^2}\psi(x) + V(x)\psi(x) = E\psi(x)$$

This is the only case the Schrödinger equation is an ordinary differential equation, rather than a partial differential equation. The general solutions are always of the form:

$$\Psi(x,t) = \psi(x)e^{-iEt/\hbar}.$$

For N particles in one dimension, the Hamiltonian is:

$$\hat{H} = \sum_{n=1}^{N} \frac{\hat{p}_n^2}{2m_n} + V(x_1, x_2, \cdots x_N), \quad \hat{p}_n = -i\hbar \frac{\partial}{\partial x_n}$$

where the position of particle n is x_n. The corresponding Schrödinger equation is:

$$-\frac{\hbar^2}{2} \sum_{n=1}^{N} \frac{1}{m_n} \frac{\partial^2}{\partial x_n^2} \psi(x_1, x_2, \cdots x_N) + V(x_1, x_2, \cdots x_N)\psi(x_1, x_2, \cdots x_N) = E\psi(x_1, x_2, \cdots x_N).$$

so the general solutions have the form:

$$\Psi(x_1, x_2, \cdots x_N, t) = e^{-iEt/\hbar} \psi(x_1, x_2 \cdots x_N)$$

For non-interacting distinguishable particles, the potential of the system only influences each particle separately, so the total potential energy is the sum of potential energies for each particle:

$$V(x_1, x_2, \cdots x_N) = \sum_{n=1}^{N} V(x_n).$$

and the wavefunction can be written as a product of the wavefunctions for each particle:

$$\Psi(x_1, x_2, \cdots x_N, t) = e^{-iEt/\hbar} \prod_{n=1}^{N} \psi(x_n),$$

For non-interacting identical particles, the potential is still a sum, but wavefunction is a bit more complicated – it is a sum over the permutations of products of the separate wavefunctions to account for particle exchange. In general for interacting particles, the above decompositions are *not* possible.

Free Particle

For no potential, $V = 0$, so the particle is free and the equation reads:

$$-E\psi = \frac{\hbar^2}{2m} \frac{d^2\psi}{dx^2}$$

which has oscillatory solutions for $E > 0$ (the C_n are arbitrary constants):

$$\psi_E(x) = C_1 e^{i\sqrt{2mE/\hbar^2}\,x} + C_2 e^{-i\sqrt{2mE/\hbar^2}\,x}$$

and exponential solutions for $E < 0$

$$\psi_{-|E|}(x) = C_1 e^{\sqrt{2m|E|/\hbar^2}\,x} + C_2 e^{-\sqrt{2m|E|/\hbar^2}\,x}.$$

The exponentially growing solutions have an infinite norm, and are not physical. They are not allowed in a finite volume with periodic or fixed boundary conditions.

Constant Potential

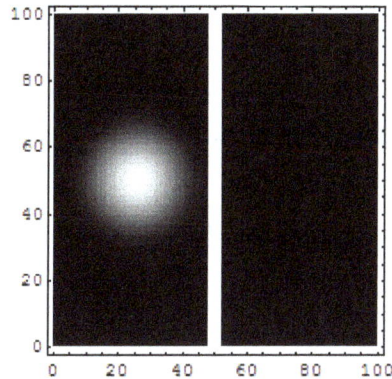

For a constant potential, $V = V_0$, the solution is oscillatory for $E > V_0$ and exponential for $E < V_0$, corresponding to energies that are allowed or disallowed in classical mechanics. Oscillatory solutions have a classically allowed energy and correspond to actual classical motions, while the exponential solutions have a disallowed energy and describe a small amount of quantum bleeding into the classically disallowed region, due to quantum tunneling. If the potential V_0 grows to infinity, the motion is classically confined to a finite region. Viewed far enough away, every solution is reduced to an exponential; the condition that the exponential is decreasing restricts the energy levels to a discrete set, called the allowed energies.

Harmonic Oscillator

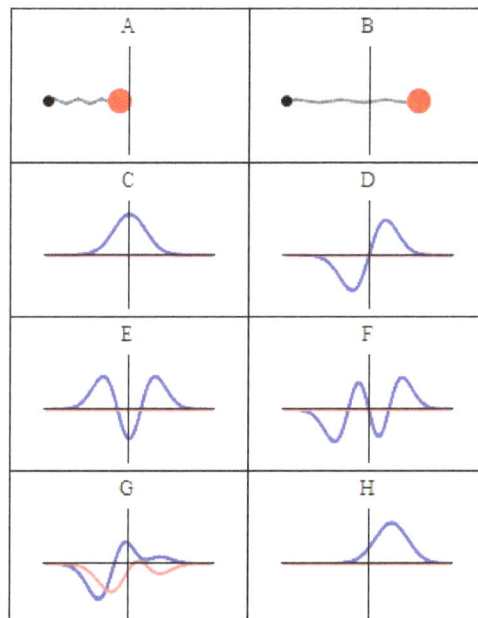

A harmonic oscillator in classical mechanics (A–B) and quantum mechanics (C–H). In (A–B), a ball, attached to a spring, oscillates back and forth. (C–H) are six solutions to the Schrödinger Equation for this situation. The horizontal axis is position, the vertical axis is the real part (blue) or imaginary part (red) of the wavefunction. Stationary states, or energy eigenstates, which are solutions to the time-independent Schrödinger equation, are shown in C, D, E, F, but not G or H.

The Schrödinger equation for this situation is

$$E\psi = -\frac{\hbar^2}{2m}\frac{d^2}{dx^2}\psi + \frac{1}{2}m\omega^2 x^2 \psi$$

It is a notable quantum system to solve for; since the solutions are exact (but complicated – in terms of Hermite polynomials), and it can describe or at least approximate a wide variety of other systems, including vibrating atoms, molecules, and atoms or ions in lattices, and approximating other potentials near equilibrium points. It is also the basis of perturbation methods in quantum mechanics.

There is a family of solutions – in the position basis they are

$$\psi_n(x) = \sqrt{\frac{1}{2^n n!}} \cdot \left(\frac{m\omega}{\pi\hbar}\right)^{1/4} \cdot e^{-\frac{m\omega x^2}{2\hbar}} \cdot H_n\left(\sqrt{\frac{m\omega}{\hbar}}x\right)$$

where $n = 0,1,2,...$, and the functions H_n are the Hermite polynomials.

Three-dimensional Examples

The extension from one dimension to three dimensions is straightforward, all position and momentum operators are replaced by their three-dimensional expressions and the partial derivative with respect to space is replaced by the gradient operator.

The Hamiltonian for one particle in three dimensions is:

$$\hat{H} = \frac{\hat{\mathbf{p}} \cdot \hat{\mathbf{p}}}{2m} + V(\mathbf{r}), \quad \hat{\mathbf{p}} = -i\hbar\nabla$$

generating the equation:

$$-\frac{\hbar^2}{2m}\nabla^2\psi(\mathbf{r}) + V(\mathbf{r})\psi(\mathbf{r}) = E\psi(\mathbf{r})$$

with stationary state solutions of the form:

$$\Psi(\mathbf{r},t) = \psi(\mathbf{r})e^{-iEt/\hbar}$$

where the position of the particle is r. Two useful coordinate systems for solving the Schrödinger equation are Cartesian coordinates so that r = (x, y, z) and spherical polar coordinates so that r = (r, θ, φ), although other orthogonal coordinates are useful for solving the equation for systems with certain geometric symmetries.

For N particles in three dimensions, the Hamiltonian is:

$$\hat{H} = \sum_{n=1}^{N}\frac{\hat{\mathbf{p}}_n \cdot \hat{\mathbf{p}}_n}{2m_n} + V(\mathbf{r}_1, \mathbf{r}_2, \cdots \mathbf{r}_N), \quad \hat{\mathbf{p}}_n = -i\hbar\nabla_n$$

where the position of particle n is r_n and the gradient operators are partial derivatives with respect to the particle's position coordinates. In Cartesian coordinates, for particle n, the position vector is $r_n = (x_n, y_n, z_n)$ while the gradient and Laplacian operator are respectively:

$$\nabla_n = \mathbf{e}_x \frac{\partial}{\partial x_n} + \mathbf{e}_y \frac{\partial}{\partial y_n} + \mathbf{e}_z \frac{\partial}{\partial z_n}, \quad \nabla_n^2 = \nabla_n \cdot \nabla_n = \frac{\partial^2}{\partial x_n^2} + \frac{\partial^2}{\partial y_n^2} + \frac{\partial^2}{\partial z_n^2}$$

The Schrödinger equation is:

$$-\frac{\hbar^2}{2} \sum_{n=1}^{N} \frac{1}{m_n} \nabla_n^2 \Psi(\mathbf{r}_1, \mathbf{r}_2, \cdots \mathbf{r}_N) + V(\mathbf{r}_1, \mathbf{r}_2, \cdots \mathbf{r}_N) \Psi(\mathbf{r}_1, \mathbf{r}_2, \cdots \mathbf{r}_N) = E\Psi(\mathbf{r}_1, \mathbf{r}_2, \cdots \mathbf{r}_N)$$

with stationary state solutions:

$$\Psi(\mathbf{r}_1, \mathbf{r}_2 \cdots \mathbf{r}_N, t) = e^{-iEt/\hbar} \psi(\mathbf{r}_1, \mathbf{r}_2 \cdots \mathbf{r}_N)$$

Again, for non-interacting distinguishable particles the potential is the sum of particle potentials

$$V(\mathbf{r}_1, \mathbf{r}_2, \cdots \mathbf{r}_N) = \sum_{n=1}^{N} V(\mathbf{r}_n)$$

and the wavefunction is a product of the particle wavefuntions

$$\Psi(\mathbf{r}_1, \mathbf{r}_2 \cdots \mathbf{r}_N, t) = e^{-iEt/\hbar} \prod_{n=1}^{N} \psi(\mathbf{r}_n).$$

For non-interacting identical particles, the potential is a sum but the wavefunction is a sum over permutations of products. The previous two equations do not apply to interacting particles.

Hydrogen Atom

This form of the Schrödinger equation can be applied to the hydrogen atom:

$$E\psi = -\frac{\hbar^2}{2\mu} \nabla^2 \psi - \frac{e^2}{4\pi\varepsilon_0 r} \psi$$

where e is the electron charge, r is the position of the electron ($r = |r|$ is the magnitude of the position), the potential term is due to the Coulomb interaction, wherein ε_0 is the electric constant (permittivity of free space) and

$$\mu = \frac{m_e m_p}{m_e + m_p}$$

is the 2-body reduced mass of the hydrogen nucleus (just a proton) of mass m_p and the electron of mass m_e. The negative sign arises in the potential term since the proton and electron are oppositely charged. The reduced mass in place of the electron mass is used since the electron and proton together orbit each other about a common centre of mass, and constitute a two-body problem to solve. The motion of the electron is of principle interest here, so the equivalent one-body problem is the motion of the electron using the reduced mass.

The wavefunction for hydrogen is a function of the electron's coordinates, and in fact can be separated into functions of each coordinate. Usually this is done in spherical polar coordinates:

$$\psi(r,\theta,\phi) = R(r)Y_\ell^m(\theta,\phi) = R(r)\Theta(\theta)\Phi(\phi)$$

where R are radial functions and $Ym\ell(\theta, \varphi)$ are spherical harmonics of degree ℓ and order m. This is the only atom for which the Schrödinger equation has been solved for exactly. Multi-electron atoms require approximative methods. The family of solutions are:

$$\psi_{n\ell m}(r,\theta,\phi) = \sqrt{\left(\frac{2}{na_0}\right)^3 \frac{(n-\ell-1)!}{2n[(n+\ell)!]}} e^{-r/na_0}\left(\frac{2r}{na_0}\right)^\ell L_{n-\ell-1}^{2\ell+1}\left(\frac{2r}{na_0}\right)\cdot Y_\ell^m(\theta,\phi)$$

where:

- $a_0 = \dfrac{4\pi\varepsilon_0\hbar^2}{m_e e^2}$ is the Bohr radius,

- $L_{n-\ell-1}^{2\ell+1}(\cdots)$ are the generalized Laguerre polynomials of degree $n-\ell-1$.

- n, ℓ, m are the principal, azimuthal, and magnetic quantum numbers respectively: which take the values:

$$= 1,2,3,\ldots\ell = 0,1,2,\ldots,n-1 m = -\ell,\ldots,\ell$$

Two-electron Atoms or Ions

The equation for any two-electron system, such as the neutral helium atom (He, $Z = 2$), the negative hydrogen ion (H⁻, $Z = 1$), or the positive lithium ion (Li⁺, $Z = 3$) is:

$$E\psi = -\hbar^2\left[\frac{1}{2\mu}(\nabla_1^2 + \nabla_2^2) + \frac{1}{M}\nabla_1\cdot\nabla_2\right]\psi + \frac{e^2}{4\pi\varepsilon_0}\left[\frac{1}{r_{12}} - Z\left(\frac{1}{r_1} + \frac{1}{r_2}\right)\right]\psi$$

where r_1 is the position of one electron ($r_1 = |r_1|$ is its magnitude), r_2 is the position of the other electron ($r_2 = |r_2|$ is the magnitude), $r_{12} = |r_{12}|$ is the magnitude of the separation between them given by

$$|\mathbf{r}_{12}| = |\mathbf{r}_2 - \mathbf{r}_1|$$

μ is again the two-body reduced mass of an electron with respect to the nucleus of mass M, so this time

$$\mu = \frac{m_e M}{m_e + M}$$

and Z is the atomic number for the element (not a quantum number).

The cross-term of two laplacians

$$\frac{1}{M} \nabla_1 \cdot \nabla_2$$

is known as the *mass polarization term*, which arises due to the motion of atomic nuclei. The wavefunction is a function of the two electron's positions:

$$\psi = \psi(\mathbf{r}_1, \mathbf{r}_2).$$

There is no closed form solution for this equation.

Time Dependent

This is the equation of motion for the quantum state. In the most general form, it is written:

$$i\hbar \frac{\partial}{\partial t} \Psi = \hat{H} \Psi.$$

and the solution, the wavefunction, is a function of all the particle coordinates of the system and time. Following are specific cases.

For one particle in one dimension, the Hamiltonian

$$\hat{H} = \frac{\hat{p}^2}{2m} + V(x,t), \quad \hat{p} = -i\hbar \frac{\partial}{\partial x}$$

generates the equation:

$$i\hbar \frac{\partial}{\partial t} \Psi(x,t) = -\frac{\hbar^2}{2m} \frac{\partial^2}{\partial x^2} \Psi(x,t) + V(x,t)\Psi(x,t)$$

For N particles in one dimension, the Hamiltonian is:

$$\hat{H} = \sum_{n=1}^{N} \frac{\hat{p}_n^2}{2m_n} + V(x_1, x_2, \cdots x_N, t), \quad \hat{p}_n = -i\hbar \frac{\partial}{\partial x_n}$$

where the position of particle n is x_n, generating the equation:

$$i\hbar \frac{\partial}{\partial t} \Psi(x_1, x_2 \cdots x_N, t) = -\frac{\hbar^2}{2} \sum_{n=1}^{N} \frac{1}{m_n} \frac{\partial^2}{\partial x_n^2} \Psi(x_1, x_2 \cdots x_N, t) + V(x_1, x_2 \cdots x_N, t)\Psi(x_1, x_2 \cdots x_N, t).$$

For one particle in three dimensions, the Hamiltonian is:

$$\hat{H} = \frac{\hat{\mathbf{p}} \cdot \hat{\mathbf{p}}}{2m} + V(\mathbf{r},t), \quad \hat{\mathbf{p}} = -i\hbar\nabla$$

generating the equation:

$$i\hbar\frac{\partial}{\partial t}\Psi(\mathbf{r},t) = -\frac{\hbar^2}{2m}\nabla^2\Psi(\mathbf{r},t) + V(\mathbf{r},t)\Psi(\mathbf{r},t)$$

For N particles in three dimensions, the Hamiltonian is:

$$\hat{H} = \sum_{n=1}^{N}\frac{\hat{\mathbf{p}}_n \cdot \hat{\mathbf{p}}_n}{2m_n} + V(\mathbf{r}_1,\mathbf{r}_2,\cdots\mathbf{r}_N,t), \quad \hat{\mathbf{p}}_n = -i\hbar\nabla_n$$

where the position of particle n is $\mathbf{r}_n$, generating the equation:

$$i\hbar\frac{\partial}{\partial t}\Psi(\mathbf{r}_1,\mathbf{r}_2,\cdots\mathbf{r}_N,t) = -\frac{\hbar^2}{2}\sum_{n=1}^{N}\frac{1}{m_n}\nabla_n^2\Psi(\mathbf{r}_1,\mathbf{r}_2,\cdots\mathbf{r}_N,t) + V(\mathbf{r}_1,\mathbf{r}_2,\cdots\mathbf{r}_N,t)\Psi(\mathbf{r}_1,\mathbf{r}_2,\cdots\mathbf{r}_N,t)$$

This last equation is in a very high dimension, so the solutions are not easy to visualize.

Solution Methods

General techniques:	Methods for special cases:
<ul><li>Perturbation theory</li><li>The variational method</li><li>Quantum Monte Carlo methods</li><li>Density functional theory</li><li>The WKB approximation and semi-classical expansion</li></ul>	<ul><li>List of quantum-mechanical systems with analytical solutions</li><li>Hartree–Fock method and post Hartree–Fock methods</li></ul>

Properties

The Schrödinger equation has the following properties: some are useful, but there are shortcomings. Ultimately, these properties arise from the Hamiltonian used, and solutions to the equation.

Linearity

In the development above, the Schrödinger equation was made to be linear for generality, though this has other implications. If two wave functions ψ_1 and ψ_2 are solutions, then so is any linear combination of the two:

$$\psi = a\psi_1 + b\psi_2$$

where a and b are any complex numbers (the sum can be extended for any number of wavefunctions). This property allows superpositions of quantum states to be solutions of the Schrödinger equation. Even more generally, it holds that a general solution to the Schrödinger equation can be found by taking a weighted sum over all single state solutions achievable. For example, consider a wave function $\Psi(x, t)$ such that the wave function is a product of two functions: one time independent, and one time dependent. If states of definite energy found using the time independent Schrödinger equation are given by $\psi_E(x)$ with amplitude A_n and time dependent phase factor is given by

$$e^{-iE_n t/\hbar},$$

then a valid general solution is

$$\Psi(x,t) = \sum_n A_n \psi_{E_n}(x) e^{-iE_n t/\hbar}.$$

Additionally, the ability to scale solutions allows one to solve for a wave function without normalizing it first. If one has a set of normalized solutions ψ_n, then

$$\Psi = \sum_n A_n \psi_n$$

can be normalized by ensuring that

$$\sum_n |A_n|^2 = 1.$$

This is much more convenient than having to verify that

$$\int_{-\infty}^{\infty} |\Psi(x)|^2 \, dx = \int_{-\infty}^{\infty} \Psi(x)\Psi^*(x) \, dx = 1.$$

Real Energy Eigenstates

For the time-independent equation, an additional feature of linearity follows: if two wave functions ψ_1 and ψ_2 are solutions to the time-independent equation with the same energy E, then so is any linear combination:

$$\hat{H}(a\psi_1 + b\psi_2) = a\hat{H}\psi_1 + b\hat{H}\psi_2 = E(a\psi_1 + b\psi_2).$$

Two different solutions with the same energy are called *degenerate.*

In an arbitrary potential, if a wave function ψ solves the time-independent equation, so does its complex conjugate, denoted ψ^*. By taking linear combinations, the real and imaginary parts of ψ are each solutions. If there is no degeneracy they can only differ by a factor.

In the time-dependent equation, complex conjugate waves move in opposite directions. If $\Psi(x, t)$ is one solution, then so is $\Psi(x, -t)$. The symmetry of complex conjugation is called time-reversal symmetry.

Space and Time Derivatives

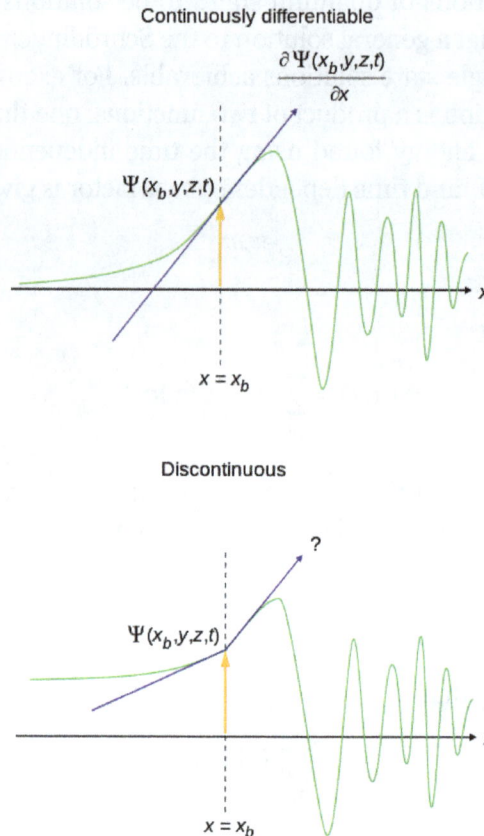

Continuously differentiable

Discontinuous

Continuity of the wavefunction and its first spatial derivative (in the x direction, y and z coordinates not shown), at some time t.

The Schrödinger equation is first order in time and second in space, which describes the time evolution of a quantum state (meaning it determines the future amplitude from the present).

Explicitly for one particle in 3-dimensional Cartesian coordinates – the equation is

$$i\hbar\frac{\partial\Psi}{\partial t} = -\frac{\hbar^2}{2m}\left(\frac{\partial^2\Psi}{\partial x^2} + \frac{\partial^2\Psi}{\partial y^2} + \frac{\partial^2\Psi}{\partial z^2}\right) + V(x,y,z,t)\Psi.$$

The first time partial derivative implies the initial value (at $t = 0$) of the wavefunction

$$\Psi(x,y,z,0)$$

is an arbitrary constant. Likewise – the second order derivatives with respect to space implies the wavefunction *and* its first order spatial derivatives

$$\Psi(x_b,y_b,z_b,t)\frac{\partial}{\partial x}\Psi(x_b,y_b,z_b,t) \quad \frac{\partial}{\partial y}\Psi(x_b,y_b,z_b,t) \quad \frac{\partial}{\partial z}\Psi(x_b,y_b,z_b,t)$$

are all arbitrary constants at a given set of points, where x_b, y_b, z_b are a set of points describing boundary b (derivatives are evaluated at the boundaries). Typically there are one or two boundaries, such as the step potential and particle in a box respectively.

As the first order derivatives are arbitrary, the wavefunction can be a continuously differentiable function of space, since at any boundary the gradient of the wavefunction can be matched.

On the contrary, wave equations in physics are usually *second order in time*, notable are the family of classical wave equations and the quantum Klein–Gordon equation.

Local Conservation of Probability

The Schrödinger equation is consistent with probability conservation. Multiplying the Schrödinger equation on the right by the complex conjugate wavefunction, and multiplying the wavefunction to the left of the complex conjugate of the Schrödinger equation, and subtracting, gives the continuity equation for probability:

$$\frac{\partial}{\partial t}\rho(\mathbf{r},t)+\nabla\cdot\mathbf{j}=0,$$

where

$$\rho=|\Psi|^2=\Psi^*(\mathbf{r},t)\Psi(\mathbf{r},t)$$

is the probability density (probability per unit volume, * denotes complex conjugate), and

$$\mathbf{j}=\frac{1}{2m}\left(\Psi^*\hat{\mathbf{p}}\Psi-\Psi\hat{\mathbf{p}}\Psi^*\right)$$

is the probability current (flow per unit area).

Hence predictions from the Schrödinger equation do not violate probability conservation.

Positive Energy

If the potential is bounded from below, meaning there is a minimum value of potential energy, the eigenfunctions of the Schrödinger equation have energy which is also bounded from below. This can be seen most easily by using the variational principle, as follows.

For any linear operator $\hat{A}$ bounded from below, the eigenvector with the smallest eigenvalue is the vector ψ that minimizes the quantity

$$\langle\psi\,|\,\hat{A}\,|\,\psi\rangle$$

over all ψ which are normalized. In this way, the smallest eigenvalue is expressed through the variational principle. For the Schrödinger Hamiltonian $\hat{H}$ bounded from below, the smallest eigenvalue is called the ground state energy. That energy is the minimum value of

$$\langle\psi\,|\,\hat{H}\,|\,\psi\rangle=\int\psi^*(\mathbf{r})\left[-\frac{\hbar^2}{2m}\nabla^2\psi(\mathbf{r})+V(\mathbf{r})\psi(\mathbf{r})\right]d^3\mathbf{r}=\int\left[\frac{\hbar^2}{2m}|\nabla\psi|^2+V(\mathbf{r})|\psi|^2\right]d^3\mathbf{r}=\langle\hat{H}\rangle$$

(using integration by parts). Due to the complex modulus of ψ^2 (which is positive definite), the right hand side always greater than the lowest value of $V(x)$. In particular, the ground state energy is positive when $V(x)$ is everywhere positive.

For potentials which are bounded below and are not infinite over a region, there is a ground state which minimizes the integral above. This lowest energy wavefunction is real and positive definite – meaning the wavefunction can increase and decrease, but is positive for all positions. It physically cannot be negative: if it were, smoothing out the bends at the sign change (to minimize the wavefunction) rapidly reduces the gradient contribution to the integral and hence the kinetic energy, while the potential energy changes linearly and less quickly. The kinetic and potential energy are both changing at different rates, so the total energy is not constant, which can't happen (conservation). The solutions are consistent with Schrödinger equation if this wavefunction is positive definite.

The lack of sign changes also shows that the ground state is nondegenerate, since if there were two ground states with common energy E, not proportional to each other, there would be a linear combination of the two that would also be a ground state resulting in a zero solution.

Analytic Continuation to Diffusion

The above properties (positive definiteness of energy) allow the analytic continuation of the Schrödinger equation to be identified as a stochastic process. This can be interpreted as the Huygens–Fresnel principle applied to De Broglie waves; the spreading wavefronts are diffusive probability amplitudes. For a free particle (not subject to a potential) in a random walk, substituting $\tau = it$ into the time-dependent Schrödinger equation gives:

$$\frac{\partial}{\partial \tau} X(\mathbf{r}, \tau) = \frac{\hbar}{2m} \nabla^2 X(\mathbf{r}, \tau), \quad X(\mathbf{r}, \tau) = \Psi(\mathbf{r}, \tau / i)$$

which has the same form as the diffusion equation, with diffusion coefficient $\hbar/2m$. In that case, the diffusivity yields the De Broglie relation in accordance with the Markov process.

Relativistic Quantum Mechanics

Relativistic quantum mechanics is obtained where quantum mechanics and special relativity simultaneously apply. In general, one wishes to build relativistic wave equations from the relativistic energy–momentum relation

$$E^2 = (pc)^2 + (m_0 c^2)^2,$$

instead of classical energy equations. The Klein–Gordon equation and the Dirac equation are two such equations. The Klein–Gordon equation,

$$\frac{1}{c^2} \frac{\partial^2}{\partial t^2} \psi - \nabla^2 \psi + \frac{m^2 c^2}{\hbar^2} \psi = 0., ,$$

was the first such equation to be obtained, even before the non-relativistic one, and applies to massive spinless particles. The Dirac equation arose from taking the "square root" of the Klein–Gordon equation by factorizing the entire relativistic wave operator into a product of two operators – one of these is the operator for the entire Dirac equation.

The general form of the Schrödinger equation remains true in relativity, but the Hamiltonian is less obvious. For example, the Dirac Hamiltonian for a particle of mass m and electric charge q in an electromagnetic field (described by the electromagnetic potentials φ and A) is:

$$\hat{H}_{Dirac} = \gamma^0 \left[c\gamma \cdot \left(\hat{p} - qA \right) + mc^2 + \gamma^0 q\phi \right],$$

in which the $\gamma = (\gamma^1, \gamma^2, \gamma^3)$ and γ^0 are the Dirac gamma matrices related to the spin of the particle. The Dirac equation is true for all spin-$\frac{1}{2}$ particles, and the solutions to the equation are 4-component spinor fields with two components corresponding to the particle and the other two for the antiparticle.

For the Klein–Gordon equation, the general form of the Schrödinger equation is inconvenient to use, and in practice the Hamiltonian is not expressed in an analogous way to the Dirac Hamiltonian. The equations for relativistic quantum fields can be obtained in other ways, such as starting from a Lagrangian density and using the Euler–Lagrange equations for fields, or use the representation theory of the Lorentz group in which certain representations can be used to fix the equation for a free particle of given spin (and mass).

In general, the Hamiltonian to be substituted in the general Schrödinger equation is not just a function of the position and momentum operators (and possibly time), but also of spin matrices. Also, the solutions to a relativistic wave equation, for a massive particle of spin s, are complex-valued $2(2s + 1)$-component spinor fields.

Quantum Field Theory

The general equation is also valid and used in quantum field theory, both in relativistic and non-relativistic situations. However, the solution ψ is no longer interpreted as a "wave", but should be interpreted as an operator acting on states existing in a Fock space.

First Order Form

The Schrödinger equation can also be derived from a first order form similar to the manner in which the Klein-Gordon equation can be derived from the Dirac equation. In 1D the first order equation is given by

$$-i\partial_z \psi = (i\eta\partial_t + \eta^\dagger m)\psi$$

This equation allows for the inclusion of spin in non-relativistic quantum mechanics. Squaring the above equation yields the Schrödinger equation in 1D. The matrices obey the following properties

$$\eta^2 = 0$$

$$(\eta^\dagger)^2 = 0$$

$$\{\eta, \eta^{\dagger}\} = 2I$$

The 3 dimensional version of the equation is given by

$$-i\gamma_i \partial_i \psi = (i\eta \partial_t + \eta^{\dagger} m)\psi$$

Here $\eta = (\gamma_0 + i\gamma_5)/\sqrt{2}$ is a 4×4 nilpotent matrix and γ_i are the Dirac gamma matrices ($i = 1, 2, 3$). The Schrödinger equation in 3D can be obtained by squaring the above equation. In the non-relativistic limit $E - m \simeq E'$ and $E + m \simeq 2m$, the above equation can be derived from the Dirac equation.

Schrödinger Field

In quantum mechanics and quantum field theory, a Schrödinger field, named after Erwin Schrödinger, is a quantum field which obeys the Schrödinger equation. While any situation described by a Schrödinger field can also be described by a many-body Schrödinger equation for identical particles, the field theory is more suitable for situations where the particle number changes.

A Schrödinger field is also the classical limit of a quantum Schrödinger field, a classical wave which satisfies the Schrödinger equation. Unlike the quantum mechanical wavefunction, if there are interactions between the particles the equation will be nonlinear. These nonlinear equations describe the classical wave limit of a system of interacting identical particles.

The path integral of a Schrödinger field is also known as a coherent state path integral, because the field itself is an annihilation operator whose eigenstates can be thought of as coherent states of the harmonic oscillations of the field modes.

Schrödinger fields are useful for describing Bose–Einstein condensation, the Bogolyubov–de Gennes equation of superconductivity, superfluidity, and many-body theory in general. They are also a useful alternative formalism for nonrelativistic quantum mechanics.

A Schrödinger field is the nonrelativistic limit of a Klein–Gordon field.

Summary

A Schrödinger field is a quantum field whose quanta obey the Schrödinger equation. In the classical limit, it can be understood as the quantized wave equation of a Bose Einstein condensate or a superfluid.

Free Field

A Schrödinger field has the free field Lagrangian

$$L = \psi^{\dagger} \left(i \frac{\partial}{\partial t} + \frac{\nabla^2}{2m} \right) \psi.$$

When ψ is a complex valued field in a path integral, or equivalently an operator with canonical commutation relations, it describes a collection of identical nonrelativistic bosons. When ψ is a grassmann valued field, or equivalently an operator with canonical anticommutation relations, the field describes identical fermions.

External Potential

If the particles interact with an external potential $V(x)$, the interaction makes a local contribution to the action:

$$S = \int_{xt} \psi^\dagger \left(i\frac{\partial}{\partial t} + \frac{\nabla^2}{2m} \right)\psi - \psi^\dagger(x)\psi(x)V(x).$$

If the ordinary Schrödinger equation for V has known energy eigenstates $\phi_i(x)$ with energies E_i, then the field in the action can be rotated into a diagonal basis by a mode expansion:

$$\psi(x) = \sum_i \psi_i \phi_i(x).$$

The action becomes:

$$S = \int_t \sum_i \psi_i^\dagger \left(i\frac{\partial}{\partial t} - E_i \right)\psi_i$$

which is the position-momentum path integral for a collection of independent Harmonic oscillators.

To see the equivalence, note that decomposed into real and imaginary parts the action is:

$$S = \int_t \sum_i 2\psi_r \frac{d\psi_i}{dt} - E_i(\psi_r^2 + \psi_i^2)$$

after an integration by parts. Integrating over ψ_r gives the action

$$S = \int_t \sum_i \frac{1}{E_i} \left(\frac{d\psi_i}{dt} \right)^2 - E_i\psi_i^2$$

which, rescaling ψ_i, is a harmonic oscillator action with frequency E_i.

Pair Potential

When the particles interact with a pair potential $V(x_1, x_2)$,, the interaction is a nonlocal contribution to the action:

$$S = \int_{xt} \psi^\dagger \left(i\frac{\partial}{\partial t} + \frac{\nabla^2}{3m} \right)\psi - \int_{xy} \psi^\dagger(y)\psi^\dagger(x)V(x, y)\psi(x)\psi(y).$$

A pair-potential is the non-relativistic limit of a relativistic field coupled to electrodynamics. Ignoring the propagating degrees of freedom, the interaction between nonrelativistic electrons is the coulomb repulsion. In 2+1 dimensions, this is:

$$V(x, y) = \frac{j^2}{|y - x|}.$$

When coupled to an external potential to model classical positions of nuclei, a Schrödinger field with this pair potential describes nearly all of condensed matter physics. The exceptions are effects like superfluidity, where the quantum mechanical interference of nuclei is important, and inner shell electrons where the electron motion can be relativistic.

Nonlinear Schrödinger Equation

A special case of a delta-function interaction $V(x_1, x_2)$ is widely studied, and is known as the non-linear Schrödinger equation. Because the interactions always happen when two particles occupy the same point, the action for the nonlinear Schrödinger equation is local:

$$S = \int_x \psi^\dagger \left(i\frac{\partial}{\partial t} + \frac{\nabla^2}{2m} \right) \psi + \lambda \psi^\dagger \psi^\dagger \psi \psi$$

The interaction strength λ requires renormalization in dimensions higher than 2 and in two dimensions it has logarithmic divergence. In any dimensions, and even with power-law divergence, the theory is well defined. If the particles are fermions, the interaction vanishes.

Many-body Potentials

The potentials can include many-body contributions. The interacting Lagrangian is then:

$$L_i = \int_x \psi^\dagger(x_1)\psi^\dagger(x_2)\cdots\psi^\dagger(x_n)V(x_1, x_2, \ldots, x_n)\psi(x_1)\psi(x_2)\cdots\psi(x_n).$$

These types of potentials are important in some effective descriptions of close-packed atoms. Higher order interactions are less and less important.

Canonical Formalism

The canonical momentum association with the field ψ is

$$\Pi(x) = i\psi^\dagger.$$

The canonical commutation relations are like an independent harmonic oscillator at each point:

$$[\psi(x), \psi^\dagger(y)] = \delta(x - y).$$

The field Hamiltonian is

$$H = S - \int \Pi(x) \frac{d}{dt} \psi = \int \frac{|\nabla \psi|^2}{2m} + \int_{xy} \psi^\dagger(x)\psi^\dagger(y)V(x,y)\psi(x)\psi(y)$$

and the field equation for any interaction is a nonlinear and nonlocal version of the Schrödinger equation. For pairwise interactions:

$$i\frac{\partial}{\partial t}\psi = -\frac{\nabla^2}{2m}\psi + \left(\int_y V(x,y)\psi^\dagger(y)\psi(y)\right)\psi(x).$$

Perturbation Theory

The expansion in Feynman diagrams is called many-body perturbation theory. The propagator is

$$G(k) = \frac{1}{i\omega - \dfrac{k^2}{2m}}.$$

The interaction vertex is the Fourier transform of the pair-potential. In all the interactions, the number of incoming and outgoing lines is equal.

Exposition

Identical Particles

The many body Schrödinger equation for identical particles describes the time evolution of the many-body wavefunction $\psi(x_1, x_2...x_N)$ which is the probability amplitude for N particles to have the listed positions. The Schrödinger equation for ψ is:

$$i\frac{d}{dt}\psi = \left(\frac{\nabla_1^2}{2m} + \frac{\nabla_2^2}{2m} + \cdots + \frac{\nabla_N^2}{2m} + V(x_1, x_2, \ldots, x_N)\right)\psi$$

with Hamiltonian

$$H = \frac{p_1^2}{2m} + \frac{p_2^2}{2m} + \cdots + \frac{p_N^2}{2m} + V(x_1, \ldots, x_N).$$

Since the particles are indistinguishable, the wavefunction has some symmetry under switching positions. Either

1. $\psi(x_1, x_2, \ldots) = \psi(x_2, x_1, \ldots)$ for bosons, ,

2. $\psi(x_1, x_2, \ldots) = -\psi(x_2, x_1, \ldots)$ for fermions .

Since the particles are indistinguishable, the potential V must be unchanged under permutations. If

$$V(x_1, \ldots, x_N) = V_1(x_1) + V_2(x_2) + \cdots + V_N(x_N)$$

then it must be the case that $V_1 = V_2 = \cdots = V_N$. If

$$V(x_1 \ldots, x_N) = V_{1,2}(x_1, x_2) + V_{1,3}(x_2, x_3) + V_{2,3}(x_1, x_2)$$

then $V_{1,2} = V_{1,3} = V_{2,3}$ and so on.

In the Schrödinger equation formalism, the restrictions on the potential are ad-hoc, and the classical wave limit is hard to reach. It also has limited usefulness if a system is open to the environment, because particles might coherently enter and leave.

Nonrelativistic Fock Space

A Schrödinger field is defined by extending the Hilbert space of states to include configurations with arbitrary particle number. A nearly complete basis for this set of states is the collection:

$$| N; x_1, \ldots, x_N \rangle$$

labeled by the total number of particles and their position. An arbitrary state with particles at separated positions is described by a superposition of states of this form.

$$\psi_0 | 0 \rangle + \int_x \psi_1(x) | 1; x \rangle + \int_{x_1 x_2} \psi_2(x_1, x_2) | 2; x_1 x_2 \rangle + \ldots$$

In this formalism, keep in mind that any two states whose positions can be permuted into each other are really the same, so the integration domains need to avoid double counting. Also keep in mind that the states with more than one particle at the same point have not yet been defined. The quantity ψ_0 is the amplitude that no particles are present, and its absolute square is the probability that the system is in the vacuum.

In order to reproduce the Schrödinger description, the inner product on the basis states should be

$$\langle 1; x_1 | 1; y_1 \rangle = \delta(x_1 - y_1)$$

$$\langle 2; x_1 x_2 | 2; y_1 y_2 \rangle = \delta(x_1 - y_1)\delta(x_2 - y_2) \pm \delta(x_1 - y_2)\delta(x_2 - y_1)$$

and so on. Since the discussion is nearly formally identical for bosons and fermions, although the physical properties are different, from here on the particles will be bosons.

There are natural operators in this Hilbert space. One operator, called $\psi^\dagger(x)$, is the operator which introduces an extra particle at x. It is defined on each basis state:

$$\psi^\dagger(x) | N; x_1 \ldots x_n \rangle = | N+1; x_1, \ldots, x_n, x \rangle$$

with slight ambiguity when a particle is already at x.

Another operator removes a particle at x, and is called ψ. This operator is the conjugate of the operator $\psi^\dagger$. Because $\psi^\dagger$ has no matrix elements which connect to states with no particle at x, ψ must give zero when acting on such a state.

$$\psi(x)\,|\,N;x_1...,x_N\rangle = \delta(x-x_1)\,|\,N-1;x_2...,x_N\rangle + \delta(x-x_2)\,|\,N-1;x_1,x_3...,x_N\rangle + ...$$

The position basis is an inconvenient way to understand coincident particles because states with a particle localized at one point have infinite energy, so intuition is difficult. In order to see what happens when two particles are at exactly the same point, it is mathematically simplest either to make space into a discrete lattice, or to Fourier transform the field in a finite volume.

The operator

$$\psi^\dagger(k) = \int_x e^{-ikx}\psi^\dagger(x)$$

creates a superposition of one particle states in a plane wave state with momentum k, in other words, it produces a new particle with momentum k. The operator

$$\psi(k) = \int_x e^{ikx}\psi(x)$$

annihilates a particle with momentum k.

If the potential energy for interaction of infinitely distant particles vanishes, the fourier transformed operators in infinite volume create states which are noninteracting. The states are infinitely spread out, and the chance that the particles are nearby is zero.

The matrix elements for the operators between non-coincident points reconstructs the matrix elements of the Fourier transform between all modes:

1. $\psi^\dagger(k)\psi^\dagger(k') - \psi^\dagger(k')\psi^\dagger(k) = 0$

2. $\psi(k)\psi(k') - \psi(k')\psi(k) = 0$

3. $\psi(k)\psi^\dagger(k') - \psi(k')\psi^\dagger(k) = \delta(k-k')$

where the delta function is either the Dirac delta function or the Kronecker delta, depending on whether the volume is infinite or finite.

The commutation relations now determine the operators completely, and when the spatial volume is finite, there are no conceptual hurdle to understand coinciding momenta because momenta are discrete. In a discrete momentum basis, the basis states are:

$$|\,n_1,n_2,...n_k\rangle$$

where the n's are the number of particles at each momentum. For fermions and anyons, the number of particles at any momentum is always either zero or one. The operators ψ_k have harmonic-oscillator like matrix elements between states, independent of the interaction:

$$\psi^\dagger(k)\,|..,n_k,...\rangle = \sqrt{n_k+1}\,|...,n_k+1,...\rangle$$

$$\psi(k)\,|...,n_k,...\rangle = \sqrt{n_k}\,|...,n_k-1,...\rangle$$

So that the operator

$$\sum_k \psi^\dagger(k)\psi(k) = \int_x \psi^\dagger(x)\psi(x)$$

counts the total number of particles.

Now it is easy to see that the matrix elements of $\psi(x)$ and $\psi^\dagger(x)$ have harmonic oscillator commutation relations too.

1. $[\psi(x), \psi(y)] = [\psi^\dagger(x), \psi^\dagger(y)] = 0$

2. $[\psi(x), \psi^\dagger(y)] = \delta(x - y)$

So that there really is no difficulty with coincident particles in position space.

The operator $\psi^\dagger(x)\psi(x)$ which removes and replaces a particle, acts as a sensor to detect if a particle is present at x. The operator $\psi^\dagger \nabla \psi$ acts to multiply the state by the gradient of the many body wavefunction. The operator

$$H = \int_x \psi^\dagger(x)\frac{\nabla^2}{2m}\psi(x)$$

acts to reproduce the right hand side of the Schrödinger equation when acting on any basis state, so that

$$\psi^\dagger i \frac{d}{dt}\psi = \psi^\dagger \frac{-\nabla^2}{2m}\psi$$

holds as an operator equation. Since this is true for an arbitrary state, it is also true without the $\psi^\dagger$.

$$i\frac{\partial}{\partial t}\psi = \frac{-\nabla^2}{2m}\psi \ .$$

To add interactions, add nonlinear terms in the field equations. The field form automatically ensures that the potentials obey the restrictions from symmetry.

Field Hamiltonian

The field Hamiltonian which reproduces the equations of motion is

$$H = \frac{\nabla \psi^\dagger \nabla \psi}{2m}$$

The Heisenberg equations of motion for this operator reproduces the equation of motion for the field.

To find the classical field Lagrangian, apply a Legendre transform to the classical limit of the Hamiltonian.

$$L = \psi^\dagger \left(i \frac{\partial}{\partial t} + \frac{\nabla^2}{2m} \right) \psi$$

Although this is correct classically, the quantum mechanical transformation is not completely conceptually straightforward because the path integral is over eigenvalues of operators ψ which are not hermitian and whose eigenvalues are not orthogonal. The path integral over field states therefore seems naively to be overcounting. This is not the case, because the time derivative term in L includes the overlap between the different field states.

Nonlinear Schrödinger Equation

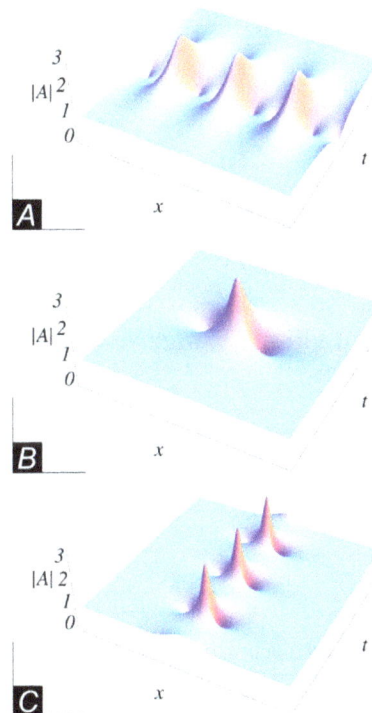

Absolute value of the complex envelope of exact analytical breather solutions of the nonlinear Schrödinger (NLS) equation in nondimensional form. (A) The Akhmediev breather; (B) the Peregrine breather; (C) the Kuznetsov–Ma breather.

In theoretical physics, the (one-dimensional) nonlinear Schrödinger equation (NLSE) is a nonlinear variation of the Schrödinger equation. It is a classical field equation whose principal applications are to the propagation of light in nonlinear optical fibers and planar waveguides and to Bose-Einstein condensates confined to highly anisotropic cigar-shaped traps, in the mean-field regime. Additionally, the equation appears in the studies of small-amplitude gravity waves on the surface of deep inviscid (zero-viscosity) water; the Langmuir waves in hot plasmas; the propagation of plane-diffracted wave beams in the focusing regions of the ionosphere; the propagation of Davydov's alpha-helix solitons, which are responsible for energy transport along molecular chains; and many others. More generally, the NLSE appears as one of universal equations that describe the evolution of slowly varying packets of quasi-monochromatic waves in weakly nonlinear media that have dispersion. Unlike the linear Schrödinger equation, the NLSE never describes the time evolution of a quantum state (except hypothetically, as in some early attempts, in the 1970s, to explain the quantum measurement process). The 1D NLSE is an example of an integrable model.

In quantum mechanics, the 1D NLSE is a special case of the classical nonlinear Schrödinger field, which in turn is a classical limit of a quantum Schrödinger field. Conversely, when the classical Schrödinger field is canonically quantized, it becomes a quantum field theory (which is linear, despite the fact that it is called "quantum *nonlinear* Schrödinger equation") that describes bosonic point particles with delta-function interactions — the particles either repel or attract when they are at the same point. In fact, when the number of particles is finite, this quantum field theory is equivalent to the Lieb–Liniger model. Both the quantum and the classical 1D nonlinear Schrödinger equations are integrable. Of special interest is the limit of infinite strength repulsion, in which case the Lieb–Liniger model becomes the Tonks–Girardeau gas (also called the hard-core Bose gas, or impenetrable Bose gas). In this limit, the bosons may, by a change of variables that is a continuum generalization of the Jordan–Wigner transformation, be transformed to a system one-dimensional noninteracting spinless fermions.

The nonlinear Schrödinger equation is a simplified 1+1-dimensional form of the Ginzburg–Landau equation introduced in 1950 in their work on superconductivity, and was written down explicitly by R. Y. Chiao, E. Garmire, and C. H. Townes (1964, equation (5)) in their study of optical beams.

Multi-dimensional version replaces the second spatial derivative by the Laplacian. In more than one dimension, the equation is not integrable, it allows for a collapse and wave turbulence

Equation

The nonlinear Schrödinger equation is a nonlinear partial differential equation, applicable to classical and quantum mechanics.

Classical Equation

The classical field equation (in dimensionless form) is:

> **Nonlinear Schrödinger equation** *(Classical field theory)*
>
> $$i\partial_t \psi = -\frac{1}{2}\partial_x^2 \psi + \kappa |\psi|^2 \psi$$

for the complex field $\psi(x,t)$.

This equation arises from the Hamiltonian

$$H = \int dx \left[\frac{1}{2}|\partial_x \psi|^2 + \frac{\kappa}{2}|\psi|^4 \right]$$

with the Poisson brackets

$$\{\psi(x),\psi(y)\} = \{\psi^*(x),\psi^*(y)\} = 0$$

$$\{\psi^*(x),\psi(y)\} = i\delta(x-y).$$

Unlike its linear counterpart, it never describes the time evolution of a quantum state.

The case with negative κ is called focusing and allows for bright soliton solutions (localized in space, and having spatial attenuation towards infinity) as well as breather solutions. It can be solved exactly by use of the inverse scattering transform, as shown by Zakharov & Shabat (1972) . The other case, with κ positive, is the defocusing NLS which has dark soliton solutions (having constant amplitude at infinity, and a local spatial dip in amplitude).

Quantum Mechanics

To get the quantized version, simply replace the Poisson brackets by commutators

$$[\psi(x),\psi(y)] = [\psi^*(x),\psi^*(y)] = 0$$
$$[\psi^*(x),\psi(y)] = -\delta(x-y)$$

and normal order the Hamiltonian

$$H = \int dx \left[\frac{1}{2}\partial_x\psi^\dagger\partial_x\psi + \frac{\kappa}{2}\psi^\dagger\psi^\dagger\psi\psi \right].$$

The quantum version was solved by Bethe ansatz by Lieb and Liniger. Thermodynamics was described by Chen Nin Yang. Quantum correlation functions also were evaluated by Korepin in 1993. The model has higher conservation laws - Davies and Korepin in 1989 expressed them in terms of local fields.

Solving the Equation

The nonlinear Schrödinger equation is integrable in 1d: Zakharov and Shabat (1972) solved it with the inverse scattering transform. The corresponding linear system of equations is known as the Zakharov–Shabat system:

$$\phi_x = J\phi\Lambda + U\phi$$
$$\phi_t = 2J\phi\Lambda^2 + 2U\phi\Lambda + (JU^2 - JU_x)\phi,$$

where

$$\Lambda = \begin{pmatrix} \lambda_1 & 0 \\ 0 & \lambda_2 \end{pmatrix}, \quad J = i\sigma_z = \begin{pmatrix} i & 0 \\ 0 & -i \end{pmatrix}, \quad U = i\begin{pmatrix} 0 & q \\ r & 0 \end{pmatrix}.$$

The nonlinear Schrödinger equation arises as compatibility condition of the Zakharov–Shabat system:

$$\phi_{xt} = \phi_{tx} \quad \Rightarrow \quad U_t = -JU_{xx} + 2JU^2U \quad \Leftrightarrow \quad \begin{cases} iq_t = q_{xx} + 2qrq \\ ir_t = -r_{xx} - 2qrr. \end{cases}$$

By setting $q = r^*$ or $q = -r^*$ the nonlinear Schrödinger equation with attractive or repulsive inter-action is obtained.

An alternative approach uses the Zakharov–Shabat system directly and employs the following Darboux transformation:

$$\phi \rightarrow \phi[1] = \phi\Lambda - \sigma\phi$$
$$U \rightarrow U[1] = U + [J, \sigma]$$
$$\sigma = \varphi\Omega\varphi^{-1}$$

which leaves the system invariant.

Here, φ is another invertible matrix solution (different from ϕ) of the Zakharov–Shabat system with spectral parameter Ω:

$$\varphi_x = J\varphi\Omega + U\varphi$$
$$\varphi_t = 2J\varphi\Omega^2 + 2U\varphi\Omega + (JU^2 - JU_x)\varphi.$$

Starting from the trivial solution $U = 0$ and iterating, one obtains the solutions with n solitons.

Computational solutions are found using a variety of methods, like the split-step method.

Galilean Invariance

The nonlinear Schrödinger equation is Galilean invariant in the following sense:

Given a solution $\psi(x, t)$ a new solution can be obtained by replacing x with $x + vt$ everywhere in $\psi(x, t)$ and by appending a phase factor of

$$\psi(x,t) \mapsto \psi_{[v]}(x,t) = \psi(x+vt,t)\, e^{-iv(x+vt/2)}.$$

The Nonlinear Schrödinger Equation in Fiber Optics

In optics, the nonlinear Schrödinger equation occurs in the Manakov system, a model of wave propagation in fiber optics. The function ψ represents a wave and the nonlinear Schrödinger equation describes the propagation of the wave through a nonlinear medium. The second-order deriv-ative represents the dispersion, while the κ term represents the nonlinearity. The equation models many nonlinearity effects in a fiber, including but not limited to self-phase modulation, four-wave mixing, second harmonic generation, stimulated Raman scattering, etc.

The Nonlinear Schrödinger Equation in Water Waves

For water waves, the nonlinear Schrödinger equation describes the evolution of the envelope of modulated wave groups. In a paper in 1968, Vladimir E. Zakharov describes the Hamiltonian structure of water waves. In the same paper Zakharov shows, that for slowly modulated wave groups, the wave amplitude satisfies the nonlinear Schrödinger equation, approximately. The value of the nonlinearity parameter κ depends on the relative water depth. For deep water, with the water depth large compared to the wave length of the water waves, κ is negative and envelope solitons may occur.

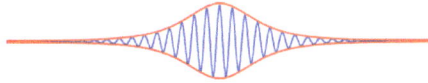

A hyperbolic secant (sech) envelope soliton for surface waves on deep water.
Blue line: water waves.
Red line: envelope soliton.

For shallow water, with wavelengths longer than 4.6 times the water depth, the nonlinearity parameter κ is positive and *wave groups* with *envelope* solitons do not exist. Note, that in shallow water *surface-elevation* solitons or waves of translation do exist, but they are not governed by the nonlinear Schrödinger equation.

The nonlinear Schrödinger equation is thought to be important for explaining the formation of rogue waves.

The complex field ψ, as appearing in the nonlinear Schrödinger equation, is related to the amplitude and phase of the water waves. Consider a slowly modulated carrier wave with water surface elevation η of the form:

$$\eta = a(x_0,t_0) \cos\left[k_0 x_0 - \omega_0 t_0 - \theta(x_0,t_0)\right],$$

where $a(x_0, t_0)$ and $\theta(x_0, t_0)$ are the slowly modulated amplitude and phase. Further ω_0 and k_0 are the (constant) angular frequency and wavenumber of the carrier waves, which have to satisfy the dispersion relation $\omega_0 = \Omega(k_0)$. Then

$$\psi = a \exp\left(i\theta\right).$$

So its modulus $|\psi|$ is the wave amplitude a, and its argument $\arg(\psi)$ is the phase θ.

The relation between the physical coordinates (x_0, t_0) and the (x, t) coordinates, as used in the nonlinear Schrödinger equation given above, is given by:

$$x = k_0\left[x_0 - \Omega'(k_0)\, t_0\right], \quad t = k_0^2\left[-\Omega''(k_0)\right] t_0$$

Thus (x, t) is a transformed coordinate system moving with the group velocity $\Omega'(k_0)$ of the carrier waves, The dispersion-relation curvature $\Omega''(k_0)$ is always negative for water waves under the action of gravity.

For waves on the water surface of deep water, the coefficients of importance for the nonlinear Schrödinger equation are:

$$\kappa = -2k_0^2, \quad \Omega(k_0) = \sqrt{gk_0} = \omega_0$$

so

$$\Omega'(k_0) = \frac{1}{2}\frac{\omega_0}{k_0}, \quad \Omega''(k_0) = -\frac{1}{4}\frac{\omega_0^3}{k_0^3}$$

where g is the acceleration due to gravity at the Earth's surface.

In the original (x_0, t_0) coordinates the nonlinear Schrödinger equation for water waves reads:

$$i\partial_{t_0} A + i\Omega'(k_0)\partial_{x_0} A + \tfrac{1}{2}\Omega''(k_0)\partial_{x_0 x_0} A - v|A|^2 A = 0,$$

with $A = \psi^*$ (i.e. the complex conjugate of Ψ) and $v = k_0^2 \Omega''(k_0)$. So $v = \tfrac{1}{2}\omega_0 k_0^2$ for deep water waves.

Gauge Equivalent Counterpart

NLSE (1) is gauge equivalent to the following isotropic Landau-Lifshitz equation (LLE) or Heisenberg ferromagnet equation

$$\vec{S}_t = \vec{S} \wedge \vec{S}_{xx}.$$

Note that this equation admits several integrable and non-integrable generalizations in 2 + 1 dimensions like the Ishimori equation and so on.

Relation to Vortices

Hasimoto (1972) showed that the work of da Rios (1906) on vortex filaments is closely related to the nonlinear Schrödinger equation. Subsequently Salman (2013) used this correspondence to show that breather solutions can also arise for a vortex filament.

Fractional Schrödinger Equation

The fractional Schrödinger equation is a fundamental equation of fractional quantum mechanics. It was discovered by Nick Laskin (1999) as a result of extending the Feynman path integral, from the Brownian-like to Lévy-like quantum mechanical paths. The term *fractional Schrödinger equation* was coined by Nick Laskin.

Fundamentals

The fractional Schrödinger equation in the form originally obtained by Nick Laskin is:

$$i\hbar \frac{\partial \psi(\mathbf{r}, t)}{\partial t} = D_\alpha (-\hbar^2 \Delta)^{\alpha/2} \psi(\mathbf{r}, t) + V(\mathbf{r}, t)\psi(\mathbf{r}, t)$$

- r is the 3-dimensional position vector,
- $\hbar$ is the reduced Planck constant,

- $\psi(r, t)$ is the wavefunction, which is the quantum mechanical probability amplitude for the particle to have a given position r at any given time t,

- $V(r, t)$ is a potential energy,

- $\Delta = \partial^2/\partial r^2$ is the Laplace operator.

Further,

- D_α is a scale constant with physical dimension $[D_\alpha] = [\text{energy}]^{1-\alpha}\cdot[\text{length}]^{\alpha}[\text{time}]^{-\alpha}$, at $\alpha = 2$, $D_2 = 1/2m$, where m is a particle mass,

- the operator $(-\hbar^2\Delta)^{\alpha/2}$ is the 3-dimensional fractional quantum Riesz derivative defined by;

$$(-\hbar^2\Delta)^{\alpha/2}\psi(\mathbf{r},t) = \frac{1}{(2\pi\hbar)^3}\int d^3p\, e^{i\frac{\mathbf{pr}}{\hbar}}|\mathbf{p}|^\alpha\, \varphi(\mathbf{p},t),$$

Here, the wave functions in the position and momentum spaces; $\psi(\mathbf{r},t)$ and $\varphi(\mathbf{p},t)$ are related each other by the 3-dimensional Fourier transforms:

$$\psi(\mathbf{r},t) = \frac{1}{(2\pi\hbar)^3}\int d^3p\, e^{i\mathbf{p}\cdot\mathbf{r}/\hbar}\varphi(\mathbf{p},t), \qquad \varphi(\mathbf{p},t) = \int d^3r\, e^{-i\mathbf{p}\cdot\mathbf{r}/\hbar}\psi(\mathbf{r},t).$$

The index α in the fractional Schrödinger equation is the Lévy index, $1 < \alpha \le 2$. Thus, the fractional Schrödinger equation includes a space derivative of fractional order α instead of the second order ($\alpha = 2$) space derivative in the standard Schrödinger equation. Thus, the fractional Schrödinger equation is a fractional differential equation in accordance with modern terminology. This is the main point of the term *fractional Schrödinger equation* or a more general term fractional quantum mechanics. At $\alpha = 2$ fractional Schrödinger equation becomes the well-known Schrödinger equation.

The fractional Schrödinger equation has the following operator form

$$i\hbar\frac{\partial\psi(\mathbf{r},t)}{\partial t} = \hat{H}_\alpha\psi(\mathbf{r},t)$$

where the fractional Hamilton operator $\hat{H}_\alpha$ is given by

$$\hat{H}_\alpha = D_\alpha(-\hbar^2\Delta)^{\alpha/2} + V(\mathbf{r},t).$$

The Hamilton operator, $\hat{H}_\alpha$ corresponds to the classical mechanics Hamiltonian function introduced by Nick Laskin

$$H_\alpha(\mathbf{p},\mathbf{r}) = D_\alpha|\mathbf{p}|^\alpha + V(\mathbf{r},t),$$

where p and r are the momentum and the position vectors respectively.

Time-independent Fractional Schrödinger Equation

The special case when the Hamiltonian H_α is independent of time

$$H_\alpha = D_\alpha(-\hbar^2\Delta)^{\alpha/2} + V(\mathbf{r}),$$

is of great importance for physical applications. It is easy to see that in this case there exist the special solution of the fractional Schrödinger equation

$$\psi(\mathbf{r},t) = e^{-(i/\hbar)Et}\phi(\mathbf{r}),$$

where $\phi(\mathbf{r})$ satisfies

$$H_\alpha\phi(\mathbf{r}) = E\phi(\mathbf{r}),$$

or

$$D_\alpha(-\hbar^2\Delta)^{\alpha/2}\phi(\mathbf{r}) + V(\mathbf{r})\phi(\mathbf{r}) = E\phi(\mathbf{r}).$$

This is the time-independent fractional Schrödinger equation.

Thus, we see that the wave function $\psi(\mathbf{r},t)$ oscillates with a definite frequency. In classical physics the frequency corresponds to the energy. Therefore, the quantum mechanical state has a definite energy E. The probability to find a particle at $\mathbf{r}$ is the absolute square of the wave function $|\psi(\mathbf{r},t)|^2$. Because of time-independent fractional Schrödinger equation this is equal to $|\phi(\mathbf{r})|^2$ and does not depend upon the time. That is, the probability of finding the particle at $\mathbf{r}$ is independent of the time. One can say that the system is in a stationary state. In other words, there is no variation in the probabilities as a function of time.

Probability Current Density

The conservation law of fractional quantum mechanical probability has been discovered for the first time by D.A.Tayurskii and Yu.V. Lysogorski

$$\frac{\partial\rho(\mathbf{r},t)}{\partial t} + \nabla\cdot\mathbf{j}(\mathbf{r},t) + K(\mathbf{r},t) = 0,$$

where $\rho(\mathbf{r},t) = \psi^*(\mathbf{r},t)\psi(\mathbf{r},t)$ is the quantum mechanical probability density and the vector $\mathbf{j}(\mathbf{r},t)$ can be called by the fractional probability current density vector

$$\mathbf{j}(\mathbf{r},t) = \frac{D_\alpha\hbar}{i}\left(\psi^*(\mathbf{r},t)(-\hbar^2\Delta)^{\alpha/2-1}\nabla\psi(\mathbf{r},t) - \psi(\mathbf{r},t)(-\hbar^2\Delta)^{\alpha/2-1}\nabla\psi^*(\mathbf{r},t)\right),$$

and

$$K(\mathbf{r},t) = \frac{D_\alpha\hbar}{i}\left(\nabla\psi(\mathbf{r},t)(-\hbar^2\Delta)^{\alpha/2-1}\nabla\psi^*(\mathbf{r},t) - (\nabla\psi^*(\mathbf{r},t)(-\hbar^2\Delta)^{\alpha/2-1}\nabla\psi(\mathbf{r},t)\right),$$

here we use the notation: $\nabla = \partial/\partial\mathbf{r}$.

It has been found in Ref. that there are quantum physical conditions when the new term $K(\mathbf{r},t)$ is negligible and we come to the continuity equation for quantum probability current and quantum density:

$$\frac{\partial \rho(\mathbf{r},t)}{\partial t} + \nabla \cdot \mathbf{j}(\mathbf{r},t) = 0.$$

Introducing the momentum operator $\hat{\mathbf{p}} = \dfrac{\hbar}{i}\dfrac{\partial}{\partial \mathbf{r}}$ we can write the vector $\mathbf{j}$ in the form

$$\mathbf{j} = D_{\alpha}\left(\psi(\hat{\mathbf{p}}^2)^{\alpha/2-1}\hat{\mathbf{p}}\psi^* + \psi^*(\hat{\mathbf{p}}^{*2})^{\alpha/2-1}\hat{\mathbf{p}}^*\psi\right).$$

This is fractional generalization of the well-known equation for probability current density vector of standard quantum mechanics.

Velocity Operator

The quantum mechanical velocity operator $\hat{\mathbf{v}}$ is defined as follows:

$$\hat{\mathbf{v}} = \frac{i}{\hbar}(H_{\alpha}\hat{\mathbf{r}} - \hat{\mathbf{r}}H_{\alpha}),$$

Straightforward calculation results in

$$\hat{\mathbf{v}} = \alpha D_{\alpha}\,|\hat{\mathbf{p}}^2|^{\alpha/2-1}\,\hat{\mathbf{p}}.$$

Hence,

$$\mathbf{j} = \frac{1}{\alpha}\left(\psi\hat{\mathbf{v}}\psi^* + \psi^*\hat{\mathbf{v}}\psi\right), \qquad 1 < \alpha \le 2.$$

To get the probability current density equal to 1 (the current when one particle passes through unit area per unit time) the wave function of a free particle has to be normalized as

$$\psi(\mathbf{r},t) = \sqrt{\frac{\alpha}{2\mathrm{v}}}\exp\left[\frac{i}{\hbar}(\mathbf{p}\cdot\mathbf{r} - Et)\right], \qquad E = D_{\alpha}\,|\mathbf{p}|^{\alpha}, \qquad 1 < \alpha \le 2,$$

where v is the particle velocity, $\mathrm{v} = \alpha D_{\alpha}p^{\alpha-1}$.

Then we have

$$\mathbf{j} = \frac{\mathbf{v}}{\mathrm{v}}, \qquad \mathbf{v} = \alpha D_{\alpha}\,|\mathbf{p}^2|^{\frac{\alpha}{2}-1}\,\mathbf{p},$$

that is, the vector $\mathbf{j}$ is indeed the unit vector.

Physical Applications

Fractional Bohr Atom

When $V(\mathbf{r})$ is the potential energy of hydrogenlike atom,

$$V(\mathbf{r}) = -\frac{Ze^2}{|\mathbf{r}|},$$

where e is the electron charge and Z is the atomic number of the hydrogenlike atom, (so Ze is the nuclear charge of the atom), we come to following fractional eigenvalue problem,

$$D_\alpha(-\hbar^2\Delta)^{\alpha/2}\phi(\mathbf{r}) - \frac{Ze^2}{|\mathbf{r}|}\phi(\mathbf{r}) = E\phi(\mathbf{r}).$$

This eigenvalue problem has first been introduced and solved by Nick Laskin in.

Using the first Niels Bohr postulate yields

$$\alpha D_\alpha\left(\frac{n\hbar}{a_n}\right)^\alpha = \frac{Ze^2}{a_n},$$

and it gives us the equation for the Bohr radius of the fractional hydrogenlike atom

$$a_n = a_0 n^{\alpha/(\alpha-1)}.$$

Here a_0 is the fractional Bohr radius (the radius of the lowest, $n = 1$, Bohr orbit) defined as,

$$a_0 = \left(\frac{\alpha D_\alpha\hbar^\alpha}{Ze^2}\right)^{1/(\alpha-1)}.$$

The energy levels of the fractional hydrogenlike atom are given by

$$E = (1-\alpha)E\ n^{-\alpha/(\alpha-1)}, \qquad 1 < \alpha \le 2,$$

where E_0 is the binding energy of the electron in the lowest Bohr orbit that is, the energy required to put it in a state with $E = 0$ corresponding to $n = \infty$,

$$E_0 = \left(\frac{Ze^2}{\alpha D_\alpha^{1/\alpha}\hbar}\right)^{\alpha/(\alpha-1)}.$$

The energy $(\alpha - 1)E_0$ divided by $\hbar c$, $(\alpha - 1)E_0/\hbar c$, can be considered as fractional generalization of the Rydberg constant of standard quantum mechanics. For $\alpha = 2$ and $Z = 1$ the formula $(\alpha - 1)E_0/\hbar c$ is transformed into

$$\mathrm{Ry} = me^4/2\hbar^3 c,,$$

which is the well-known expression for the Rydberg formula.

According to the second Niels Bohr postulate, the frequency of radiation ω associated with the transition, say, for example from the orbit m to the orbit n, is,

$$\omega = \frac{(1-\alpha)E_0}{\hbar}\left[\frac{1}{n^{\frac{\alpha}{\alpha-1}}} - \frac{1}{m^{\frac{\alpha}{\alpha-1}}}\right].$$

The above equations are fractional generalization of the Bohr model. In the special Gaussian case, when ($a = 2$) those equations give us the well-known results of the Bohr model.

The Infinite Potential Well

A particle in a one-dimensional well moves in a potential field $V(x)$, which is zero for $-a \le x \le a$ and which is infinite elsewhere,

$$V(x) = \infty, \qquad x < -a \qquad \text{(i)}$$

$$V(x) = 0, \quad -a \le x \le a \qquad \text{(ii)}$$

$$V(x) = \infty, \qquad x > a \qquad \text{(iii)}$$

It is evident *a priori* that the energy spectrum will be discrete. The solution of the fractional Schrödinger equation for the stationary state with well-defined energy E is described by a wave function $\psi(x),$, which can be written as

$$\psi(x,t) = \left(-i\frac{Et}{\hbar}\right)\phi(x),$$

where $\phi(x),$ is now time independent. In regions (i) and (iii), the fractional Schrödinger equation can be satisfied only if we take $\phi(x) = 0$. In the middle region (ii), the time-independent fractional Schrödinger equation is.

$$D_\alpha(\hbar\nabla)^\alpha \phi(x) = E\phi(x).$$

This equation defines the wave functions and the energy spectrum within region (ii), while outside of the region (ii), x<-a and x>a, the wave functions are zero. The wave function $\phi(x)$ has to be continuous everywhere, thus we impose the boundary conditions $\phi(-a) = \phi(a) = 0$ for the solutions of the *time-independent fractional Schrödinger equation*. Then the solution in region (ii) can be written as

$$\phi(x) = A\exp(ikx) + B\exp(-ikx).$$

To satisfy the boundary conditions we have to choose

$$A = -B \exp(-i2ka),$$

and

$$\sin(2ka) = 0.$$

It follows from the last equation that

$$2ka = n\pi.$$

Then the even ($\phi_n^{\text{even}}(-x) = \phi_n^{\text{even}}(x)$ under reflection $x \to -x$) solution of the time-independent fractional Schrödinger equation $\phi^{\text{even}}(x)$ in the infinite potential well is

$$\phi_n^{\text{even}}(x) = \frac{1}{\sqrt{a}} \cos\left[\frac{n\pi x}{2a}\right], \quad n = 1, 3, 5, \dots.$$

The odd ($\phi_n^{\text{odd}}(-x) = -\phi_n^{\text{odd}}(x)$ under reflection $x \to -x$) solution of the time-independent fractional Schrödinger equation $\phi^{\text{even}}(x)$ in the infinite potential well is

$$\phi_n^{\text{odd}}(x) = \frac{1}{\sqrt{a}} \sin\left[\frac{n\pi x}{2a}\right], \quad n = 2, 4, 6, \dots.$$

The solutions $\phi^{\text{even}}(x)$ and $\phi^{\text{odd}}(x)$ have the property that

$$\int_{-a}^{a} dx\, \phi_m^{\text{even}}(x)\phi_n^{\text{even}}(x) = \int_{-a}^{a} dx\, \phi_m^{\text{odd}}(x)\phi_n^{\text{odd}}(x) = \delta_{mn},$$

where δ_{mn} is the Kronecker symbol and

$$\int_{-a}^{a} dx\, \phi_m^{\text{even}}(x)\phi_n^{\text{odd}}(x) = 0.$$

The eigenvalues of the particle in an infinite potential well are

$$E_n = D_\alpha \left(\frac{\pi\hbar}{2a}\right)^\alpha n^\alpha, \qquad n = 1, 2, 3\dots, \qquad 1 < \alpha \leq 2.$$

It is obvious that in the Gaussian case ($\alpha = 2$) above equations are ö transformed into the standard quantum mechanical equations for a particle in a box

The state of the lowest energy, the ground state, in the infinite potential well is represented by the $\phi_n^{\text{even}}(x)$ at $n=1$,

$$\phi_{\text{ground}}(x) \equiv \phi_1^{\text{even}}(x) = \frac{1}{\sqrt{a}} \cos\left(\frac{\pi x}{2a}\right),$$

and its energy is

$$E_{\text{ground}} = D_\alpha \left(\frac{\pi \hbar}{2a}\right)^\alpha.$$

Fractional Quantum Oscillator

Fractional quantum oscillator introduced by Nick Laskin is the fractional quantum mechanical model with the Hamiltonian operator $H_{\alpha,\beta}$ defined as

$$H_{\alpha,\beta} = D_\alpha(-\hbar^2 \Delta)^{\alpha/2} + q^2 |\mathbf{r}|^\beta, \quad 1 < \alpha \le 2, \quad 1 < \beta \le 2,,,$$

where q is interaction constant.

The fractional Schrödinger equation for the wave function $\psi(\mathbf{r}, t)$ of the fractional quantum oscillator is,

$$i\hbar \frac{\partial \psi(\mathbf{r}, t)}{\partial t} = D_\alpha(-\hbar^2 \Delta)^{\alpha/2} \psi(\mathbf{r}, t) + q^2 |\mathbf{r}|^\beta \psi(\mathbf{r}, t)$$

Aiming to search for solution in form

$$\psi(\mathbf{r}, t) = e^{-iEt/\hbar} \phi(\mathbf{r}),$$

we come to the time-independent fractional Schrödinger equation,

$$D_\alpha(-\hbar^2 \Delta)^{\alpha/2} \phi(\mathbf{r}, t) + q^2 |\mathbf{r}|^\beta \phi(\mathbf{r}, t) = E\phi(\mathbf{r}, t).$$

The Hamiltonian $H_{\alpha,\beta}$ is the fractional generalization of the 3D quantum harmonic oscillator Hamiltonian of standard quantum mechanics.

Energy Levels of the 1D Fractional Quantum Oscillator in Semiclassical Approximation

The energy levels of 1D fractional quantum oscillator with the Hamiltonian function $H_\alpha = D_\alpha |p|^\alpha + q^2 |x|^\beta$ were found in semiclassical approximation.

We set the total energy equal to E, so that

$$E = D_\alpha |p|^\alpha + q^2 |x|^\beta,$$

whence

$$|p| = \left(\frac{1}{D_\alpha}(E - q^2 |x|^\beta)\right)^{1/\alpha}.$$

At the turning points $p = 0$. Hence, the classical motion is possible in the range $|x| \le (E/q^2)^{1/\beta}$..

A routine use of the Bohr-Sommerfeld quantization rule yields

$$2\pi\hbar(n+\frac{1}{2}) = \oint pdx = 4\int_0^{x_m} pdx = 4\int_0^{x_m} D_\alpha^{-1/\alpha}(E - q^2 |x|^\beta)^{1/\alpha} dx,$$

where the notation $\oint$ means the integral over one complete period of the classical motion and $x_m = (E/q^2)^{1/\beta}$ is the turning point of classical motion.

To evaluate the integral in the right hand we introduce a new variable $y = x(E/q^2)^{-1/\beta}$.. Then we have

$$\int_0^{x_m} D_\alpha^{-1/\alpha}(E - q^2 |x|^\beta)^{1/\alpha} dx = \frac{1}{D_\alpha^{1/\alpha} q^{2/\beta}} E^{\frac{1}{\alpha}+\frac{1}{\beta}} \int_0^1 dy(1-y^\beta)^{1/\alpha}.$$

The integral over dy can be expressed in terms of the Beta-function,

$$\int_0^1 dy(1-y^\beta)^{1/\alpha} = \frac{1}{\beta}\int_0^1 dz z^{\frac{1}{\beta}-1}(1-z)^{\frac{1}{\alpha}} = \frac{1}{\beta}B\left(\frac{1}{\beta},\frac{1}{\alpha}+1\right).$$

Therefore,

$$2\pi\hbar(n+\frac{1}{2}) = \frac{4}{D_\alpha^{1/\alpha} q^{2/\beta}} E^{\frac{1}{\alpha}+\frac{1}{\beta}} \frac{1}{\beta}B\left(\frac{1}{\beta},\frac{1}{\alpha}+1\right).$$

The above equation gives the energy levels of stationary states for the 1D fractional quantum oscillator,

$$E_n = \left(\frac{\pi\hbar\beta D_\alpha^{1/\alpha} q^{2/\beta}}{2B(\frac{1}{\beta},\frac{1}{\alpha}+1)}\right)^{\frac{\alpha\beta}{\alpha+\beta}} \left(n+\frac{1}{2}\right)^{\frac{\alpha\beta}{\alpha+\beta}}.$$

This equation is generalization of the well-known energy levels equation of the standard quantum harmonic oscillator and is transformed into it at $\alpha = 2$ and $\beta = 2$. It follows from this equation that

at $\frac{1}{\alpha}+\frac{1}{\beta} = 1$ the energy levels are equidistant. When $1 < \alpha \le 2$ and $1 < \ \le 2$ the equidistant energy

levels can be for $\alpha = 2$ and $\beta = 2$ only. It means that the only standard quantum harmonic oscillator has an equidistant energy spectrum.

Fractional Quantum Mechanics in Solid State Systems

The effective mass of states in solid state systems can depend on the wave vector k, i.e. formally one considers m=m(k). Polariton Bose-Einstein condensate modes are examples of states in solid state systems with mass sensitive to variations and locally in k fractional quantum mechanics is experimentally feasible .

References

- Richard Liboff (2002). Introductory Quantum Mechanics (4th ed.). Addison Wesley. ISBN 0-8053-8714-5.

- Malomed, Boris (2005), "Nonlinear Schrödinger Equations", in Scott, Alwyn, Encyclopedia of Nonlinear Science, New York: Routledge, pp. 639–643

- L.D. Landau and E.M. Lifshitz, Quantum mechanics (Non-relativistic Theory), Vol.3, Third Edition, Course of Theoretical Physics, Butterworth-Heinemann, Oxford, 2003

- J. Klafter; S.C. Lim; R. Metzler (2012). Fractional Dynamics: Recent Advances. World Scientific. p. 426. ISBN 981-434-059-6.

Path Integrals in Quantum Mechanics

Path integral formulation is the description of quantum theory that helps in the generalization of the action principle of classical mechanics. The topics elucidated in this chapter are relation between Schrödinger's equation and the path integral formulation of quantum mechanics, propagator and the Feynman diagram. The chapter serves as a source to understand the path integrals in quantum mechanics.

Path Integral Formulation

The path integral formulation of quantum mechanics is a description of quantum theory which generalizes the action principle of classical mechanics. It replaces the classical notion of a single, unique classical trajectory for a system with a sum, or functional integral, over an infinity of quantum mechanically possible trajectories to compute a quantum amplitude.

This formulation has proven crucial to the subsequent development of theoretical physics, because manifest Lorentz covariance (time and space components of quantities enters equations in the same way) is easier to achieve than in the operator formalism of canonical quantization. Unlike previous methods, the path-integral allows a physicist to easily change coordinates between very different canonical descriptions of the same quantum system. Another advantage is that it is in practice easier to guess the correct form of the Lagrangian of a theory, which naturally enters the path integrals, than the Hamiltonian. Possible downsides of the approach include that unitarity (this is related to conservation of probability; the probabilities of all physically possible outcomes must add up to one) of the S-matrix is obscure in the formulation. The path-integral approach has been proved to be equivalent to the other formalisms of quantum mechanics and quantum field theory. Thus, by *deriving* either approach from the other, problems associated with one or the other approach (as exemplified by Lorentz covariance or unitarity) go away.

The path integral also relates quantum and stochastic processes, and this provided the basis for the grand synthesis of the 1970s which unified quantum field theory with the statistical field theory of a fluctuating field near a second-order phase transition. The Schrödinger equation is a diffusion equation with an imaginary diffusion constant, and the path integral is an analytic continuation of a method for summing up all possible random walks.

The basic idea of the path integral formulation can be traced back to Norbert Wiener, who introduced the Wiener integral for solving problems in diffusion and Brownian motion. This idea was extended to the use of the Lagrangian in quantum mechanics by P. A. M. Dirac in his 1933 paper. The complete method was developed in 1948 by Richard Feynman. Some preliminaries were worked out earlier in his doctoral work under the supervision of John Archibald Wheeler. The original motivation stemmed from the desire to obtain a quantum-mechanical formulation for the Wheeler–Feynman absorber theory using a Lagrangian (rather than a Hamiltonian) as a starting point.

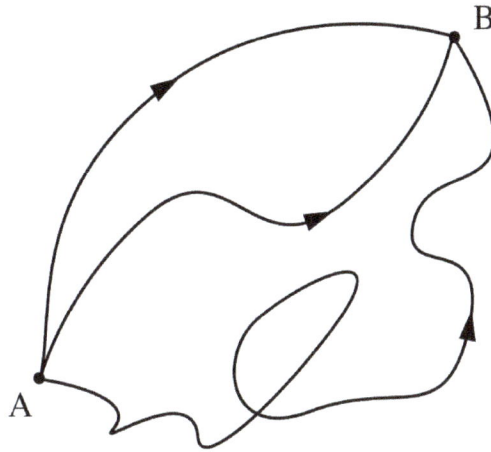

These are just three of the paths that contribute to the quantum amplitude for a particle moving from point A at some time t_0 to point B at some other time t_1.

Quantum Action Principle

In quantum mechanics, as in classical mechanics, the Hamiltonian is the generator of time-trans-lations. This means that the state at a slightly later time differs from the state at the current time by the result of acting with the Hamiltonian operator (multiplied by the negative imaginary unit, $-i$). For states with a definite energy, this is a statement of the de Broglie relation between frequency and energy, and the general relation is consistent with that plus the superposition principle.

The Hamiltonian in classical mechanics is derived from a Lagrangian, which is a more fundamen-tal quantity relative to special relativity. The Hamiltonian indicates how to march forward in time, but the time is different in different reference frames. So the Hamiltonian is different in different frames, and this type of symmetry is not apparent in the original formulation of quantum mechan-ics.

The Hamiltonian is a function of the position and momentum at one time, and it determines the position and momentum a little later. The Lagrangian is a function of the position now and the position a little later (or, equivalently for infinitesimal time separations, it is a function of the posi-tion and velocity). The relation between the two is by a Legendre transform, and the condition that determines the classical equations of motion (the Euler–Lagrange equations) is that the action is an extremum.

In quantum mechanics, the Legendre transform is hard to interpret, because the motion is not over a definite trajectory. In classical mechanics, with discretization in time, the Legendre trans-form becomes,

$$\epsilon H = p(t)\big(q(t+\epsilon)-q(t)\big)-\epsilon L$$

and

$$p = \frac{\partial L}{\partial \dot{q}}$$

where the partial derivative with respect to q holds $q(t + \varepsilon)$ fixed. The inverse Legendre transform is:

$$\epsilon L = \epsilon p \dot{q} - \epsilon H$$

where

$$\dot{q} = \frac{\partial H}{\partial p}$$

and the partial derivative now is with respect to p at fixed q.

In quantum mechanics, the state is a superposition of different states with different values of q, or different values of p, and the quantities p and q can be interpreted as noncommuting operators. The operator p is only definite on states that are indefinite with respect to q. So consider two states separated in time and act with the operator corresponding to the Lagrangian:

$$e^{i\left(p\left(q(t+\epsilon)-q(t)\right)-\epsilon H(p,q)\right)}$$

If the multiplications implicit in this formula are reinterpreted as *matrix* multiplications, the first factor is

$$e^{-ipq(t)}$$

And if this is also interpreted as a matrix multiplication, the sum over all states integrates over all $q(t)$, and so it takes the Fourier transform in $q(t)$, to change basis to $p(t)$. That is the action on the Hilbert space – change basis to **p** at time **t**.

Next comes:

$$e^{-i\epsilon H(p,q)}$$

or evolve an infinitesimal time into the future.

Finally, the last factor in this interpretation is

$$e^{ipq(t+\epsilon)}$$

which means change basis back to **q** at a later time.

This is not very different from just ordinary time evolution: the H factor contains all the dynamical information – it pushes the state forward in time. The first part and the last part are just Fourier transforms to change to a pure q basis from an intermediate p basis.

"...we see that the integrand in (11) must be of the form $e^{iF/h}$ where F is a function of q_T, q_1, q_2,... q_m, q_t, which remains finite as h tends to zero. Let us now picture one of the intermediate qs, say q_k, as varying continuously while the other ones are fixed. Owing to the smallness of h, we shall then in general have F/h varying extremely rapidly. This means that $e^{iF/h}$ will vary periodically with a very high frequency about the value zero, as a result of which its integral will be practically zero. The only important part in the domain of integration of q_k is thus that for which a comparatively large variation in q_k produces only a very small variation in F. This part is the neighbourhood of a point for which F is stationary with respect to small variations in q_k. We can apply this argument to each of the variables of integration ....and obtain the result that the only important part in the domain of integration is that for which F is stationary for small variations in all intermediate qs. ...We see that F has for its classical analogue $\int t\, T\, L\, dt$, which is just the action function which classical mechanics requires to be stationary for small variations in all the intermediate qs. This shows the way in which equation (11) goes over into classical results when h becomes extremely small."

Dirac (1933), p. 69

Another way of saying this is that since the Hamiltonian is naturally a function of p and q, exponentiating this quantity and changing basis from p to q at each step allows the matrix element of H to be expressed as a simple function along each path. This function is the quantum analog of the classical action. This observation is due to Paul Dirac.

Dirac further noted that one could square the time-evolution operator in the S representation

$$e^{i\epsilon S}$$

and this gives the time evolution operator between time t and time $t + 2\epsilon$. While in the H representation the quantity that is being summed over the intermediate states is an obscure matrix element, in the S representation it is reinterpreted as a quantity associated to the path. In the limit that one takes a large power of this operator, one reconstructs the full quantum evolution between two states, the early one with a fixed value of $q(0)$ and the later one with a fixed value of $q(t)$. The result is a sum over paths with a phase which is the quantum action. Crucially, Dirac identified in this paper the deep quantum mechanical reason for the principle of least action controlling the classical limit.

Feynman's Interpretation

Dirac's work did not provide a precise prescription to calculate the sum over paths, and he did not show that one could recover the Schrödinger equation or the canonical commutation relations from this rule. This was done by Feynman. That is, the classical path arises naturally in the classical limit.

Feynman showed that Dirac's quantum action was, for most cases of interest, simply equal to the classical action, appropriately discretized. This means that the classical action is the phase acquired by quantum evolution between two fixed endpoints. He proposed to recover all of quantum mechanics from the following postulates:

1. The probability for an event is given by the modulus length squared of a complex number called the "probability amplitude".

2. The probability amplitude is given by adding together the contributions of all paths in configuration space.

3. The contribution of a path is proportional to $e^{iS/\hbar}$, where S is the action given by the time integral of the Lagrangian along the path.

In order to find the overall probability amplitude for a given process, then, one adds up, or integrates, the amplitude of the 3rd postulate over the space of *all* possible paths of the system in between the initial and final states, including those that are absurd by classical standards. In calculating the probability amplitude for a single particle to go from one space-time coordinate to another, it is correct to include paths in which the particle describes elaborate curlicues, curves in which the particle shoots off into outer space and flies back again, and so forth. The path integral assigns to all these amplitudes *equal weight* but varying phase, or argument of the complex number. Contributions from paths wildly different from the classical trajectory may be suppressed by interference.

Feynman showed that this formulation of quantum mechanics is equivalent to the canonical approach to quantum mechanics when the Hamiltonian is at most quadratic in the momentum. An amplitude computed according to Feynman's principles will also obey the Schrödinger equation for the Hamiltonian corresponding to the given action.

The path integral formulation of quantum field theory represents the transition amplitude (corresponding to the classical correlation function) as a weighted sum of all possible histories of the system from the initial to the final state. A Feynman diagram is a graphical representation of a perturbative contribution to the transition amplitude.

Concrete Formulation

Feynman's postulates can be interpreted as follows:

Time-Slicing Definition

For a particle in a smooth potential, the path integral is approximated by zigzag paths, which in one dimension is a product of ordinary integrals. For the motion of the particle from position x_a at time t_a to x_b at time t_b, the time sequence

$$t_a = t_0 < t_1 < \ldots < t_{n-1} < t_n < t_{n+1} = t_b$$

can be divided up into $n + 1$ smaller segments $t_j - t_{j-1}$, where $j = 1,\ldots,n + 1$, of fixed duration

$$\epsilon = \Delta t = \frac{t_b - t_a}{n+1}.$$

This process is called *time-slicing*.

An approximation for the path integral can be computed as proportional to

$$\int_{-\infty}^{+\infty} \ldots \int_{-\infty}^{+\infty} \exp\left(\frac{i}{\hbar} \int_{t_a}^{t_b} L\big(x(t), v(t), t\big)dt \right) dx_0 \ldots dx_n$$

where $L(x,v,t)$ is the Lagrangian of the one-dimensional system with position variable $x(t)$ and velocity $v = \dot{x}(t)$ considered, and dx_j corresponds to the position at the jth time step, if the time integral is approximated by a sum of n terms.

In the limit $n \rightarrow \infty$, this becomes a functional integral, which, apart from a nonessential factor, is directly the product of the probability amplitudes $\langle x_b, t_b | x_a, t_a \rangle$ (more precisely, since one must work with a continuous spectrum, the respective densities) to find the quantum mechanical parti-cle at t_a in the initial state x_a and at t_b in the final state x_b.

Actually L is the classical Lagrangian of the one-dimensional system considered, also

$$L(x,\dot{x},t) = p \cdot \dot{x} - H(x,p,t),$$

where H is the Hamiltonian,

$$p = \frac{\partial L}{\partial \dot{x}},$$

and the abovementioned "zigzagging" corresponds to the appearance of the terms:

$$\exp\left(\frac{i}{\hbar}\epsilon \sum_{j=1}^{n+1} L\left(\tilde{x}_j, \frac{x_j - x_{j-1}}{\epsilon}, j\right)\right)$$

In the Riemannian sum approximating the time integral, which are finally integrated over x_1 to x_n with the integration measure $dx_1...dx_n$, $\tilde{x}_j$ is an arbitrary value of the interval corresponding to j, e.g. its center, $x_j + x_{j-1}/2$.

Thus, in contrast to classical mechanics, not only does the stationary path contribute, but actually all virtual paths between the initial and the final point also contribute.

Feynman's time-sliced approximation does not, however, exist for the most important quan-tum-mechanical path integrals of atoms, due to the singularity of the Coulomb potential e^2/r at the origin. Only after replacing the time t by another path-dependent pseudo-time parameter

$$s = \int \frac{dt}{r(t)}$$

the singularity is removed and a time-sliced approximation exists, that is exactly integrable, since it can be made harmonic by a simple coordinate transformation, as discovered in 1979 by İsmail Hakkı Duru and Hagen Kleinert. The combination of a path-dependent time transformation and a coordinate transformation is an important tool to solve many path integrals and is called gener-ically the Duru–Kleinert transformation.

Free Particle

The path integral representation gives the quantum amplitude to go from point x to point y as an integral over all paths. For a free particle action (for simplicity let $m = 1$, $\hbar = 1$):

$$S = \int \frac{\dot{x}^2}{2} dt$$

the integral can be evaluated explicitly.

To do this, it is convenient to start without the factor i in the exponential, so that large deviations are suppressed by small numbers, not by cancelling oscillatory contributions.

$$K(x-y;T) = \int_{x(0)=x}^{x(T)=y} \exp\left(-\int_0^T \frac{\dot{x}^2}{2} dt\right) Dx$$

Splitting the integral into time slices:

$$K(x,y;T) = \int_{x(0)=x}^{x(T)=y} \Pi_t \exp\left(-\frac{1}{2}\left(\frac{x(t+\epsilon)-x(t)}{\epsilon}\right)^2 \epsilon\right) Dx$$

where the Dx is interpreted as a finite collection of integrations at each integer multiple of ϵ. Each factor in the product is a Gaussian as a function of $x(t + \epsilon)$ centered at $x(t)$ with variance ϵ. The multiple integrals are a repeated convolution of this Gaussian G_ϵ with copies of itself at adjacent times.

$$K(x-y;T) = G_\epsilon * G_\epsilon ... * G_\epsilon$$

Where the number of convolutions is T/ϵ. The result is easy to evaluate by taking the Fourier transform of both sides, so that the convolutions become multiplications.

$$\tilde{K}(p;T) = \tilde{G}_\epsilon(p)^{T/\epsilon}$$

The Fourier transform of the Gaussian G is another Gaussian of reciprocal variance:

$$\tilde{G}_\epsilon(p) = e^{-\frac{\epsilon p^2}{2}}$$

and the result is:

$$\tilde{K}(p;T) = e^{-\frac{T p^2}{2}}$$

The Fourier transform gives K, and it is a Gaussian again with reciprocal variance:

$$K(x-y;T) \propto e^{-\frac{(x-y)^2}{2T}}$$

The proportionality constant is not really determined by the time slicing approach, only the ratio of values for different endpoint choices is determined. The proportionality constant should be chosen to ensure that between each two time-slices the time evolution is quantum-mechanically unitary, but a more illuminating way to fix the normalization is to consider the path integral as a description of a stochastic process.

The result has a probability interpretation. The sum over all paths of the exponential factor can be seen as the sum over each path of the probability of selecting that path. The probability is the product over each segment of the probability of selecting that segment, so that each segment is probabilistically independently chosen. The fact that the answer is a Gaussian spreading linearly in time is the central limit theorem, which can be interpreted as the first historical evaluation of a statistical path integral.

The probability interpretation gives a natural normalization choice. The path integral should be defined so that:

$$\int K(x-y;T)dy = 1$$

This condition normalizes the Gaussian, and produces a Kernel which obeys the diffusion equation:

$$\frac{d}{dt}K(x;T) = \frac{\nabla^2}{2}K$$

For oscillatory path integrals, ones with an i in the numerator, the time-slicing produces convolved Gaussians, just as before. Now, however, the convolution product is marginally singular since it requires careful limits to evaluate the oscillating integrals. To make the factors well defined, the easiest way is to add a small imaginary part to the time increment ε. This is closely related to Wick rotation. Then the same convolution argument as before gives the propagation kernel:

$$K(x-y;T) \propto e^{\frac{i(x-y)^2}{2T}}$$

Which, with the same normalization as before (not the sum-squares normalization – this function has a divergent norm), obeys a free Schrödinger equation

$$\frac{d}{dt}K(x;T) = i\frac{\nabla^2}{2}K$$

This means that any superposition of Ks will also obey the same equation, by linearity. Defining

$$\psi_t(y) = \int \psi_0(x)K(x-y;t)dx = \int \psi_0(x)\int_{x(0)=x}^{x(t)=y} e^{iS}Dx$$

then ψ_t obeys the free Schrödinger equation just as K does:

$$i\frac{\partial}{\partial t}\psi_t = -\frac{\nabla^2}{2}\psi_t$$

Simple Harmonic Oscillator

The Lagrangian for the simple harmonic oscillator is

$$\mathcal{L} = \tfrac{1}{2}m\dot{x}^2 - \tfrac{1}{2}m\omega^2 x^2.$$

Write its trajectory $x(t)$ as the classical trajectory plus some perturbation, $x(t) = x_c(t) + \delta x(t)$ and the action as $S = S_c + \delta S$. The classical trajectory can be written as:

$$x_c(t) = x_i \frac{\sin\omega(t_f - t)}{\sin\omega(t_f - t_i)} + x_f \frac{\sin\omega(t - t_i)}{\sin\omega(t_f - t_i)}.$$

This trajectory corresponds to the classical action:

$$S_c = \int_{t_i}^{t_f} \mathcal{L}\,dt = \int_{t_i}^{t_f} \left(\tfrac{1}{2}m\dot{x}^2 - \tfrac{1}{2}m\omega^2 x^2 \right) dt = \frac{1}{2}m\omega \left(\frac{(x_i^2 + x_f^2)\cos\omega(t_f - t_i) - 2x_i x_f}{\sin\omega(t_f - t_i)} \right)$$

Next, expand the non-classical contribution to the action δS, as a Fourier series, which gives

$$S = S_c + \sum_{n=1}^{\infty} \tfrac{1}{2}a_n^2 \frac{m}{2}\left(\frac{(n\pi)^2}{t_f - t_i} - \omega^2(t_f - t_i) \right)$$

This means the propagator is

$$K(x_f, t_f; x_i, t_i) = Qe^{\frac{iS_c}{\hbar}} \prod_{j=1}^{\infty} \frac{j\pi}{\sqrt{2}} \int da_j \exp\left(\frac{i}{2\hbar}a_j^2 \frac{m}{2}\left(\frac{(j\pi)^2}{t_f - t_i} - \omega^2(t_f - t_i) \right) \right) = e^{\frac{iS_c}{\hbar}} Q \prod_{j=1}^{\infty} \left(1 - \left(\frac{\omega(t_f - t_i)}{j\pi} \right)^2 \right)^{-\frac{1}{2}}$$

for some normalization

$$Q = \sqrt{\frac{m}{2\pi i\hbar(t_f - t_i)}}.$$

Using the infinite product representation of the sinc function

$$\prod_{j=1}^{\infty}\left(1 - \frac{x^2}{j^2} \right) = \frac{\sin\pi x}{\pi x},$$

the propagator can be written as

$$K(x_f, t_f; x_i, t_i) = Qe^{\frac{iS_c}{\hbar}} \sqrt{\frac{\omega(t_f - t_i)}{\sin\omega(t_f - t_i)}} = e^{\frac{iS_c}{\hbar}} \sqrt{\frac{m\omega}{2\pi i\hbar\sin\omega(t_f - t_i)}}.$$

Let $T = t_f - t_i$. We can write our propagator in terms of energy eigenstates as,

$$K(x_f,t_f;x_i,t_i) = \left(\frac{m\omega}{2\pi i\hbar \sin \omega T}\right)^{\frac{1}{2}} \exp\left(\frac{i}{\hbar}\tfrac{1}{2}m\omega \frac{\left(x_i^2+x_f^2\right)\cos \omega T - 2x_i x_f}{\sin \omega T}\right) = \sum_{n=0}^{\infty} \exp\left(-\frac{iE_n T}{\hbar}\right)\psi_n(x_f)^*\psi_n(x_i).$$

Using the identities $i \sin \omega T = 1/2e^{i\omega T}(1 - e^{-2i\omega T})$ and $\cos \omega T = 1/2e^{i\omega T}(1 + e^{-2i\omega T})$,

$$K(x_f,t_f;x_i,t_i) = \left(\frac{m\omega}{\pi\hbar}\right)^{\frac{1}{2}} e^{-\frac{i\omega T}{2}}\left(1-e^{-2i\omega T}\right)^{-\frac{1}{2}}\exp\left(-\frac{m\omega}{2\hbar}\left(\left(x_i^2+x_f^2\right)\frac{1+e^{-2i\omega T}}{1-e^{-2i\omega T}} - \frac{4x_i x_f e^{-i\omega T}}{1-e^{-2i\omega T}}\right)\right)$$

We can absorb all the terms after the first $e^{-i\omega T/2}$ into $R(T)$, thereby giving,

$$K(x_f,t_f;x_i,t_i) = \left(\frac{m\omega}{\pi\hbar}\right)^{\frac{1}{2}} e^{-\frac{i\omega T}{2}} \cdot R(T).$$

We can expand $R(T)$ in the powers of $e^{-i\omega T}$. All the terms in that expansion get multiplied by the $e^{-i\omega T/2}$ factor in the front and so we get terms that look like

$$e^{-\frac{i\omega T}{2}}e^{-in\omega T} = e^{-i\omega T\left(\frac{1}{2}+n\right)} \quad \text{for } n = 0,1,2,\dots.$$

Comparing that to the eigenstate expansion, we get the energy spectrum for simple harmonic oscillator,

$$E_n = \left(n+\tfrac{1}{2}\right)\hbar\omega.$$

The Schrödinger Equation

The path integral reproduces the Schrödinger equation for the initial and final state even when a potential is present. This is easiest to see by taking a path-integral over infinitesimally separated times.

$$\psi(y;t+\epsilon) = \int_{-\infty}^{\infty} \psi(x;t) \int_{x(t)=x}^{x(t+\epsilon)=y} e^{i\int_t^{t+\epsilon}\left(\frac{\dot{x}^2}{2}-V(x)\right)dt} Dx(t)dx \qquad (1)$$

Since the time separation is infinitesimal and the cancelling oscillations become severe for large values of $\dot{x}$, the path integral has most weight for y close to x. In this case, to lowest order the potential energy is constant, and only the kinetic energy contribution is nontrivial. The exponential of the action is

$$e^{-i\epsilon V(x)}e^{i\frac{\dot{x}^2}{2}\epsilon}$$

The first term rotates the phase of $\psi(x)$ locally by an amount proportional to the potential energy. The second term is the free particle propagator, corresponding to i times a diffusion process. To lowest order in ϵ they are additive; in any case one has with (1):

$$\psi(y;t+\epsilon) \approx \int \psi(x;t) e^{-i\epsilon V(x)} e^{\frac{i(x-y)^2}{2\epsilon}} \, dx.$$

As mentioned, the spread in ψ is diffusive from the free particle propagation, with an extra infinitesimal rotation in phase which slowly varies from point to point from the potential:

$$\frac{\partial \psi}{\partial t} = i \cdot \left(\tfrac{1}{2} \nabla^2 - V(x) \right) \psi$$

and this is the Schrödinger equation. Note that the normalization of the path integral needs to be fixed in exactly the same way as in the free particle case. An arbitrary continuous potential does not affect the normalization, although singular potentials require careful treatment.

Equations of Motion

Since the states obey the Schrödinger equation, the path integral must reproduce the Heisenberg equations of motion for the averages of x and $\dot{x}$ variables, but it is instructive to see this directly. The direct approach shows that the expectation values calculated from the path integral reproduce the usual ones of quantum mechanics.

Start by considering the path integral with some fixed initial state

$$\int \psi_0(x) \int_{x(0)=x} e^{iS(x,\dot{x})} Dx$$

Now note that $x(t)$ at each separate time is a separate integration variable. So it is legitimate to change variables in the integral by shifting: $x(t) = u(t) + \epsilon(t)$ where $\epsilon(t)$ is a different shift at each time but $\epsilon(0) = \epsilon(T) = 0$, since the endpoints are not integrated:

$$\int \psi_0(x) \int_{u(0)=x} e^{iS(u+\epsilon, \dot{u}+\dot{\epsilon})} Du$$

The change in the integral from the shift is, to first infinitesimal order in ϵ:

$$\int \psi_0(x) \int_{u(0)=x} \left(\int \frac{\partial S}{\partial u}\epsilon + \frac{\partial S}{\partial \dot{u}}\dot{\epsilon} \, dt \right) e^{iS} Du$$

which, integrating by parts in t, gives:

$$\int \psi_0(x) \int_{u(0)=x} -\left(\int \left(\frac{d}{dt}\frac{\partial S}{\partial \dot{u}} - \frac{\partial S}{\partial u} \right) \epsilon(t) dt \right) e^{iS} Du$$

But this was just a shift of integration variables, which doesn't change the value of the integral for any choice of $\epsilon(t)$. The conclusion is that this first order variation is zero for an arbitrary initial state and at any arbitrary point in time:

$$\left\langle \psi_0 \left| \frac{\delta S}{\delta x}(t) \right| \psi_0 \right\rangle = 0$$

this is the Heisenberg equation of motion.

If the action contains terms which multiply $\dot{x}$ and x, at the same moment in time, the manipulations above are only heuristic, because the multiplication rules for these quantities is just as non-commuting in the path integral as it is in the operator formalism.

Stationary Phase Approximation

If the variation in the action exceeds $\hbar$ by many orders of magnitude, we typically have destructive phase interference other than in the vicinity of those trajectories satisfying the Euler–Lagrange equation, which is now reinterpreted as the condition for constructive phase interference. This can be shown using the method of stationary phase applied to the propagator. As $\hbar$ decreases, the exponential in the integral oscillates rapidly in the complex domain for any change in the action. Thus, in the limit that $\hbar$ goes to zero, only points where the classical action does not vary contribute to the propagator.

Canonical Commutation Relations

The formulation of the path integral does not make it clear at first sight that the quantities x and p do not commute. In the path integral, these are just integration variables and they have no obvious ordering. Feynman discovered that the non-commutativity is still present.

To see this, consider the simplest path integral, the brownian walk. This is not yet quantum mechanics, so in the path-integral the action is not multiplied by i:

$$S = \int \left(\frac{dx}{dt} \right)^2 dt$$

The quantity $x(t)$ is fluctuating, and the derivative is defined as the limit of a discrete difference.

$$\frac{dx}{dt} = \frac{x(t+\epsilon) - x(t)}{\epsilon}$$

Note that the distance that a random walk moves is proportional to $\sqrt{t}$, so that:

$$x(t+\epsilon) - x(t) \approx \sqrt{\epsilon}$$

This shows that the random walk is not differentiable, since the ratio that defines the derivative diverges with probability one.

The quantity $x\dot{x}$ is ambiguous, with two possible meanings:

$$[1] = x\frac{dx}{dt} = x(t)\frac{x(t+\epsilon) - x(t)}{\epsilon}$$

$$[2] = x\frac{dx}{dt} = x(t+\epsilon)\frac{x(t+\epsilon)-x(t)}{\epsilon}$$

In elementary calculus, the two are only different by an amount which goes to 0 as ϵ goes to 0. But in this case, the difference between the two is not 0:

$$[2] - [1] = \frac{\left(x(t+\epsilon)-x(t)\right)}{\epsilon} \approx \frac{\epsilon}{\epsilon}$$

give a name to the value of the difference for any one random walk:

$$\frac{\left(x(t+\epsilon)-x(t)\right)^2}{\epsilon} = f(t)$$

and note that $f(t)$ is a rapidly fluctuating statistical quantity, whose average value is 1, i.e. a normalized "Gaussian process". The fluctuations of such a quantity can be described by a statistical Lagrangian

$$\mathcal{L} = (f(t)-1)^2,$$

and the equations of motion for f derived from extremizing the action S corresponding to L just set it equal to 1. In physics, such a quantity is "equal to 1 as an operator identity". In mathematics, it "weakly converges to 1". In either case, it is 1 in any expectation value, or when averaged over any interval, or for all practical purpose.

Defining the time order to *be* the operator order:

$$[x,\dot{x}] = x\frac{dx}{dt} - \frac{dx}{dt}x = 1$$

This is called the Itō lemma in stochastic calculus, and the (euclideanized) canonical commutation relations in physics.

For a general statistical action, a similar argument shows that

$$\left[x, \frac{\partial S}{\partial \dot{x}}\right] = 1$$

and in quantum mechanics, the extra imaginary unit in the action converts this to the canonical commutation relation,

$$[x,p] = i$$

Particle in Curved Space

For a particle in curved space the kinetic term depends on the position and the above time slicing cannot be applied, this being a manifestation of the notorious operator ordering problem in

Schrödinger quantum mechanics. One may, however, solve this problem by transforming the time-sliced flat-space path integral to curved space using a multivalued coordinate transformation (nonholonomic mapping explained here).

The Path Integral and the Partition Function

The path integral is just the generalization of the integral above to all quantum mechanical problems—

$$Z = \int e^{\frac{iS[x]}{\hbar}} Dx \quad \text{where } S[x] = \int_0^T L[x(t)]dt$$

is the action of the classical problem in which one investigates the path starting at time $t = 0$ and ending at time $t = T$, and Dx denotes integration over all paths. In the classical limit, $S[x] \gg \hbar$, the path of minimum action dominates the integral, because the phase of any path away from this fluctuates rapidly and different contributions cancel.

The connection with statistical mechanics follows. Considering only paths which begin and end in the same configuration, perform the Wick rotation $it = \tau$, i.e., make time imaginary, and integrate over all possible beginning-ending configurations. The path integral now resembles the partition function of statistical mechanics defined in a canonical ensemble with inverse temperature proportional to imaginary time, $1/T = k_B\tau/\hbar$. Strictly speaking, though, this is the partition function for a statistical field theory.

Clearly, such a deep analogy between quantum mechanics and statistical mechanics cannot be dependent on the formulation. In the canonical formulation, one sees that the unitary evolution operator of a state is given by

$$|\alpha;t\rangle = e^{-\frac{iHt}{\hbar}} |\alpha;0\rangle$$

where the state α is evolved from time $t = 0$. If one makes a Wick rotation here, and finds the amplitude to go from any state, back to the same state in (imaginary) time iT is given by

$$Z = \text{Tr}\left[e^{\frac{-HT}{\hbar}} \right]$$

which is precisely the partition function of statistical mechanics for the same system at temperature quoted earlier. One aspect of this equivalence was also known to Erwin Schrödinger who remarked that the equation named after him looked like the diffusion equation after Wick rotation.

Measure Theoretic Factors

Sometimes (e.g. a particle moving in curved space) we also have measure-theoretic factors in the functional integral.

$$\int \mu[x] e^{iS[x]} Dx$$

This factor is needed to restore unitarity.

For instance, if

$$S = \int \left(\frac{m}{2} g_{ij} \dot{x}^i \dot{x}^j - V(x) \right) dt ,$$

then it means that each spatial slice is multiplied by the measure $\sqrt{g}$. This measure cannot be expressed as a functional multiplying the Dx measure because they belong to entirely different classes.

Quantum Field Theory

The path integral formulation was very important for the development of quantum field theory. Both the Schrödinger and Heisenberg approaches to quantum mechanics single out time, and are not in the spirit of relativity. For example, the Heisenberg approach requires that scalar field operators obey the commutation relation

$$[\phi(x), \partial_t \phi(y)] = i \delta^3(x - y)$$

for x and y two simultaneous spatial positions, and this is not a relativistically invariant concept. The results of a calculation *are* covariant, but the symmetry is not apparent in intermediate stages. If naive field theory calculations did not produce infinite answers in the continuum limit, this would not have been such a big problem – it would just have been a bad choice of coordinates. But the lack of symmetry means that the infinite quantities must be cut off, and the bad coordinates make it nearly impossible to cut off the theory without spoiling the symmetry. This makes it difficult to extract the physical predictions, which require a careful limiting procedure.

The problem of lost symmetry also appears in classical mechanics, where the Hamiltonian formulation also superficially singles out time. The Lagrangian formulation makes the relativistic invariance apparent. In the same way, the path integral is manifestly relativistic. It reproduces the Schrödinger equation, the Heisenberg equations of motion, and the canonical commutation relations and shows that they are compatible with relativity. It extends the Heisenberg type operator algebra to operator product rules which are new relations difficult to see in the old formalism.

Further, different choices of canonical variables lead to very different seeming formulations of the same theory. The transformations between the variables can be very complicated, but the path integral makes them into reasonably straightforward changes of integration variables. For these reasons, the Feynman path integral has made earlier formalisms largely obsolete.

The price of a path integral representation is that the unitarity of a theory is no longer self-evident, but it can be proven by changing variables to some canonical representation. The path integral itself also deals with larger mathematical spaces than is usual, which requires more careful mathematics not all of which has been fully worked out. The path integral historically was not immediately accepted, partly because it took many years to incorporate fermions properly. This required physicists to invent an entirely new mathematical object – the Grassmann variable – which also allowed changes of variables to be done naturally, as well as allowing constrained quantization.

The integration variables in the path integral are subtly non-commuting. The value of the product of two field operators at what looks like the same point depends on how the two points are ordered in space and time. This makes some naive identities fail.

The Propagator

In relativistic theories, there is both a particle and field representation for every theory. The field representation is a sum over all field configurations, and the particle representation is a sum over different particle paths.

The nonrelativistic formulation is traditionally given in terms of particle paths, not fields. There, the path integral in the usual variables, with fixed boundary conditions, gives the probability amplitude for a particle to go from point x to point y in time T.

$$K(x, y; T) = \langle y; T \mid x; 0 \rangle = \int_{x(0)=x}^{x(T)=y} e^{iS[x]} Dx$$

This is called the propagator. Superposing different values of the initial position x with an arbitrary initial state $\psi_0(x)$ constructs the final state.

$$\psi_T(y) = \int_x \psi_0(x) K(x, y; T) dx = \int^{x(T)=y} \psi_0(x(0)) e^{iS[x]} Dx$$

For a spatially homogeneous system, where $K(x, y)$ is only a function of $(x - y)$, the integral is a convolution, the final state is the initial state convolved with the propagator.

$$\psi_T = \psi_0 * K(; T)$$

For a free particle of mass m, the propagator can be evaluated either explicitly from the path integral or by noting that the Schrödinger equation is a diffusion equation in imaginary time and the solution must be a normalized Gaussian:

$$K(x, y; T) \propto e^{\frac{im(x-y)^2}{2T}}$$

Taking the Fourier transform in $(x - y)$ produces another Gaussian:

$$K(p; T) = e^{\frac{iTp^2}{2m}}$$

and in p-space the proportionality factor here is constant in time, as will be verified in a moment. The Fourier transform in time, extending $K(p; T)$ to be zero for negative times, gives Green's function, or the frequency space propagator:

$$G_F(p, E) = \frac{-i}{E - \dfrac{\vec{p}^2}{2m} + i\epsilon}$$

Which is the reciprocal of the operator which annihilates the wavefunction in the Schrödinger equation, which wouldn't have come out right if the proportionality factor weren't constant in the p-space representation.

The infinitesimal term in the denominator is a small positive number which guarantees that the inverse Fourier transform in E will be nonzero only for future times. For past times, the inverse Fourier transform contour closes toward values of E where there is no singularity. This guarantees that K propagates the particle into the future and is the reason for the subscript on G. The infinitesimal term can be interpreted as an infinitesimal rotation toward imaginary time.

It is also possible to reexpress the nonrelativistic time evolution in terms of propagators which go toward the past, since the Schrödinger equation is time-reversible. The past propagator is the same as the future propagator except for the obvious difference that it vanishes in the future, and in the Gaussian t is replaced by $-t$. In this case, the interpretation is that these are the quantities to convolve the final wavefunction so as to get the initial wavefunction.

$$G_B(p, E) = \frac{-i}{-E - \dfrac{i\vec{p}^2}{2m} + i\epsilon}$$

Given the nearly identical only change is the sign of E and ε, the parameter E in Green's function can either be the energy if the paths are going toward the future, or the negative of the energy if the paths are going toward the past.

For a nonrelativistic theory, the time as measured along the path of a moving particle and the time as measured by an outside observer are the same. In relativity, this is no longer true. For a relativistic theory the propagator should be defined as the sum over all paths which travel between two points in a fixed proper time, as measured along the path. These paths describe the trajectory of a particle in space and in time.

$$K(x - y, T) = \int_{x(0)=x}^{x(T)=y} e^{i\int_0^T \sqrt{\dot{x}^2} - \alpha \, d\tau}$$

The integral above is not trivial to interpret, because of the square root. Fortunately, there is a heuristic trick. The sum is over the relativistic arclength of the path of an oscillating quantity, and like the nonrelativistic path integral should be interpreted as slightly rotated into imaginary time. The function $K(x - y, \tau)$ can be evaluated when the sum is over paths in Euclidean space.

$$K(x - y, T) = e^{-\alpha T} \int_{x(0)=x}^{x(T)=y} e^{-L}$$

This describes a sum over all paths of length T of the exponential of minus the length. This can be given a probability interpretation. The sum over all paths is a probability average over a path constructed step by step. The total number of steps is proportional to T, and each step is less likely the longer it is. By the central limit theorem, the result of many independent steps is a Gaussian of variance proportional to T.

$$K(x-y,T) = e^{-\alpha T} e^{-\frac{(x-y)^2}{T}}$$

The usual definition of the relativistic propagator only asks for the amplitude is to travel from x to y, after summing over all the possible proper times it could take.

$$K(x-y) = \int_0^\infty K(x-y,T) W(T) dT$$

Where $W(T)$ is a weight factor, the relative importance of paths of different proper time. By the translation symmetry in proper time, this weight can only be an exponential factor, and can be absorbed into the constant α.

$$K(x-y) = \int_0^\infty e^{-\frac{(x-y)^2}{T} - \alpha T} dT$$

This is the Schwinger representation. Taking a Fourier transform over the variable $(x - y)$ can be done for each value of T separately, and because each separate T contribution is a Gaussian, gives whose Fourier transform is another Gaussian with reciprocal width. So in p-space, the propagator can be reexpressed simply:

$$K(p) = \int_0^\infty e^{-Tp^2 - T\alpha} dT = \frac{1}{p^2 + \alpha}$$

Which is the Euclidean propagator for a scalar particle. Rotating p_0 to be imaginary gives the usual relativistic propagator, up to a factor of $-i$ and an ambiguity which will be clarified below.

$$K(p) = \frac{i}{p_0^2 - \vec{p}^2 - m^2}$$

This expression can be interpreted in the nonrelativistic limit, where it is convenient to split it by partial fractions:

$$2p_0 K(p) = \frac{i}{p_0 - \sqrt{\vec{p}^2 + m^2}} + \frac{i}{p_0 + \sqrt{\vec{p}^2 + m^2}}$$

For states where one nonrelativistic particle is present, the initial wavefunction has a frequency distribution concentrated near $p_0 = m$. When convolving with the propagator, which in p space just means multiplying by the propagator, the second term is suppressed and the first term is enhanced. For frequencies near $p_0 = m$, the dominant first term has the form:

$$2m K_{NR}(p) = \frac{i}{(p_0 - m) - \frac{\vec{p}^2}{2m}}$$

This is the expression for the nonrelativistic Green's function of a free Schrödinger particle.

The second term has a nonrelativistic limit also, but this limit is concentrated on frequencies which are negative. The second pole is dominated by contributions from paths where the proper time and the coordinate time are ticking in an opposite sense, which means that the second term is to be interpreted as the antiparticle. The nonrelativistic analysis shows that with this form the antiparticle still has positive energy.

The proper way to express this mathematically is that, adding a small suppression factor in proper time, the limit where $t \to -\infty$ of the first term must vanish, while the $t \to +\infty$ limit of the second term must vanish. In the Fourier transform, this means shifting the pole in p_0 slightly, so that the inverse Fourier transform will pick up a small decay factor in one of the time directions:

$$K(p) = \frac{i}{p_0 - \sqrt{\vec{p}^2 + m^2} + i\epsilon} + \frac{i}{p_0 - \sqrt{\vec{p}^2 + m^2} - i\epsilon}$$

Without these terms, the pole contribution could not be unambiguously evaluated when taking the inverse Fourier transform of p_0. The terms can be recombined:

$$K(p) = \frac{i}{p^2 - m^2 + i\epsilon}$$

Which when factored, produces opposite sign infinitesimal terms in each factor. This is the mathematically precise form of the relativistic particle propagator, free of any ambiguities. The ε term introduces a small imaginary part to the $\alpha = m^2$, which in the Minkowski version is a small exponential suppression of long paths.

So in the relativistic case, the Feynman path-integral representation of the propagator includes paths which go backwards in time, which describe antiparticles. The paths which contribute to the relativistic propagator go forward and backwards in time, and the interpretation of this is that the amplitude for a free particle to travel between two points includes amplitudes for the particle to fluctuate into an antiparticle, travel back in time, then forward again.

Unlike the nonrelativistic case, it is impossible to produce a relativistic theory of local particle propagation without including antiparticles. All local differential operators have inverses which are nonzero outside the light cone, meaning that it is impossible to keep a particle from travelling faster than light. Such a particle cannot have a Green's function which is only nonzero in the future in a relativistically invariant theory.

Functionals of Fields

However, the path integral formulation is also extremely important in *direct* application to quantum field theory, in which the "paths" or histories being considered are not the motions of a single particle, but the possible time evolutions of a field over all space. The action is referred to technically as a functional of the field: $S[\phi]$ where the field $\phi(x^\mu)$ is itself a function of space and time, and the square brackets are a reminder that the action depends on all the field's values everywhere, not just some particular value. *One* such given function $\phi(x^\mu)$ of spacetime is called a *field configuration*. In principle, one integrates Feynman's amplitude over the class of all possible field configurations.

Much of the formal study of QFT is devoted to the properties of the resulting functional integral, and much effort (not yet entirely successful) has been made toward making these functional integrals mathematically precise.

Such a functional integral is extremely similar to the partition function in statistical mechanics. Indeed, it is sometimes *called* a partition function, and the two are essentially mathematically identical except for the factor of i in the exponent in Feynman's postulate 3. Analytically continuing the integral to an imaginary time variable (called a Wick rotation) makes the functional integral even more like a statistical partition function, and also tames some of the mathematical difficulties of working with these integrals.

Expectation Values

In quantum field theory, if the action is given by the functional S of field configurations (which only depends locally on the fields), then the time ordered vacuum expectation value of polynomially bounded functional F, $\langle F \rangle$, is given by

$$\langle F \rangle = \frac{\int \mathcal{D}\phi F[\phi] e^{iS[\phi]}}{\int \mathcal{D}\phi e^{iS[\phi]}}$$

The symbol $\int \mathcal{D}\phi$ here is a concise way to represent the infinite-dimensional integral over all possible field configurations on all of spacetime. As stated above, the unadorned path integral in the denominator ensures proper normalization.

As a Probability

Strictly speaking the only question that can be asked in physics is: *"What fraction of states satisfying condition A also satisfy condition B?"* The answer to this is a number between 0 and 1 which can be interpreted as a conditional probability which is written as $P(B|A)$. In terms of path integration, since $P(B|A) = P(A \cap B)/P(A)$ this means:

$$P(B \mid A) = \frac{\displaystyle\sum_{F \subset A \cap B} \left| \int \mathcal{D}\phi O_{in}[\phi] e^{iS[\phi]} F[\phi] \right|^2}{\displaystyle\sum_{F \subset A} \left| \int \mathcal{D}\phi O_{in}[\phi] e^{iS[\phi]} F[\phi] \right|^2}$$

where the functional $O_{in}[\phi]$ is the superposition of all incoming states that could lead to the states we are interested in. In particular this could be a state corresponding to the state of the Universe just after the Big Bang although for actual calculation this can be simplified using heuristic methods. Since this expression is a quotient of path integrals it is naturally normalised.

Schwinger–Dyson Equations

Since this formulation of quantum mechanics is analogous to classical action principles, one might expect that identities concerning the action in classical mechanics would have quantum counterparts derivable from a functional integral. This is often the case.

In the language of functional analysis, we can write the Euler–Lagrange equations as

$$\frac{\delta S[\phi]}{\delta \phi} = 0$$

(the left-hand side is a functional derivative; the equation means that the action is stationary under small changes in the field configuration). The quantum analogues of these equations are called the Schwinger–Dyson equations.

If the functional measure $D\phi$ turns out to be translationally invariant (we'll assume this for the rest of this article, although this does not hold for, let's say nonlinear sigma models) and if we assume that after a Wick rotation

$$e^{iS[\phi]},$$

which now becomes

$$e^{-H[\phi]}$$

for some H, goes to zero faster than a reciprocal of any polynomial for large values of φ, we can integrate by parts (after a Wick rotation, followed by a Wick rotation back) to get the following Schwinger–Dyson equations for the expectation:

$$\left\langle \frac{\delta F[\phi]}{\delta \phi} \right\rangle = -i \left\langle F[\phi] \frac{\delta S[\phi]}{\delta \phi} \right\rangle$$

for any polynomially-bounded functional F.

$$\left\langle F_{,i} \right\rangle = -i \left\langle F S_{,i} \right\rangle$$

in the deWitt notation.

These equations are the analog of the on shell EL equations. The time ordering is taken before the time derivatives inside the $S_{,i}$.

If J (called the source field) is an element of the dual space of the field configurations (which has at least an affine structure because of the assumption of the translational invariance for the functional measure), then, the generating functional Z of the source fields is defined to be:

$$Z[J] = \int D\phi e^{i(S[\phi]+\langle J,\phi \rangle)}.$$

Note that

$$\frac{\delta^n Z}{\delta J(x_1)\ldots\delta J(x_n)}[J] = i^n Z[J] \langle \phi(x_1)\ldots\phi(x_n) \rangle_J$$

or

$$Z^{,i_1\cdots i_n}[J] = i^n Z[J]\left\langle \phi^{i_1} \cdots \phi^{i_n} \right\rangle_J,$$

where

$$\langle F \rangle_J = \frac{\int \mathcal{D}\phi F[\phi] e^{i(S[\phi] + \langle J, \phi \rangle)}}{\int \mathcal{D}\phi e^{i(S[\phi] + \langle J, \phi \rangle)}}.$$

Basically, if $\mathcal{D}\varphi\, e^{iS[\varphi]}$ is viewed as a functional distribution (this shouldn't be taken too literally as an interpretation of QFT, unlike its Wick rotated statistical mechanics analogue, because we have time ordering complications here!), then $\langle \varphi(x_1) \cdots \varphi(x_n) \rangle$ are its moments and Z is its Fourier transform.

If F is a functional of φ, then for an operator K, $F[K]$ is defined to be the operator which substitutes K for φ. For example, if

$$F[\phi] = \frac{\partial^{k_1}}{\partial x_1^{k_1}} \phi(x_1) \cdots \frac{\partial^{k_n}}{\partial x_n^{k_n}} \phi(x_n)$$

and G is a functional of J, then

$$F\left[-i\frac{\delta}{\delta J}\right] G[J] = (-i)^n \frac{\partial^{k_1}}{\partial x_1^{k_1}} \frac{\delta}{\delta J(x_1)} \cdots \frac{\partial^{k_n}}{\partial x_n^{k_n}} \frac{\delta}{\delta J(x_n)} G[J].$$

Then, from the properties of the functional integrals

$$\left\langle \frac{\delta S}{\delta \phi(x)}[\phi] + J(x) \right\rangle_J = 0$$

we get the "master" Schwinger–Dyson equation:

$$\frac{\delta S}{\delta \phi(x)}\left[-i\frac{\delta}{\delta J}\right] Z[J] + J(x)Z[J] = 0$$

or

$$S_{,i}[-i\partial]Z + J_i Z = 0.$$

If the functional measure is not translationally invariant, it might be possible to express it as the product $M[\varphi]\,\mathcal{D}\varphi$ where M is a functional and $\mathcal{D}\varphi$ is a translationally invariant measure. This is true, for example, for nonlinear sigma models where the target space is diffeomorphic to $\mathbb{R}^n$. However, if the target manifold is some topologically nontrivial space, the concept of a translation does not even make any sense.

In that case, we would have to replace the S in this equation by another functional

$$\hat{S} = S - i\ln(M)$$

If we expand this equation as a Taylor series about $J = 0$, we get the entire set of Schwinger–Dyson equations.

Localization

The path integrals are usually thought of as being the sum of all paths through an infinite space-time. However, in Local quantum field theory we would restrict everything to lie within a finite *causally complete* region, for example inside a double light-cone. This gives a more mathematically precise and physically rigorous definition of quantum field theory.

Ward–Takahashi Identities

Now how about the on shell Noether's theorem for the classical case? Does it have a quantum analog as well? Yes, but with a caveat. The functional measure would have to be invariant under the one parameter group of symmetry transformation as well.

Let's just assume for simplicity here that the symmetry in question is local (not local in the sense of a gauge symmetry, but in the sense that the transformed value of the field at any given point under an infinitesimal transformation would only depend on the field configuration over an arbitrarily small neighborhood of the point in question). Let's also assume that the action is local in the sense that it is the integral over spacetime of a Lagrangian, and that

$$Q[\mathcal{L}(x)] = \partial_\mu f^\mu(x)$$

for some function f where f only depends locally on φ (and possibly the spacetime position).

If we don't assume any special boundary conditions, this would not be a "true" symmetry in the true sense of the term in general unless $f = 0$ or something. Here, Q is a derivation which generates the one parameter group in question. We could have antiderivations as well, such as BRST and supersymmetry.

Let's also assume

$$\int \mathcal{D}\phi Q[F][\phi] = 0$$

for any polynomially-bounded functional F. This property is called the invariance of the measure. And this does not hold in general.

Then,

$$\int \mathcal{D}\phi Q\left[Fe^{iS}\right][\phi] = 0,$$

which implies

$$\langle Q[F]\rangle + i\left\langle F\int_{\partial V} f^\mu \, ds_\mu\right\rangle = 0$$

where the integral is over the boundary. This is the quantum analog of Noether's theorem.

Now, let's assume even further that Q is a local integral

$$Q = \int d^d x\, q(x)$$

where

$$q(x)[\phi(y)] = \delta^{(d)}(X - y)Q[\phi(y)]$$

so that

$$q(x)[S] = \partial_\mu j^\mu(x)$$

where

$$j^\mu(x) = f^\mu(x) - \frac{\partial}{\partial(\partial_\mu \phi)}\mathcal{L}(x)Q[\phi]$$

(this is assuming the Lagrangian only depends on ϕ and its first partial derivatives! More general Lagrangians would require a modification to this definition!). Note that we're NOT insisting that $q(x)$ is the generator of a symmetry (i.e. we are *not* insisting upon the gauge principle), but just that Q is. And we also assume the even stronger assumption that the functional measure is locally invariant:

$$\int \mathcal{D}\phi\, q(x)[F][\phi] = 0.$$

Then, we would have

$$\langle q(x)[F]\rangle + i\langle Fq(x)[S]\rangle = \langle q(x)[F]\rangle + i\left\langle F\partial_\mu j^\mu(x)\right\rangle = 0.$$

Alternatively,

$$q(x)[S]\left[-i\frac{\delta}{\delta J}\right]Z[J] + J(x)Q[\phi(x)]\left[-i\frac{\delta}{\delta J}\right]Z[J] = \partial_\mu j^\mu(x)\left[-i\frac{\delta}{\delta J}\right]Z[J] + J(x)Q[\phi(x)]\left[-i\frac{\delta}{\delta J}\right]Z[J] = 0.$$

The above two equations are the Ward–Takahashi identities.

Now for the case where $f = 0$, we can forget about all the boundary conditions and locality assumptions. We'd simply have

$$\langle Q[F]\rangle = 0.$$

Alternatively,

$$\int d^d x\, J(x)Q[\phi(x)]\left[-i\frac{\delta}{\delta J}\right]Z[J] = 0.$$

The Need for Regulators and Renormalization

Path integrals as they are defined here require the introduction of regulators. Changing the scale of the regulator leads to the renormalization group. In fact, renormalization is the major obstruction to making path integrals well-defined.

The Path Integral in Quantum-Mechanical Interpretation

In one interpretation of quantum mechanics, the "sum over histories" interpretation, the path integral is taken to be fundamental and reality is viewed as a single indistinguishable "class" of paths which all share the same events. For this interpretation, it is crucial to understand what exactly an event is. The sum over histories method gives identical results to canonical quantum mechanics, and Sinha and Sorkin claim the interpretation explains the Einstein–Podolsky–Rosen paradox without resorting to nonlocality.

Someadvocates of interpretations of quantum mechanics emphasizing decoherence have attempted to make more rigorous the notion of extracting a classical-like "coarse-grained" history from the space of all possible histories.

Quantum Gravity

Whereas in quantum mechanics the path integral formulation is fully equivalent to other formulations, it may be that it can be extended to quantum gravity, which would make it different from the Hilbert space model. Feynman had some success in this direction and his work has been extended by Hawking and others. Approaches that use this method include causal dynamical triangulations and spinfoam models.

Quantum Tunneling

Quantum tunnelling can be modeled by using the path integral formation to determine the action of the trajectory through a potential barrier. Using the WKB approximation, the tunneling rate (Γ) can be determined to be of the form

$$\Gamma = A_\mathrm{o} \exp\left(-\frac{S_\mathrm{eff}}{\hbar}\right)$$

with the effective action S_eff and pre-exponential factor A_o. This form is specifically useful in a dissipative system, in which the systems and surroundings must be modeled together. Using the Langevin equation to model Brownian motion, the path integral formation can be used to determine an effective action and pre-exponential model to see the effect of dissipation on tunnelling. From this model, tunneling rates of macroscopic systems (at finite temperatures) can be predicted.

Relation between Schrödinger's Equation and the Path Integral Formulation of Quantum Mechanics

This article relates the Schrödinger equation with the path integral formulation of quantum mechanics using a simple nonrelativistic one-dimensional single-particle Hamiltonian composed of kinetic and potential energy.

Background

Schrödinger's Equation

Schrödinger's equation, in bra–ket notation, is

$$i\hbar\frac{d}{dt}|\psi\rangle = \hat{H}|\psi\rangle$$

where $\hat{H}$ is the Hamiltonian operator. We have assumed for simplicity that there is only one spatial dimension.

The Hamiltonian operator can be written

$$\hat{H} = \frac{\hat{p}^2}{2m} + V(\hat{q})$$

where $V(\hat{q})$ is the potential energy, m is the mass and we have assumed for simplicity that there is only one spatial dimension q.

The formal solution of the equation is

$$|\psi(t)\rangle = \exp\left(-\frac{i}{\hbar}\hat{H}t\right)|q_0\rangle \equiv \exp\left(-\frac{i}{\hbar}\hat{H}t\right)|0\rangle$$

where we have assumed the initial state is a free-particle spatial state $|q_0\rangle$.

The transition probability amplitude for a transition from an initial state $|0\rangle$ to a final free-particle spatial state $|F\rangle$ at time T is

$$\langle F|\psi(t)\rangle = \left\langle F\left|\exp\left(-\frac{i}{\hbar}\hat{H}T\right)\right|0\right\rangle.$$

Path Integral Formulation

The path integral formulation states that the transition amplitude is simply the integral of the quantity

$$\exp\left(\frac{i}{\hbar}S\right)$$

over all possible paths from the initial state to the final state. Here S is the classical action.

The reformulation of this transition amplitude, originally due to Dirac and conceptualized by Feynman, forms the basis of the path integral formulation.

From Schrödinger's Equation to the Path Integral Formulation

Note: the following derivation is heuristic (it is valid in cases in which the potential, $V(q)$, commutes with the momentum, p). Following Feynman, this derivation can be made rigorous by writing the momentum, p, as the product of mass, m, and a difference in position at two points, x_a and x_b, separated by a time difference, δt, thus quantizing distance.

$$p = m\left(\frac{x_b - x_a}{\delta t}\right)$$

Note 2: There are two errata on page 11 in Zee, both of which are corrected here.

We can divide the time interval $[0, T]$ into N segments of length

$$\delta t = \frac{T}{N}.$$

The transition amplitude can then be written

$$\left\langle F \left| \exp\left(-\frac{i}{\hbar}\hat{H}T\right)\right| 0 \right\rangle = \left\langle F \left| \exp\left(-\frac{i}{\hbar}\hat{H}\delta t\right)\exp\left(-\frac{i}{\hbar}\hat{H}\delta t\right)\cdots\exp\left(-\frac{i}{\hbar}\hat{H}\delta t\right)\right| 0 \right\rangle.$$

We can insert the identity matrix

$$I = \int dq\, |q\rangle\langle q|$$

$N - 1$ times between the exponentials to yield

$$\left\langle F \left| \exp\left(-\frac{i}{\hbar}\hat{H}T\right)\right| 0 \right\rangle = \left(\prod_{j=1}^{N-1}\int dq_j\right)\left\langle F \left| \exp\left(-\frac{i}{\hbar}\hat{H}\delta t\right)\right| q_{N-1}\right\rangle\left\langle q_{N-1} \left| \exp\left(-\frac{i}{\hbar}\hat{H}\delta t\right)\right| q_{N-2}\right\rangle\cdots\left\langle q_1 \left| \exp\left(-\frac{i}{\hbar}\hat{H}\delta t\right)\right| 0\right\rangle.$$

Each individual transition probability can be written

$$\left\langle q_{j+1} \left| \exp\left(-\frac{i}{\hbar}\hat{H}\delta t\right)\right| q_j \right\rangle = \left\langle q_{j+1} \left| \exp\left(-\frac{i}{\hbar}\frac{\hat{p}^2}{2m}\delta t\right)\exp\left(-\frac{i}{\hbar}V(q_j)\delta t\right)\right| q_j \right\rangle.$$

We can insert the identity

$$I = \int \frac{dp}{2\pi}\, |p\rangle\langle p|$$

into the amplitude to yield

$$
\left\langle q_{j+1} \left| \exp\left(-\frac{i}{\hbar}\hat{H}\delta t\right)\right| q_j \right\rangle
$$

$$
= \exp\left(-\frac{i}{\hbar}V(q_j)\delta t\right)\int\frac{dp}{2\pi}\left\langle q_{j+1}\left|\exp\left(-\frac{i}{\hbar}\frac{p^2}{2m}\delta t\right)\right| p\right\rangle\langle p | q_j\rangle
$$

$$
= \exp\left(-\frac{i}{\hbar}V(q_j)\delta t\right)\int\frac{dp}{2\pi}\exp\left(-\frac{i}{\hbar}\frac{p^2}{2m}\delta t\right)\langle q_{j+1} | p\rangle\langle p | q_j\rangle
$$

$$
= \exp\left(-\frac{i}{\hbar}V(q_j)\delta t\right)\int\frac{dp}{2\pi\hbar}\exp\left(-\frac{i}{\hbar}\frac{p^2}{2m}\delta t - \frac{i}{\hbar}p(q_{j+1}-q_j)\right)
$$

where we have used the fact that the free particle wave function is

$$
\langle p | q_j\rangle = \frac{\exp\left(\dfrac{i}{\hbar}pq_j\right)}{\sqrt{\hbar}}.
$$

The integral over p can be performed

$$
\left\langle q_{j+1}\left|\exp\left(-\frac{i}{\hbar}\hat{H}\delta t\right)\right| q_j\right\rangle = \left(\frac{-im}{2\pi\delta t\hbar}\right)^{\frac{1}{2}}\exp\left[\frac{i}{\hbar}\delta t\left(\frac{1}{2}m\left(\frac{q_{j+1}-q_j}{\delta t}\right)^2 - V(q_j)\right)\right]
$$

The transition amplitude for the entire time period is

$$
\left\langle F\left|\exp\left(-\frac{i}{\hbar}\hat{H}T\right)\right| 0\right\rangle = \left(\frac{-im}{2\pi\delta t\hbar}\right)^{\frac{N}{2}}\left(\prod_{j=1}^{N-1}\int dq_j\right)\exp\left[\frac{i}{\hbar}\sum_{j=0}^{N-1}\delta t\left(\frac{1}{2}m\left(\frac{q_{j+1}-q_j}{\delta t}\right)^2 - V(q_j)\right)\right].
$$

If we take the limit of large N the transition amplitude reduces to

$$
\left\langle F\left|\exp\left(-\frac{i}{\hbar}\hat{H}T\right)\right| 0\right\rangle = \int Dq(t)\exp\left[\frac{i}{\hbar}S\right]
$$

where S is the classical action given by

$$
S = \int_0^T dt L\left(q(t), \dot{q}(t)\right)
$$

and L is the classical Lagrangian given by

$$
L(q,\dot{q}) = \frac{1}{2}m\dot{q}^2 - V(q)
$$

Any possible path of the particle, going from the initial state to the final state, is approximated as a broken line and included in the measure of the integral

$$\int Dq(t) = \lim_{N \to \infty} \left(\frac{-im}{2\pi\delta t\hbar} \right)^{\frac{N}{2}} \left(\prod_{j=1}^{N-1} \int dq_j \right)$$

This expression actually defines the manner in which the path integrals are to be taken. The coefficient in front is needed to ensure that the expression has the correct dimensions, but it has no actual relevance in any physical application.

This recovers the path integral formulation from Schrödinger's equation.

Propagator

In quantum mechanics and quantum field theory, the propagator is a function that specifies the probability amplitude for a particle to travel from one place to another in a given time, or to travel with a certain energy and momentum. In Feynman diagrams, which serve to calculate the rate of collisions in quantum field theory, virtual particles contribute their propagator to the rate of the scattering event described by the respective diagram. These may also be viewed as the inverse of the wave operator appropriate to the particle, and are, therefore, often called *(causal) Green's functions* (called "*causal*" to distinguish it from the elliptic Laplacian Green's function).

Non-relativistic Propagators

In non-relativistic quantum mechanics, the propagator gives the probability amplitude for a particle to travel from one spatial point at one time to another spatial point at a later time. It is the Green's function (fundamental solution) for the Schrödinger equation. This means that, if a system has Hamiltonian H, then the appropriate propagator is a function

$$G(x,t;x',t') = \frac{1}{i\hbar} \Theta(t-t') K(x,t;x',t')$$

satisfying

$$\left(i\hbar \frac{\partial}{\partial t} - H_x \right) G(x,t;x',t') = \delta(x-x')\delta(t-t') ,$$

where H_x denotes the Hamiltonian written in terms of the x coordinates, $\delta(x)$ denotes the Dirac delta-function, $\Theta(x)$ is the Heaviside step function and $K(x, t ;x', t')$ is the kernel of the differential operator in question, often referred to as the propagator instead of G in this context, and henceforth in this article. This propagator can also be written as

$$K(x,t;x',t') = \left\langle x | \hat{U}(t,t') | x' \right\rangle ,$$

where $\hat{U}(t, t')$ is the unitary time-evolution operator for the system taking states at time t to states at time t'.

The quantum mechanical propagator may also be found by using a path integral,

$$K(x,t;x',t') = \int \exp\left[\frac{i}{\hbar}\int_t^{t'} L(\dot{q},q,t)dt\right] D[q(t)]$$

where the boundary conditions of the path integral include $q(t) = x$, $q(t') = x'$. Here L denotes the Lagrangian of the system. The paths that are summed over move only forwards in time.

In non-relativistic quantum mechanics, the propagator lets you find the state of a system given an initial state and a time interval. The new state is given by the equation

$$\psi(x,t) = \int_{-\infty}^{\infty} \psi(x',t')K(x,t;x',t')dx'.$$

If $K(x,t;x',t')$ only depends on the difference $x - x'$, this is a convolution of the initial state and the propagator. This kernel is the kernel of integral transform.

Basic Examples: Propagator of Free Particle and Harmonic Oscillator

For a time-translationally invariant system, the propagator only depends on the time difference $t - t'$, so it may be rewritten as

$$K(x,t;x',t') = K(x, x';t - t').$$

The propagator of a one-dimensional free particle, with the far-right expression obtained via saddle-point methods, is then

$$K(x,x';t) = \frac{1}{2\pi}\int_{-\infty}^{+\infty} dk\, e^{ik(x-x')} e^{-\frac{i\hbar k^2 t}{2m}} = \left(\frac{m}{2\pi i\hbar t}\right)^{\frac{1}{2}} e^{-\frac{m(x-x')^2}{2i\hbar t}}.$$

Similarly, the propagator of a one-dimensional quantum harmonic oscillator is the Mehler kernel,

$$K(x,x';t) = \left(\frac{m\omega}{2\pi i\hbar \sin \omega t}\right)^{\frac{1}{2}} \exp\left(-\frac{m\omega((x^2 + x'^2)\cos \omega t - 2xx')}{2i\hbar \sin \omega t}\right).$$

The latter may be obtained from the previous free particle result upon making use of van Kortryk's SU(2) Lie-group identity,

$$\exp\left(-\frac{it}{\hbar}\left(\frac{1}{2m}p^2 + \frac{1}{2}m\omega^2 x^2\right)\right)$$

$$= \exp\left(-\frac{im\omega}{2\hbar}\mathsf{x}^2\tan\left(\frac{\omega t}{2}\right)\right)\exp\left(-\frac{i}{2m\omega\hbar}\mathsf{p}^2\sin\left(\omega t\right)\right)\exp\left(-\frac{im\omega}{2\hbar}\mathsf{x}^2\tan\left(\frac{\omega t}{2}\right)\right),$$

valid for operators x and p satisfying the Heisenberg relation $[\mathsf{x},\mathsf{p}]=i\hbar..$

For the N-dimensional case, the propagator can be simply obtained by the product

$$K(\vec{x},\vec{x}';t)=\prod_{q=1}^{N}K(x_q,x_{q'};t)\,.$$

Relativistic Propagators

In relativistic quantum mechanics and quantum field theory the propagators are Lorentz invariant. They give the amplitude for a particle to travel between two spacetime points.

Scalar Propagator

In quantum field theory the theory of a free (non-interacting) scalar field is a useful and simple example which serves to illustrate the concepts needed for more complicated theories. It describes spin zero particles. There are a number of possible propagators for free scalar field theory. We now describe the most common ones.

Position Space

The position space propagators are Green's functions for the Klein–Gordon equation. This means they are functions $G(x,y)$ which satisfy

$$(\square_x+m^2)G(x,y)=-\delta(x-y)$$

where:

- x,y are two points in Minkowski spacetime.
- $\square_x=\frac{\partial^2}{\partial t^2}-\nabla^2$ is the d'Alembertian operator acting on the x coordinates.
- $\delta(x-y)$ is the Dirac delta-function.

(As typical in relativistic quantum field theory calculations, we use units where the speed of light, c, and Planck's reduced constant, $\hbar$, are set to unity.)

We shall restrict attention to 4-dimensional Minkowski spacetime. We can perform a Fourier transform of the equation for the propagator, obtaining

$$\left(-p^2+m^2\right)G(p)=-1.$$

This equation can be inverted in the sense of distributions noting that the equation $xf(x)=1$ has the solution,

$$f(x) = \frac{1}{x \pm i\varepsilon} = \frac{1}{x} \mp i\pi\delta(x),$$

with ε implying the limit to zero. Below, we discuss the right choice of the sign arising from causality requirements.

The solution is

$$G(x,y) = \frac{1}{(2\pi)^4} \int d^4 p \frac{e^{-ip(x-y)}}{p^2 - m^2 \pm i\varepsilon},$$

where

$$p(x-y) := p_0(x^0 - y^0) - \vec{p} \cdot (\vec{x} - \vec{y})$$

is the 4-vector inner product.

The different choices for how to deform the integration contour in the above expression lead to different forms for the propagator. The choice of contour is usually phrased in terms of the p_0 integral.

The integrand then has two poles at

$$p_0 = \pm\sqrt{\vec{p}^2 + m^2}$$

so different choices of how to avoid these lead to different propagators.

Causal Propagators

Retarded Propagator

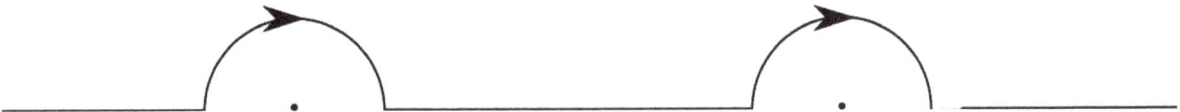

A contour going clockwise over both poles gives the causal retarded propagator. This is zero if x and y are spacelike or if $x^0 < y^0$ (i.e. if y is to the future of x).

This choice of contour is equivalent to calculating the limit,

$$G_{\text{ret}}(x,y) = \lim_{\varepsilon \to 0} \frac{1}{(2\pi)^4} \int d^4 p \frac{e^{-ip(x-y)}}{(p_0 + i\varepsilon)^2 - \vec{p}^2 - m^2} = \begin{cases} \frac{1}{2\pi}\delta(\tau_{xy}^2) - \frac{mJ_1(m\tau_{xy})}{4\pi\tau_{xy}} & y \prec x \\ 0 & \text{otherwise} \end{cases}$$

Here

$$\tau_{xy} := \sqrt{(x^0 - y^0)^2 - (\vec{x} - \vec{y})^2}$$

is the proper time from x to y and J_1 is a Bessel function of the first kind. The expression $y \prec x$ means y causally precedes x which, for Minkowski spacetime, means

$$y^0 < x^0 \text{ and } \tau_{xy}^2 \geq 0.$$

This expression can also be expressed in terms of the vacuum expectation value of the commutator of the free scalar field operator,

$$G_{\text{ret}}(x, y) = i\langle 0 | [\Phi(x), \Phi(y)] | 0 \rangle \Theta(x^0 - y^0)$$

where

$$\Theta(x) := \begin{cases} 1 & x \geq 0 \\ 0 & x < 0 \end{cases}$$

is the Heaviside step function and

$$[\Phi(x), \Phi(y)] := \Phi(x)\Phi(y) - \Phi(y)\Phi(x)$$

is the commutator.

Advanced Propagator

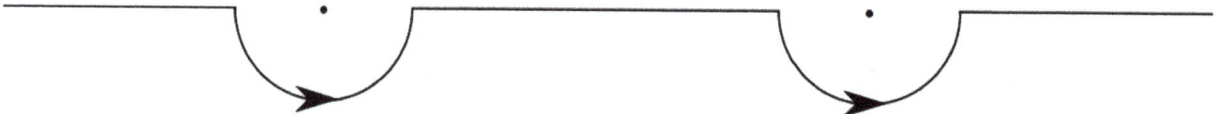

A contour going anti-clockwise under both poles gives the causal advanced propagator. This is zero if x and y are spacelike or if $x^0 > y^0$ (i.e. if y is to the past of x).

This choice of contour is equivalent to calculating the limit

$$G_{\text{adv}}(x, y) = \lim_{\varepsilon \to 0} \frac{1}{(2\pi)^4} \int d^4 p \frac{e^{-ip(x-y)}}{(p_0 - i\varepsilon)^2 - \vec{p}^2 - m^2}$$

$$= \begin{cases} -\dfrac{1}{2\pi}\delta(\tau_{xy}^2) + \dfrac{mJ_1(m\tau_{xy})}{4\pi\tau_{xy}} & x \prec y \\ 0 & \text{otherwise} \end{cases}$$

This expression can also be expressed in terms of the vacuum expectation value of the commutator of the free scalar field. In this case,

$$G_{\text{adv}}(x, y) = -i\langle 0 | [\Phi(x), \Phi(y)] | 0 \rangle \Theta(y^0 - x^0).$$

Feynman Propagator

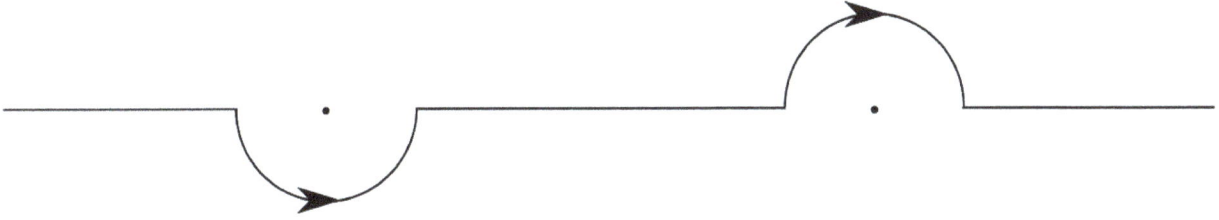

A contour going under the left pole and over the right pole gives the Feynman propagator.

This choice of contour is equivalent to calculating the limit

$$G_F(x,y) = \lim_{\varepsilon \to 0} \frac{1}{(2\pi)^4} \int d^4 p \frac{e^{-ip(x-y)}}{p^2 - m^2 + i\varepsilon} = \begin{cases} -\dfrac{1}{4\pi}\delta(s) + \dfrac{m}{8\pi\sqrt{s}}H_1^{(2)}(m\sqrt{s}) & s \geq 0 \\[2ex] -\dfrac{im}{4\pi^2\sqrt{-s}}K_1(m\sqrt{-s}) & s < 0. \end{cases}$$

Here

$$s := (x^0 - y^0)^2 - (\vec{x} - \vec{y})^2,$$

where x and y are two points in Minkowski spacetime, and the dot in the exponent is a four-vector inner product. $H_1^{(2)}$ is a Hankel function and K_1 is a modified Bessel function.

This expression can be derived directly from the field theory as the vacuum expectation value of the *time-ordered product* of the free scalar field, that is, the product always taken such that the time ordering of the spacetime points is the same,

$$G_F(x-y) = -i\langle 0 | T(\Phi(x)\Phi(y)) | 0 \rangle$$
$$= -i\left\langle 0 | \left[\Theta(x^0 - y^0)\Phi(x)\Phi(y) + \Theta(y^0 - x^0)\Phi(y)\Phi(x) \right] | 0 \right\rangle.$$

This expression is Lorentz invariant, as long as the field operators commute with one another when the points x and y are separated by a spacelike interval.

The usual derivation is to insert a complete set of single-particle momentum states between the fields with Lorentz covariant normalization, then show that the Θ functions providing the causal time ordering may be obtained by a contour integral along the energy axis if the integrand is as above (hence the infinitesimal imaginary part, to move the pole off the real line).

The propagator may also be derived using the path integral formulation of quantum theory.

Momentum Space Propagator

The Fourier transform of the position space propagators can be thought of as propagators in momentum space. These take a much simpler form than the position space propagators.

They are often written with an explicit ε term although this is understood to be a reminder about which integration contour is appropriate. This ε term is included to incorporate boundary conditions and causality.

For a 4-momentum p the causal and Feynman propagators in momentum space are:

$$\tilde{G}_{\text{ret}}(p) = \frac{1}{(p_0 + i\varepsilon)^2 - \vec{p}^2 - m^2}$$

$$\tilde{G}_{\text{adv}}(p) = \frac{1}{(p_0 - i\varepsilon)^2 - \vec{p}^2 - m^2}$$

$$\tilde{G}_F(p) = \frac{1}{p^2 - m^2 + i\varepsilon}.$$

For purposes of Feynman diagram calculations it is usually convenient to write these with an additional overall factor of $-i$ (conventions vary).

Faster than Light?

The Feynman propagator has some properties that seem baffling at first. In particular, unlike the commutator, the propagator is *nonzero* outside of the light cone, though it falls off rapidly for spacelike intervals. Interpreted as an amplitude for particle motion, this translates to the virtual particle traveling faster than light. It is not immediately obvious how this can be reconciled with causality: can we use faster-than-light virtual particles to send faster-than-light messages?

The answer is no: while in classical mechanics the intervals along which particles and causal effects can travel are the same, this is no longer true in quantum field theory, where it is commutators that determine which operators can affect one another.

So what *does* the spacelike part of the propagator represent? In QFT the vacuum is an active participant, and particle numbers and field values are related by an uncertainty principle; field values are uncertain even for particle number *zero*. There is a nonzero probability amplitude to find a significant fluctuation in the vacuum value of the field $\Phi(x)$ if one measures it locally (or, to be more precise, if one measures an operator obtained by averaging the field over a small region). Furthermore, the dynamics of the fields tend to favor spatially correlated fluctuations to some extent. The nonzero time-ordered product for spacelike-separated fields then just measures the amplitude for a nonlocal correlation in these vacuum fluctuations, analogous to an EPR correlation. Indeed, the propagator is often called a *two-point correlation function* for the free field.

Since, by the postulates of quantum field theory, all observable operators commute with each other at spacelike separation, messages can no more be sent through these correlations than they can through any other EPR correlations; the correlations are in random variables.

In terms of virtual particles, the propagator at spacelike separation can be thought of as a means of calculating the amplitude for creating a virtual particle-antiparticle pair that eventually disappear into the vacuum, or for detecting a virtual pair emerging from the vacuum. In Feynman's language, such creation and annihilation processes are equivalent to a virtual particle wandering backward and forward through time, which can take it outside of the light cone. However, no causality violation is involved.

Explanation Using Limits

This can be made clearer by writing the propagator in the following form for a massless photon,

$$G_F^\varepsilon(x,y) = \frac{\varepsilon}{(x-y)^2 + i\varepsilon^2}.$$

This is the usual definition but normalised by a factor of ε. Then the rule is that one only takes the limit $\varepsilon \to 0$ at the end of a calculation.

One sees that

$$G_F^\varepsilon(x,y) = \frac{1}{\varepsilon} \quad \text{if} \quad (x-y)^2 = 0$$

and

$$\lim_{\varepsilon \to 0} G_F^\varepsilon(x,y) = 0 \quad \text{if} \quad (x-y)^2 \neq 0$$

Hence this means a single photon will always stay on the light cone. It is also shown that the total probability for a photon at any time must be normalised by the reciprocal of the following factor:

$$\lim_{\varepsilon \to 0} \int \left| G_F^\varepsilon(0,x) \right|^2 dx^3 = \lim_{\varepsilon \to 0} \int \frac{\varepsilon^2}{(\mathbf{x}^2 - t^2)^2 + \varepsilon^4} dx^3 = 2\pi^2 |t|.$$

We see that the parts outside the light cone usually are zero in the limit and only are important in Feynman diagrams.

Propagators in Feynman Diagrams

The most common use of the propagator is in calculating probability amplitudes for particle interactions using Feynman diagrams. These calculations are usually carried out in momentum space. In general, the amplitude gets a factor of the propagator for every *internal line*, that is, every line that does not represent an incoming or outgoing particle in the initial or final state. It will also get a factor proportional to, and similar in form to, an interaction term in the theory's Lagrangian for every internal vertex where lines meet. These prescriptions are known as *Feynman rules*.

Internal lines correspond to virtual particles. Since the propagator does not vanish for combinations of energy and momentum disallowed by the classical equations of motion, we say that the virtual particles are allowed to be off shell. In fact, since the propagator is obtained by inverting the wave equation, in general it will have singularities on shell.

The energy carried by the particle in the propagator can even be *negative*. This can be interpreted simply as the case in which, instead of a particle going one way, its antiparticle is going the *other* way, and therefore carrying an opposing flow of positive energy. The propagator encompasses both possibilities. It does mean that one has to be careful about minus signs for the case of fermions, whose propagators are not even functions in the energy and momentum.

Virtual particles conserve energy and momentum. However, since they can be off shell, wherever the diagram contains a closed *loop*, the energies and momenta of the virtual particles participating in the loop will be partly unconstrained, since a change in a quantity for one particle in the loop can be balanced by an equal and opposite change in another. Therefore, every loop in a Feynman diagram requires an integral over a continuum of possible energies and momenta. In general, these integrals of products of propagators can diverge, a situation that must be handled by the process of renormalization.

Other Theories

Spin $\frac{1}{2}$

If the particle possesses spin then its propagator is in general somewhat more complicated, as it will involve the particle's spin or polarization indices. The differential equation satisfied by the propagator for a spin $\frac{1}{2}$ particle is given by

$$(i\slashed{\nabla}' - m)S_F(x',x) = I_4\delta^4(x'-x),$$

where I_4 is the unit matrix in four dimensions, and employing the Feynman slash notation. This is the Dirac equation for a delta function source in spacetime. Using the momentum representation,

$$S_F(x',x) = \int \frac{d^4p}{(2\pi)^4} \exp\left[-ip\cdot(x'-x)\right]\tilde{S}_F(p),$$

the equation becomes

$$(i\slashed{\nabla}' - m)\int \frac{d^4p}{(2\pi)^4}\tilde{S}_F(p)\exp\left[-ip\cdot(x'-x)\right]$$

$$= \int \frac{d^4p}{(2\pi)^4}(\slashed{p}-m)\tilde{S}_F(p)\exp\left[-ip\cdot(x'-x)\right]$$

$$= \int \frac{d^4p}{(2\pi)^4}I_4\exp\left[-ip\cdot(x'-x)\right]$$

$$= I_4\delta^4(x'-x),$$

where on the right hand side an integral representation of the four-dimensional delta function is used. Thus

$$(\slashed{p}-mI_4)\tilde{S}_F(p) = I_4.$$

By multiplying from the left with

$$(\not{p} + m)$$

(dropping unit matrices from the notation) and using properties of the gamma matrices,

$$\not{p}\not{p} = \frac{1}{2}(\not{p}\not{p} + \not{p}\not{p})$$

$$= \frac{1}{2}(\gamma_\mu p^\mu \gamma_\nu p^\nu + \gamma_\nu p^\nu \gamma_\mu p^\mu)$$

$$= \frac{1}{2}(\gamma_\mu \gamma_\nu + \gamma_\nu \gamma_\mu)p^\mu p^\nu$$

$$= g_{\mu\nu} p^\mu p^\nu = p_\mu p^\nu = p^2,$$

the momentum-space propagator used in Feynman diagrams for a Dirac field representing the electron in quantum electrodynamics is found to have form

$$\tilde{S}_F(p) = \frac{(\not{p} + m)}{p^2 - m^2 + i\varepsilon} = \frac{(\gamma^\mu p_\mu + m)}{p^2 - m^2 + i\varepsilon}.$$

The $i\varepsilon$ downstairs is a prescription for how to handle the poles in the complex p_0-plane. It automatically yields the Feynman contour of integration by shifting the poles appropriately. It is sometimes written

$$\tilde{S}_F(p) = \frac{1}{\gamma^\mu p_\mu - m + i\varepsilon} = \frac{1}{\not{p} - m + i\varepsilon}$$

for short. It should be remembered that this expression is just shorthand notation for $(\gamma_\mu p^\mu - m)^{-1}$. "One over matrix" is otherwise nonsensical. In position space one has

$$S_F(x - y) = \int \frac{d^4 p}{(2\pi)^4} e^{-ip\cdot(x-y)} \frac{\gamma^\mu p_\mu + m}{p^2 - m^2 + i\varepsilon} = \left(\frac{\gamma^\mu (x - y)_\mu}{|x - y|^5} + \frac{m}{|x - y|^3} \right) J_1(m |x - y|).$$

This is related to the Feynman propagator by

$$S_F(x - y) = (i\not{\partial} + m)G_F(x - y)$$

where $\not{\partial} := \gamma^\mu \partial_\mu$.

Spin 1

The propagator for a gauge boson in a gauge theory depends on the choice of convention to fix the gauge. For the gauge used by Feynman and Stueckelberg, the propagator for a photon is

$$\frac{-ig^{\mu\nu}}{p^2 + i\varepsilon}.$$

The propagator for a massive vector field can be derived from the Stueckelberg Lagrangian. The general form with gauge parameter λ reads

$$\frac{g_{\mu\nu} - \dfrac{k_\mu k_\nu}{m^2}}{k^2 - m^2 + i\varepsilon} + \frac{\dfrac{k_\mu k_\nu}{m^2}}{k^2 - \dfrac{m^2}{\lambda} + i\varepsilon}.$$

With this general form one obtains the propagator in unitary gauge for $\lambda = 0$, the propagator in Feynman or 't Hooft gauge for $\lambda = 1$ and in Landau or Lorenz gauge for $\lambda = \infty$. There are also other notations where the gauge parameter is the inverse of λ. The name of the propagator however refers to its final form and not necessarily to the value of the gauge parameter.

Unitary gauge:

$$\frac{g_{\mu\nu} - \dfrac{k_\mu k_\nu}{m^2}}{k^2 - m^2 + i\varepsilon}.$$

Feynman ('t Hooft) gauge:

$$\frac{g_{\mu\nu}}{k^2 - m^2 + i\varepsilon}.$$

Landau (Lorenz) gauge:

$$\frac{g_{\mu\nu} - \dfrac{k_\mu k_\nu}{k^2}}{k^2 - m^2 + i\varepsilon}.$$

Graviton Propagator

The graviton propagator for Minkowski space in general relativity is

$$G = \frac{P^2}{k^2} - \frac{P_s^0}{2k^2},$$

where P^2 is the transverse and traceless spin-2 projection operator and P_s^0 is a spin-0 scalar multiplet. The graviton propagator for (Anti) de Sitter space is

$$G = \frac{P^2}{k^2 + 2H^2} - \frac{P_s^0}{2(k^2 - 4H^2)},$$

where H is the Hubble constant. Note that upon taking the limit $H \to 0$, the AdS propagator reduces to the Minkowski propagator.

Related Singular Functions

The scalar propagators are Green's functions for the Klein–Gordon equation. There are related singular functions which are important in quantum field theory. We follow the notation in Bjorken and Drell. These function are most simply defined in terms of the vacuum expectation value of products of field operators.

Solutions to the Klein–Gordon Equation

Pauli–Jordan Function

The commutator of two scalar field operators defines the Pauli–Jordan function $\Delta(x-y)$ by

$$\langle 0|[\Phi(x),\Phi(y)]|0\rangle = i\Delta(x-y)$$

with

$$\Delta(x-y) = G_{adv}(x-y) - G_{ret}(x-y)$$

This satisfies

$$\Delta(x-y) = -\Delta(y-x)$$

and is zero if $(x-y)^2 < 0$.

Positive and Negative Frequency Parts (Cut Propagators)

We can define the positive and negative frequency parts of $\Delta(x-y)$, sometimes called cut propagators, in a relativistically invariant way.

This allows us to define the positive frequency part:

$$\Delta_+(x-y) = \langle 0|\Phi(x)\Phi(y)|0\rangle,$$

and the negative frequency part:

$$\Delta_-(x-y) = \langle 0|\Phi(y)\Phi(x)|0\rangle.$$

These satisfy

$$i\Delta = \Delta_+ - \Delta_-$$

and

$$(\Box_x + m^2)\Delta_\pm(x-y) = 0.$$

Auxiliary Function

The anti-commutator of two scalar field operators defines $\Delta_1(x-y)$ function by

$$\langle 0|\{\Phi(x),\Phi(y)\}|0\rangle = \Delta_1(x-y)$$

with

$$\Delta_1(x-y) = \Delta_+(x-y) + \Delta_-(x-y).$$

This satisfies $\Delta_1(x-y) = \Delta_1(y-x)$.

Green's Functions for the Klein–Gordon Equation

The retarded, advanced and Feynman propagators defined above are all Green's functions for the Klein–Gordon equation.

They are related to the singular functions by

$$G_{ret}(x-y) = -\Delta(x-y)\Theta(x_0 - y_0)$$

$$G_{adv}(x-y) = \Delta(x-y)\Theta(y_0 - x_0)$$

$$2G_F(x-y) = -i\Delta_1(x-y) + \varepsilon(x_0 - y_0)\Delta(x-y)$$

where

$$\varepsilon(x_0 - y_0) = 2\Theta(x_0 - y_0) - 1.$$

Feynman Diagram

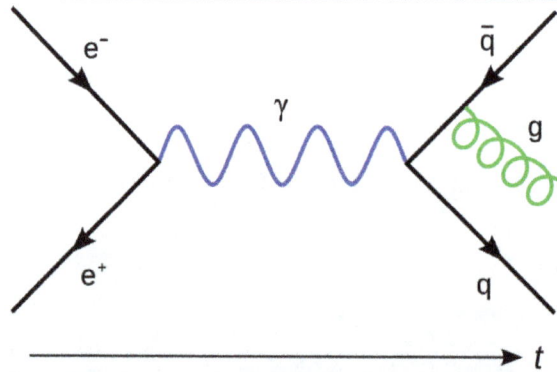

In this Feynman diagram, an electron and a positron annihilate, producing a photon (represented by the blue sine wave) that becomes a quark–antiquark pair, after which the antiquark radiates a gluon (represented by the green helix).

In theoretical physics, Feynman diagrams are pictorial representations of the mathematical expressions describing the behavior of subatomic particles. The scheme is named after its inventor, American physicist Richard Feynman, and was first introduced in 1948. The interaction of sub-atomic particles can be complex and difficult to understand intuitively. Feynman diagrams give a simple visualization of what would otherwise be a rather arcane and abstract formula. As David Kaiser writes, "since the middle of the 20th century, theoretical physicists have increasingly turned to this tool to help them undertake critical calculations", and so "Feynman diagrams have revolutionized nearly every aspect of theoretical physics". While the diagrams are applied primarily to quantum field theory, they can also be used in other fields, such as solid-state theory.

Feynman used Ernst Stueckelberg's interpretation of the positron as if it were an electron moving backward in time. Thus, antiparticles are represented as moving backward along the time axis in Feynman diagrams.

The calculation of probability amplitudes in theoretical particle physics requires the use of rather large and complicated integrals over a large number of variables. These integrals do, however, have a regular structure, and may be represented graphically as Feynman diagrams.

A Feynman diagram is a contribution of a particular class of particle paths, which join and split as described by the diagram. More precisely, and technically, a Feynman diagram is a graphical representation of a perturbative contribution to the transition amplitude or correlation function of a quantum mechanical or statistical field theory. Within the canonical formulation of quantum field theory, a Feynman diagram represents a term in the Wick's expansion of the perturbative S-matrix. Alternatively, the path integral formulation of quantum field theory represents the transition amplitude as a weighted sum of all possible histories of the system from the initial to the final state, in terms of either particles or fields. The transition amplitude is then given as the matrix element of the S-matrix between the initial and the final states of the quantum system.

Motivation and History

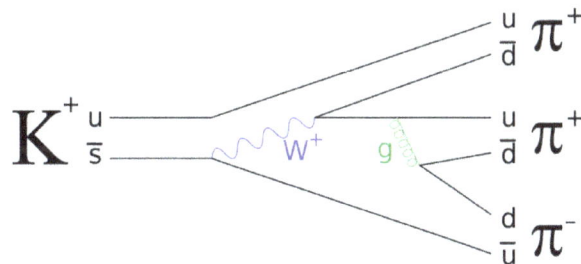

In this diagram, a kaon, made of an up and anti-strange quark, decays both weakly and strongly into three pions, with intermediate steps involving a W boson and a gluon (represented by the blue sine wave and green spiral, respectively).

When calculating scattering cross-sections in particle physics, the interaction between particles can be described by starting from a free field that describes the incoming and outgoing particles, and including an interaction Hamiltonian to describe how the particles deflect one another. The amplitude for scattering is the sum of each possible interaction history over all possible intermediate particle states. The number of times the interaction Hamiltonian acts is the order of the perturbation expansion, and the time-dependent perturbation theory for fields is known as the Dyson series. When the intermediate states at intermediate times are energy eigenstates (collections of particles with a definite momentum) the series is called old-fashioned perturbation theory.

The Dyson series can be alternatively rewritten as a sum over Feynman diagrams, where at each interaction vertex both the energy and momentum are conserved, but where the length of the energy-momentum four-vector is not equal to the mass. The Feynman diagrams are much easier to keep track of than old-fashioned terms, because the old-fashioned way treats the particle and antiparticle contributions as separate. Each Feynman diagram is the sum of exponentially many old-fashioned terms, because each internal line can separately represent either a particle or an antiparticle. In a non-relativistic theory, there are no antiparticles and there is no doubling, so each Feynman diagram includes only one term.

Feynman gave a prescription for calculating the amplitude for any given diagram from a field theory Lagrangian—the Feynman rules. Each internal line corresponds to a factor of the virtual particle's propagator; each vertex where lines meet gives a factor derived from an interaction term in the Lagrangian, and incoming and outgoing lines carry an energy, momentum, and spin.

In addition to their value as a mathematical tool, Feynman diagrams provide deep physical insight into the nature of particle interactions. Particles interact in every way available; in fact, intermediate virtual particles are allowed to propagate faster than light. The probability of each final state is then obtained by summing over all such possibilities. This is closely tied to the functional integral formulation of quantum mechanics, also invented by Feynman

The naïve application of such calculations often produces diagrams whose amplitudes are infinite, because the short-distance particle interactions require a careful limiting procedure, to include particle self-interactions. The technique of renormalization, suggested by Ernst Stueckelberg and Hans Bethe and implemented by Dyson, Feynman, Schwinger, and Tomonaga compensates for this effect and eliminates the troublesome infinities. After renormalization, calculations using Feynman diagrams match experimental results with very high accuracy.

Feynman diagram and path integral methods are also used in statistical mechanics and can even be applied to classical mechanics.

Alternative Names

Murray Gell-Mann always referred to Feynman diagrams as Stueckelberg diagrams, after a Swiss physicist, Ernst Stueckelberg, who devised a similar notation many years earlier. Stueckelberg was motivated by the need for a manifestly covariant formalism for quantum field theory, but did not provide as automated a way to handle symmetry factors and loops, although he was first to find the correct physical interpretation in terms of forward and backward in time particle paths, all without the path-integral. Historically they were sometimes called Feynman–Dyson diagrams or Dyson graphs, because when they were introduced the path integral was unfamiliar, and Freeman Dyson's derivation from old-fashioned perturbation theory was easier to follow for physicists trained in earlier methods. However, in 2006 Dyson himself stated that the diagrams should be called *Feynman diagrams* because "he taught us how to use them". This reflects historical fact: Feynman had to lobby hard for the diagrams, which confused the establishment physicists trained in equations and graphs.

Representation of Physical Reality

In their presentations of fundamental interactions, written from the particle physics perspective, Gerard 't Hooft and Martinus Veltman gave good arguments for taking the original, non-regularized Feynman diagrams as the most succinct representation of our present knowledge about the physics of quantum scattering of fundamental particles. Their motivations are consistent with the convictions of James Daniel Bjorken and Sidney Drell:

The Feynman graphs and rules of calculation summarize quantum field theory in a form in close contact with the experimental numbers one wants to understand. Although the statement of the theory in terms of graphs may imply perturbation theory, use of graphical methods in the many-body problem shows that this formalism is flexible enough to deal with phenomena of nonperturbative characters ... Some modification of the Feynman rules of calculation may well outlive the elaborate mathematical structure of local canonical quantum field theory ...

So far there are no opposing opinions. In quantum field theories the Feynman diagrams are obtained from Lagrangian by Feynman rules.

Dimensional regularization is a method for regularizing integrals in the evaluation of Feynman diagrams; it assigns values to them that are meromorphic functions of an auxiliary complex parameter d, called the dimension. Dimensional regularization writes a Feynman integral as an integral depending on the spacetime dimension d and spacetime points.

Particle-path Interpretation

A Feynman diagram is a representation of quantum field theory processes in terms of particle paths. The particle trajectories are represented by the lines of the diagram, which can be squiggly or straight, with an arrow or without, depending on the type of particle. A point where lines connect to other lines is an interaction vertex, and this is where the particles meet and interact: by emitting or absorbing new particles, deflecting one another, or changing type.

There are three different types of lines: *internal lines* connect two vertices, *incoming lines* extend from "the past" to a vertex and represent an initial state, and *outgoing lines* extend from a vertex to "the future" and represent the final state. Sometimes, the bottom of the diagram is the past and the top the future; other times, the past is to the left and the future to the right. When calculating correlation functions instead of scattering amplitudes, there is no past and future and all the lines are internal. The particles then begin and end on little x's, which represent the positions of the operators whose correlation is being calculated.

Feynman diagrams are a pictorial representation of a contribution to the total amplitude for a process that can happen in several different ways. When a group of incoming particles are to scatter off each other, the process can be thought of as one where the particles travel over all possible paths, including paths that go backward in time.

Feynman diagrams are often confused with spacetime diagrams and bubble chamber images because they all describe particle scattering. Feynman diagrams are graphs that represent the trajectories of particles in intermediate stages of a scattering process. Unlike a bubble chamber picture, only the sum of all the Feynman diagrams represent any given particle interaction; particles do not choose a particular diagram each time they interact. The law of summation is in accord with the principle of superposition—every diagram contributes to the total amplitude for the process.

Description

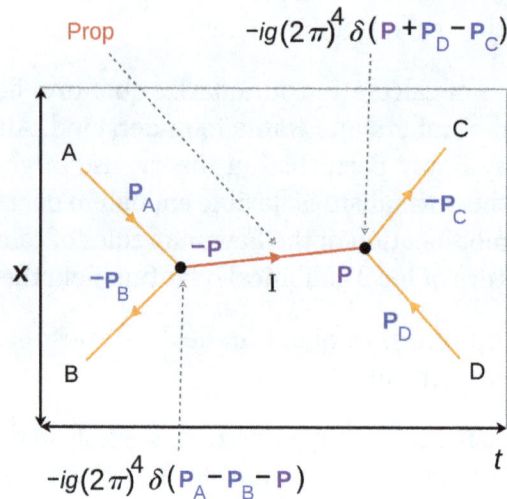

General features of the scattering process A + B → C + D:
• internal lines (red) for intermediate particles and processes, which has a propagator factor ("prop"), external lines (orange) for incoming/outgoing particles to/from vertices (black),
• at each vertex there is 4-momentum conservation using delta functions, 4-momenta entering the vertex are positive while those leaving are negative, the factors at each vertex and internal line are multiplied in the amplitude integral,
• space x and time t axes are not always shown, directions of external lines correspond to passage of time.

A Feynman diagram represents a perturbative contribution to the amplitude of a quantum transition from some initial quantum state to some final quantum state.

For example, in the process of electron-positron annihilation the initial state is one electron and one positron, the final state: two photons.

The initial state is often assumed to be at the left of the diagram and the final state at the right (although other conventions are also used quite often).

A Feynman diagram consists of points, called vertices, and lines attached to the vertices.

The particles in the initial state are depicted by lines sticking out in the direction of the initial state (e.g., to the left), the particles in the final state are represented by lines sticking out in the direction of the final state (e.g., to the right).

In QED there are two types of particles: electrons/positrons (called fermions) and photons (called gauge bosons). They are represented in Feynman diagrams as follows:

1. Electron in the initial state is represented by a solid line with an arrow pointing toward the vertex (→•).

2. Electron in the final state is represented by a line with an arrow pointing away from the vertex: (•→).

3. Positron in the initial state is represented by a solid line with an arrow pointing away from the vertex: (←•).

4. Positron in the final state is represented by a line with an arrow pointing toward the vertex: (•←).

5. Photon in the initial and the final state is represented by a wavy line (~• and •~).

In QED a vertex always has three lines attached to it: one bosonic line, one fermionic line with arrow toward the vertex, and one fermionic line with arrow away from the vertex.

The vertices might be connected by a bosonic or fermionic propagator. A bosonic propagator is represented by a wavy line connecting two vertices (•~•). A fermionic propagator is represented by a solid line (with an arrow in one or another direction) connecting two vertices, (•←•).

The number of vertices gives the order of the term in the perturbation series expansion of the transition amplitude.

Electron–positron Annihilation Example

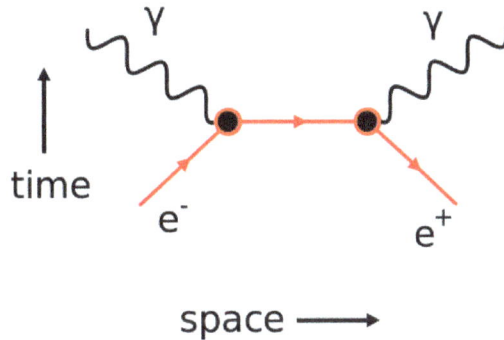

Feynman diagram of electron/positron annihilation

The electron–positron annihilation interaction:

$$e^+ e^- \rightarrow 2\gamma$$

has a contribution from the second order Feynman diagram shown adjacent:

In the initial state (at the bottom; early time) there is one electron (e^-) and one positron (e^+) and in the final state (at the top; late time) there are two photons (γ).

Canonical Quantization Formulation

The probability amplitude for a transition of a quantum system from the initial state $|i\rangle$ to the final state $|f\rangle$ is given by the matrix element

$$S_{fi} = \langle f | S | i \rangle,$$

where S is the S-matrix.

In the canonical quantum field theory the S-matrix is represented within the interaction picture by the perturbation series in the powers of the interaction Lagrangian,

$$S = \sum_{n=0}^{\infty} \frac{i^n}{n!} \int \prod_{j=1}^{n} d^4 x_j T \prod_{j=1}^{n} L_v(x_j) \equiv \sum_{n=0}^{\infty} S^{(n)},$$

where L_v is the interaction Lagrangian and T signifies the time-ordered product of operators.

A Feynman diagram is a graphical representation of a term in the Wick's expansion of the time-ordered product in the nth order term $S^{(n)}$ of the S-matrix,

$$T \prod_{j=1}^{n} L_v(x_j) = \sum_{\substack{\text{all possible} \\ \text{contractions}}} (\pm) N \prod_{j=1}^{n} L_v(x_j),$$

where N signifies the normal-product of the operators and $(\pm)$ takes care of the possible sign change when commuting the fermionic operators to bring them together for a contraction (a propagator).

Feynman Rules

The diagrams are drawn according to the Feynman rules, which depend upon the interaction Lagrangian. For the QED interaction Lagrangian

$$L_v = -g \bar{\psi} \gamma^\mu \psi A_\mu$$

describing the interaction of a fermionic field ψ with a bosonic gauge field A_μ, the Feynman rules can be formulated in coordinate space as follows:

1. Each integration coordinate x_j is represented by a point (sometimes called a vertex);

2. A bosonic propagator is represented by a wiggly line connecting two points;

3. A fermionic propagator is represented by a solid line connecting two points;

4. A bosonic field $A_{\mu(xi)}$ is represented by a wiggly line attached to the point x_i;

5. A fermionic field $\psi(x_i)$ is represented by a solid line attached to the point x_i with an arrow toward the point;

6. An anti-fermionic field $\psi(x_i)$ is represented by a solid line attached to the point x_i with an arrow away from the point;

Example: Second Order Processes in QED

The second order perturbation term in the S-matrix is

$$S^{(2)} = \frac{(ie)^2}{2!} \int d^4 x d^4 x' T \bar{\psi}(x) \gamma^\mu \psi(x) A_\mu(x) \bar{\psi}(x') \gamma^\nu \psi(x') A_\nu(x').$$

Scattering of Fermions

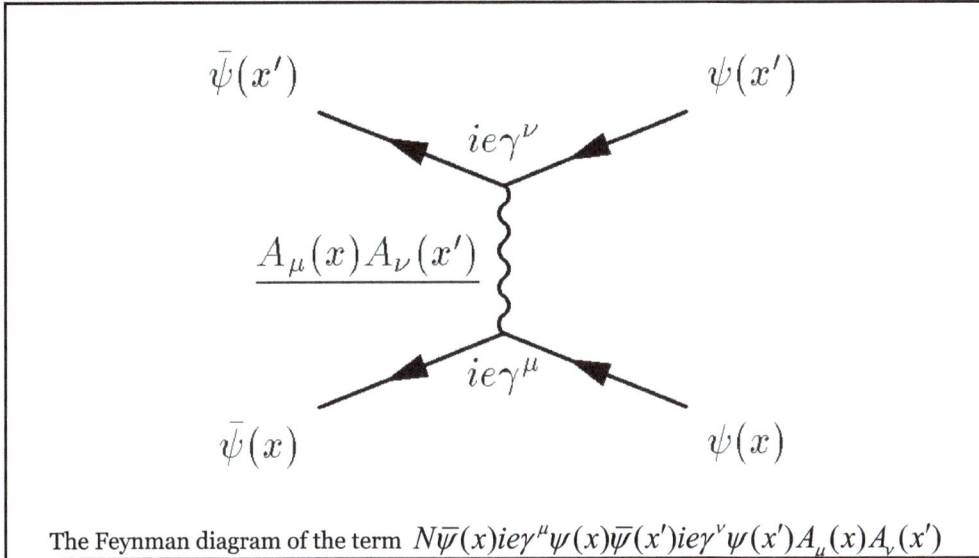

The Feynman diagram of the term $N\bar{\psi}(x)ie\gamma^\mu\psi(x)\bar{\psi}(x')ie\gamma^\nu\psi(x')A_\mu(x)A_\nu(x')$

The Wick's expansion of the integrand gives (among others) the following term

$$N\bar{\psi}(x)\gamma^\mu\psi(x)\bar{\psi}(x')\gamma^\nu\psi(x')\underline{A_\mu(x)A_\nu(x')}\,,$$

where

$$\underline{A_\mu(x)A_\nu(x')} = \int \frac{d^4k}{(2\pi)^4}\frac{-ig_{\mu\nu}}{k^2+i0}e^{-ik(x-x')}$$

is the electromagnetic contraction (propagator) in the Feynman gauge. This term is represented by the Feynman diagram at the right. This diagram gives contributions to the following processes:

1. $e^-\,e^-$ scattering (initial state at the right, final state at the left of the diagram);

2. $e^+\,e^+$ scattering (initial state at the left, final state at the right of the diagram);

3. $e^-\,e^+$ scattering (initial state at the bottom/top, final state at the top/bottom of the diagram).

Compton Scattering and Annihilation/Generation of e⁻ e⁺ pairs

Another interesting term in the expansion is

$$N\bar{\psi}(x)\gamma^\mu\underline{\psi(x)\bar{\psi}(x')}\gamma^\nu\psi(x')A_\mu(x)A_\nu(x')\,,$$

where

$$\underline{\psi(x)\bar{\psi}(x')} = \int \frac{d^4p}{(2\pi)^4}\frac{i}{\gamma p-m+i0}e^{-ip(x-x')}$$

is the fermionic contraction (propagator).

Path Integral Formulation

In a path integral, the field Lagrangian, integrated over all possible field histories, defines the probability amplitude to go from one field configuration to another. In order to make sense, the field theory should have a well-defined ground state, and the integral should be performed a little bit rotated into imaginary time, i.e. a Wick rotation.

Scalar Field Lagrangian

A simple example is the free relativistic scalar field in d dimensions, whose action integral is:

$$S = \int \tfrac{1}{2} \partial_\mu \phi \partial^\mu \phi d^d x.$$

The probability amplitude for a process is:

$$\int_A^B e^{iS} D\phi,$$

where A and B are space-like hypersurfaces that define the boundary conditions. The collection of all the $\varphi(A)$ on the starting hypersurface give the initial value of the field, analogous to the starting position for a point particle, and the field values $\varphi(B)$ at each point of the final hypersurface defines the final field value, which is allowed to vary, giving a different amplitude to end up at different values. This is the field-to-field transition amplitude.

The path integral gives the expectation value of operators between the initial and final state:

$$\int_A^B e^{iS} \phi(x_1) \cdots \phi(x_n) D\phi = \langle A | \phi(x_1) \cdots \phi(x_n) | B \rangle,$$

and in the limit that A and B recede to the infinite past and the infinite future, the only contribution that matters is from the ground state (this is only rigorously true if the path-integral is defined slightly rotated into imaginary time). The path integral can be thought of as analogous to a probability distribution, and it is convenient to define it so that multiplying by a constant doesn't change anything:

$$\frac{\int e^{iS} \phi(x_1) \cdots \phi(x_n) D\phi}{\int e^{iS} D\phi} = \langle 0 | \phi(x_1) \cdots \phi(x_n) | 0 \rangle.$$

The normalization factor on the bottom is called the *partition function* for the field, and it coincides with the statistical mechanical partition function at zero temperature when rotated into imaginary time.

The initial-to-final amplitudes are ill-defined if one thinks of the continuum limit right from the beginning, because the fluctuations in the field can become unbounded. So the path-integral can be thought of as on a discrete square lattice, with lattice spacing a and the limit $a \to 0$ should be taken carefully. If the final results do not depend on the shape of the lattice or the value of a, then the continuum limit exists.

On a Lattice

On a lattice, (i), the field can be expanded in Fourier modes:

$$\phi(x) = \int \frac{dk}{(2\pi)^d} \phi(k) e^{ik \cdot x} = \int_k \phi(k) e^{ikx}.$$

Here the integration domain is over k restricted to a cube of side length $2\pi/a$, so that large values of k are not allowed. It is important to note that the k-measure contains the factors of 2π from Fourier transforms, this is the best standard convention for k-integrals in QFT. The lattice means that fluctuations at large k are not allowed to contribute right away, they only start to contribute in the limit $a \to 0$. Sometimes, instead of a lattice, the field modes are just cut off at high values of k instead.

It is also convenient from time to time to consider the space-time volume to be finite, so that the k modes are also a lattice. This is not strictly as necessary as the space-lattice limit, because interactions in k are not localized, but it is convenient for keeping track of the factors in front of the k-integrals and the momentum-conserving delta functions that will arise.

On a lattice, (ii), the action needs to be discretized:

$$S = \sum_{\langle x,y \rangle} \tfrac{1}{2} \big(\phi(x) - \phi(y)\big)^2,$$

where $\langle x,y \rangle$ is a pair of nearest lattice neighbors x and y. The discretization should be thought of as defining what the derivative $\partial_\mu \varphi$ means.

In terms of the lattice Fourier modes, the action can be written:

$$S = \int_k \Big(\big(1 - \cos(k_1)\big) + \big(1 - \cos(k_2)\big) + \cdots + \big(1 - \cos(k_d)\big)\Big) \phi_k^* \phi^k.$$

For k near zero this is:

$$S = \int_k \tfrac{1}{2} k^2 |\phi(k)|^2.$$

Now we have the continuum Fourier transform of the original action. In finite volume, the quantity d^dk is not infinitesimal, but becomes the volume of a box made by neighboring Fourier modes, or $(2\pi/V)d$.

The field φ is real-valued, so the Fourier transform obeys:

$$\phi(k)^* = \phi(-k).$$

In terms of real and imaginary parts, the real part of $\varphi(k)$ is an even function of k, while the imaginary part is odd. The Fourier transform avoids double-counting, so that it can be written:

$$S = \int_k \tfrac{1}{2}k^2 \phi(k)\phi(-k)$$

over an integration domain that integrates over each pair $(k,-k)$ exactly once.

For a complex scalar field with action

$$S = \int \tfrac{1}{2}\partial_\mu \phi^* \partial^\mu \phi\, d^d x$$

the Fourier transform is unconstrained:

$$S = \int_k \tfrac{1}{2}k^2 \left|\phi(k)\right|^2$$

and the integral is over all k.

Integrating over all different values of $\varphi(x)$ is equivalent to integrating over all Fourier modes, because taking a Fourier transform is a unitary linear transformation of field coordinates. When you change coordinates in a multidimensional integral by a linear transformation, the value of the new integral is given by the determinant of the transformation matrix. If

$$y_i = A_{ij}x_j,$$

then

$$\det(A)\int dx_1\, dx_2 \cdots dx_n = \int dy_1\, dy_2 \cdots dy_n.$$

If A is a rotation, then

$$A^{\mathrm{T}}A = I$$

so that $\det A = \pm 1$, and the sign depends on whether the rotation includes a reflection or not.

The matrix that changes coordinates from $\varphi(x)$ to $\varphi(k)$ can be read off from the definition of a Fourier transform.

$$A_{kx} = e^{ikx}$$

and the Fourier inversion theorem tells you the inverse:

$$A_{kx}^{-1} = e^{-ikx}$$

which is the complex conjugate-transpose, up to factors of 2π. On a finite volume lattice, the determinant is nonzero and independent of the field values.

$$\det A = 1$$

and the path integral is a separate factor at each value of k.

$$\int \exp\left(\frac{i}{2}\sum_k k^2 \phi^*(k)\phi(k)\right)D\phi = \prod_k \int_{\phi_k} e^{\frac{i}{2}k^2|\phi_k|^2 d^d k}$$

The factor $d^d k$ is the infinitesimal volume of a discrete cell in k-space, in a square lattice box

$$d^d k = \left(\frac{1}{L}\right)^d,$$

where L is the side-length of the box. Each separate factor is an oscillatory Gaussian, and the width of the Gaussian diverges as the volume goes to infinity.

In imaginary time, the *Euclidean action* becomes positive definite, and can be interpreted as a probability distribution. The probability of a field having values ϕ_k is

$$e^{\int -\frac{1}{2}k^2 \phi_k^* \phi_k} = \prod_k e^{-k^2|\phi_k|^2 d^d k}.$$

The expectation value of the field is the statistical expectation value of the field when chosen according to the probability distribution:

$$\langle \phi(x_1)\cdots\phi(x_n)\rangle = \frac{\int e^{-S}\phi(x_1)\cdots\phi(x_n)D\phi}{\int e^{-S}D\phi}$$

Since the probability of ϕ_k is a product, the value of ϕ_k at each separate value of k is independently Gaussian distributed. The variance of the Gaussian is $1/k^2 d^d k$, which is formally infinite, but that just means that the fluctuations are unbounded in infinite volume. In any finite volume, the integral is replaced by a discrete sum, and the variance of the integral is V/k^2.

Monte Carlo

The path integral defines a probabilistic algorithm to generate a Euclidean scalar field configuration. Randomly pick the real and imaginary parts of each Fourier mode at wavenumber k to be a Gaussian random variable with variance $1/k^2$. This generates a configuration $\phi_C(k)$ at random, and the Fourier transform gives $\phi_C(x)$. For real scalar fields, the algorithm must generate only one of each pair $\phi(k)$, $\phi(-k)$, and make the second the complex conjugate of the first.

To find any correlation function, generate a field again and again by this procedure, and find the statistical average:

$$\langle \phi(x_1)\cdots\phi(x_n)\rangle = \lim_{|C|\to\infty} \frac{\sum_C \phi_C(x_1)\cdots\phi_C(x_n)}{|C|}$$

where $|C|$ is the number of configurations, and the sum is of the product of the field values on each configuration. The Euclidean correlation function is just the same as the correlation function in statistics or statistical mechanics. The quantum mechanical correlation functions are an analytic continuation of the Euclidean correlation functions.

For free fields with a quadratic action, the probability distribution is a high-dimensional Gaussian, and the statistical average is given by an explicit formula. But the Monte Carlo method also works well for bosonic interacting field theories where there is no closed form for the correlation functions.

Scalar Propagator

Each mode is independently Gaussian distributed. The expectation of field modes is easy to calculate:

$$\langle \phi_k \phi_{k'} \rangle = 0$$

for $k \neq k'$, since then the two Gaussian random variables are independent and both have zero mean.

$$\langle \phi_k \phi_k \rangle = \frac{V}{k^2}$$

in finite volume V, when the two k-values coincide, since this is the variance of the Gaussian. In the infinite volume limit,

$$\langle \phi(\mathrm{k})\phi(\mathrm{k}') \rangle = \delta(\mathrm{k} - \mathrm{k}')\frac{1}{\mathrm{k}^2}$$

Strictly speaking, this is an approximation: the lattice propagator is:

$$\langle \phi(k)\phi(k') \rangle = \delta(k - k')\frac{1}{2\big(d - \cos(k_1) + \cos(k_2)\cdots + \cos(k_d)\big)}$$

But near $k = 0$, for field fluctuations long compared to the lattice spacing, the two forms coincide.

It is important to emphasize that the delta functions contain factors of 2π, so that they cancel out the 2π factors in the measure for k integrals.

$$\delta(k) = (2\pi)^d \delta_D(k_1)\delta_D(k_2)\cdots\delta_D(k_d)$$

where $\delta_D(k)$ is the ordinary one-dimensional Dirac delta function. This convention for delta-functions is not universal—some authors keep the factors of 2π in the delta functions (and in the k-integration) explicit.

Equation of Motion

The form of the propagator can be more easily found by using the equation of motion for the field. From the Lagrangian, the equation of motion is:

$$\partial_\mu \partial^\mu \phi = 0$$

and in an expectation value, this says:

$$\partial_\mu \partial^\mu \langle \phi(x)\phi(y) \rangle = 0$$

Where the derivatives act on x, and the identity is true everywhere except when x and y coincide, and the operator order matters. The form of the singularity can be understood from the canonical commutation relations to be a delta-function. Defining the (Euclidean) *Feynman propagator* Δ as the Fourier transform of the time-ordered two-point function (the one that comes from the path-integral):

$$\partial \; \Delta(x) = i \; (x)$$

So that:

$$\Delta(k) = \frac{i}{k^2}$$

If the equations of motion are linear, the propagator will always be the reciprocal of the quadratic-form matrix that defines the free Lagrangian, since this gives the equations of motion. This is also easy to see directly from the path integral. The factor of i disappears in the Euclidean theory.

Wick Theorem

Because each field mode is an independent Gaussian, the expectation values for the product of many field modes obeys *Wick's theorem*:

$$\langle \phi(k_1)\phi(k_2)\cdots\phi(k_n) \rangle$$

is zero unless the field modes coincide in pairs. This means that it is zero for an odd number of φ, and for an even number of φ, it is equal to a contribution from each pair separately, with a delta function.

$$\langle \phi(k_1)\cdots\phi(k_{2n}) \rangle = \sum \prod_{i,j} \frac{\delta\left(k_i - k_j\right)}{k_i^2}$$

where the sum is over each partition of the field modes into pairs, and the product is over the pairs. For example,

$$\langle \phi(k_1)\phi(k_2)\phi(k_3)\phi(k_4) \rangle = \frac{\delta(k_1 - k_2)}{k_1^2}\frac{\delta(k_3 - k_4)}{k_3^2} + \frac{\delta(k_1 - k_3)}{k_3^2}\frac{\delta(k_2 - k_4)}{k_2^2} + \frac{\delta(k_1 - k_4)}{k_1^2}\frac{\delta(k_2 - k_3)}{k_2^2}$$

An interpretation of Wick's theorem is that each field insertion can be thought of as a dangling line, and the expectation value is calculated by linking up the lines in pairs, putting a delta function factor that ensures that the momentum of each partner in the pair is equal, and dividing by the propagator.

Higher Gaussian Moments — completing Wick's Theorem

There is a subtle point left before Wick's theorem is proved—what if more than two of the phis have the same momentum? If it's an odd number, the integral is zero; negative values cancel with the positive values. But if the number is even, the integral is positive. The previous demonstration assumed that the phis would only match up in pairs.

But the theorem is correct even when arbitrarily many of the φ are equal, and this is a notable property of Gaussian integration:

$$I = \int e^{-ax^2/2} dx = \sqrt{\frac{2\pi}{a}}$$

$$\frac{\partial^n}{\partial a^n} I = \int \frac{x^{2n}}{2^n} e^{-ax^2/2} dx = \frac{1 \cdot 3 \cdot 5 \ldots (2n-1)}{2 \cdot 2 \cdot 2 \ldots \quad \cdot 2} \sqrt{2\pi} \, a^{-\frac{2n+1}{2}}$$

Dividing by I,

$$\left\langle x^{2n} \right\rangle = \frac{\int x^{2n} e^{-ax^2/2}}{\int e^{-ax^2/2}} = 1 \cdot 3 \cdot 5 \ldots (2n-1) \frac{1}{a^n}$$

$$\left\langle x^2 \right\rangle = \frac{1}{a}$$

If Wick's theorem were correct, the higher moments would be given by all possible pairings of a list of $2n$ different x:

$$\left\langle x_1 x_2 x_3 \cdots x_{2n} \right\rangle$$

where the x are all the same variable, the index is just to keep track of the number of ways to pair them. The first x can be paired with $2n - 1$ others, leaving $2n - 2$. The next unpaired x can be paired with $2n - 3$ different x leaving $2n - 4$, and so on. This means that Wick's theorem, uncorrected, says that the expectation value of x^{2n} should be:

$$\left\langle x^{2n} \right\rangle = (2n-1) \cdot (2n-3) \ldots \cdot 5 \cdot 3 \cdot 1 \left\langle x^2 \right\rangle^n$$

and this is in fact the correct answer. So Wick's theorem holds no matter how many of the momenta of the internal variables coincide.

Interaction

Interactions are represented by higher order contributions, since quadratic contributions are always Gaussian. The simplest interaction is the quartic self-interaction, with an action:

$$S = \int \partial^\mu \phi \partial_\mu \phi + \frac{\lambda}{4!} \phi^4.$$

The reason for the combinatorial factor 4! will be clear soon. Writing the action in terms of the lattice (or continuum) Fourier modes:

$$S = \int_k k^2 |\phi(k)|^2 + \int_{k_1 k_2 k_3 k_4} \phi(k_1)\phi(k_2)\phi(k_3)\phi(k_4)\delta(k_1 + k_2 + k_3 + k_4) = S_F + X.$$

Where S_F is the free action, whose correlation functions are given by Wick's theorem. The exponential of S in the path integral can be expanded in powers of λ, giving a series of corrections to the free action.

$$e^{-S} = e^{-S_F}\left(1 + X + \frac{1}{2!}XX + \frac{1}{3!}XXX + \cdots\right)$$

The path integral for the interacting action is then a power series of corrections to the free action. The term represented by X should be thought of as four half-lines, one for each factor of $\phi(k)$. The half-lines meet at a vertex, which contributes a delta-function that ensures that the sum of the momenta are all equal.

To compute a correlation function in the interacting theory, there is a contribution from the X terms now. For example, the path-integral for the four-field correlator:

$$\langle \phi(k_1)\phi(k_2)\phi(k_3)\phi(k_4)\rangle = \frac{\int e^{-S}\phi(k_1)\phi(k_2)\phi(k_3)\phi(k_4)D\phi}{Z}$$

which in the free field was only nonzero when the momenta k were equal in pairs, is now nonzero for all values of k. The momenta of the insertions $\phi(k_i)$ can now match up with the momenta of the Xs in the expansion. The insertions should also be thought of as half-lines, four in this case, which carry a momentum k, but one that is not integrated.

The lowest-order contribution comes from the first nontrivial term $e^{-S_F}X$ in the Taylor expansion of the action. Wick's theorem requires that the momenta in the X half-lines, the $\phi(k)$ factors in X, should match up with the momenta of the external half-lines in pairs. The new contribution is equal to:

$$\lambda \frac{1}{k_1^2}\frac{1}{k_2^2}\frac{1}{k_3^2}\frac{1}{k_4^2}.$$

The 4! inside X is canceled because there are exactly 4! ways to match the half-lines in X to the external half-lines. Each of these different ways of matching the half-lines together in pairs contributes exactly once, regardless of the values of $k_{1,2,3,4}$, by Wick's theorem.

Feynman Diagrams

The expansion of the action in powers of X gives a series of terms with progressively higher number of Xs. The contribution from the term with exactly n Xs is called nth order.

The nth order terms has:

1. $4n$ internal half-lines, which are the factors of $\varphi(k)$ from the Xs. These all end on a vertex, and are integrated over all possible k.

2. external half-lines, which are the come from the $\varphi(k)$ insertions in the integral.

By Wick's theorem, each pair of half-lines must be paired together to make a *line*, and this line gives a factor of

$$\frac{\delta(k_1 + k_2)}{k_1^2}$$

which multiplies the contribution. This means that the two half-lines that make a line are forced to have equal and opposite momentum. The line itself should be labelled by an arrow, drawn parallel to the line, and labeled by the momentum in the line k. The half-line at the tail end of the arrow carries momentum k, while the half-line at the head-end carries momentum $-k$. If one of the two half-lines is external, this kills the integral over the internal k, since it forces the internal k to be equal to the external k. If both are internal, the integral over k remains.

The diagrams that are formed by linking the half-lines in the Xs with the external half-lines, representing insertions, are the Feynman diagrams of this theory. Each line carries a factor of $1/k^2$, the propagator, and either goes from vertex to vertex, or ends at an insertion. If it is internal, it is integrated over. At each vertex, the total incoming k is equal to the total outgoing k.

The number of ways of making a diagram by joining half-lines into lines almost completely cancels the factorial factors coming from the Taylor series of the exponential and the 4! at each vertex.

Loop Order

A forest diagram is one where all the internal lines have momentum that is completely determined by the external lines and the condition that the incoming and outgoing momentum are equal at each vertex. The contribution of these diagrams is a product of propagators, without any integration. A tree diagram is a connected forest diagram.

An example of a tree diagram is the one where each of four external lines end on an X. Another is when three external lines end on an X, and the remaining half-line joins up with another X, and the remaining half-lines of this X run off to external lines. These are all also forest diagrams (as every tree is a forest); an example of a forest that is not a tree is when eight external lines end on two Xs.

It is easy to verify that in all these cases, the momenta on all the internal lines is determined by the external momenta and the condition of momentum conservation in each vertex.

A diagram that is not a forest diagram is called a *loop* diagram, and an example is one where two lines of an X are joined to external lines, while the remaining two lines are joined to each other. The two lines joined to each other can have any momentum at all, since they both enter and leave the same vertex. A more complicated example is one where two Xs are joined to each other by matching the legs one to the other. This diagram has no external lines at all.

The reason loop diagrams are called loop diagrams is because the number of k-integrals that are left undetermined by momentum conservation is equal to the number of independent closed loops in the diagram, where independent loops are counted as in homology theory. The homology is re-al-valued (actually R^d valued), the value associated with each line is the momentum. The boundary operator takes each line to the sum of the end-vertices with a positive sign at the head and a negative sign at the tail. The condition that the momentum is conserved is exactly the condition that the boundary of the k-valued weighted graph is zero.

A set of k-values can be relabeled whenever there is a closed loop going from vertex to vertex, never revisiting the same vertex. Such a cycle can be thought of as the boundary of a 2-cell. The k-labelings of a graph that conserves momentum (which has zero boundary) up to redefinitions of k (up to boundaries of 2-cells) define the first homology of a graph. The number of independent momenta that are not determined is then equal to the number of independent homology loops. For many graphs, this is equal to the number of loops as counted in the most intuitive way.

Symmetry Factors

The number of ways to form a given Feynman diagram by joining together half-lines is large, and by Wick's theorem, each way of pairing up the half-lines contributes equally. Often, this completely cancels the factorials in the denominator of each term, but the cancellation is sometimes incomplete.

The uncancelled denominator is called the *symmetry factor* of the diagram. The contribution of each diagram to the correlation function must be divided by its symmetry factor.

For example, consider the Feynman diagram formed from two external lines joined to one X, and the remaining two half-lines in the X joined to each other. There are 4×3 ways to join the external half-lines to the X, and then there is only one way to join the two remaining lines to each other. The X comes divided by $4! = 4 \times 3 \times 2$, but the number of ways to link up the X half lines to make the diagram is only 4×3, so the contribution of this diagram is divided by two.

For another example, consider the diagram formed by joining all the half-lines of one X to all the half-lines of another X. This diagram is called a *vacuum bubble*, because it does not link up to any external lines. There are $4!$ ways to form this diagram, but the denominator includes a $2!$ (from the expansion of the exponential, there are two Xs) and two factors of $4!$. The contribution is multiplied by $4!/2 \times 4! \times 4! = 1/48$.

Another example is the Feynman diagram formed from two Xs where each X links up to two external lines, and the remaining two half-lines of each X are joined to each other. The number of ways to link an X to two external lines is 4×3, and either X could link up to either pair, giving an additional factor of 2. The remaining two half-lines in the two Xs can be linked to each other in two ways, so that the total number of ways to form the diagram is $4 \times 3 \times 4 \times 3 \times 2 \times 2$, while the denominator is $4! \times 4! \times 2!$. The total symmetry factor is 2, and the contribution of this diagram is divided by 2.

The symmetry factor theorem gives the symmetry factor for a general diagram: the contribution of each Feynman diagram must be divided by the order of its group of automorphisms, the number of symmetries that it has.

An automorphism of a Feynman graph is a permutation M of the lines and a permutation N of the vertices with the following properties:

1.   If a line l goes from vertex v to vertex v', then $M(l)$ goes from $N(v)$ to $N(v')$. If the line is undirected, as it is for a real scalar field, then $M(l)$ can go from $N(v')$ to $N(v)$ too.

2.   If a line l ends on an external line, $M(l)$ ends on the same external line.

3.   If there are different types of lines, $M(l)$ should preserve the type.

This theorem has an interpretation in terms of particle-paths: when identical particles are present, the integral over all intermediate particles must not double-count states that differ only by interchanging identical particles.

Proof: To prove this theorem, label all the internal and external lines of a diagram with a unique name. Then form the diagram by linking the a half-line to a name and then to the other half line.

Now count the number of ways to form the named diagram. Each permutation of the Xs gives a different pattern of linking names to half-lines, and this is a factor of $n!$. Each permutation of the half-lines in a single X gives a factor of $4!$. So a named diagram can be formed in exactly as many ways as the denominator of the Feynman expansion.

But the number of unnamed diagrams is smaller than the number of named diagram by the order of the automorphism group of the graph.

Connected Diagrams: Linked-Cluster Theorem

Roughly speaking, a Feynman diagram is called *connected* if all vertices and propagator lines are linked by a sequence of vertices and propagators of the diagram itself. If one views it as an undirected graph it is connected. The remarkable relevance of such diagrams in QFTs is due to the fact that they are sufficient to determine the quantum partition function $Z[J]$. More precisely, connected Feynman diagrams determine

$$iW[J] \equiv \ln Z[J].$$

To see this, one should recall that

$$Z[J] \propto \sum_k D_k$$

with D_k constructed from some (arbitrary) Feynman diagram that can be thought to consist of several connected components C_i. If one encounters n_i (identical) copies of a component C_i within the Feynman diagram D_k one has to include a *symmetry factor* $n_i!$. However, in the end each contribution of a Feynman diagram D_k to the partition function has the generic form

$$\prod_i \frac{C_i^{n_i}}{n_i!}$$

where i labels the (infinitely) many connected Feynman diagrams possible.

A scheme to successively create such contributions from the D_k to $Z[J]$ is obtained by

$$\left(\frac{1}{0!}+\frac{C_1}{1!}+\frac{C_1^2}{2!}+\cdots\right)\left(1+C_2+\frac{1}{2}C_2^2+\cdots\right)\cdots$$

and therefore yields

$$Z[J]\propto\prod_i\sum_{n_i=0}^{\infty}\frac{C_i^{n_i}}{n_i!}=\exp\sum_i C_i\propto\exp W[J].$$

To establish the *normalization* Z_0 = exp W = 1 one simply calculates all connected *vacuum diagrams*, i.e., the diagrams without any *sources J* (sometimes referred to as *external legs* of a Feynman diagram).

Vacuum Bubbles

An immediate consequence of the linked-cluster theorem is that all vacuum bubbles, diagrams without external lines, cancel when calculating correlation functions. A correlation function is given by a ratio of path-integrals:

$$\langle\phi_1(x_1)\cdots\phi_n(x_n)\rangle=\frac{\int e^{-S}\phi_1(x_1)\cdots\phi_n(x_n)D\phi}{\int e^{-S}D\phi}.$$

The top is the sum over all Feynman diagrams, including disconnected diagrams that do not link up to external lines at all. In terms of the connected diagrams, the numerator includes the same contributions of vacuum bubbles as the denominator:

$$\int e^{-S}\phi_1(x_1)\cdots\phi_n(x_n)D\phi=\left(\sum E_i\right)\left(\exp\left(\sum_i C_i\right)\right).$$

Where the sum over E diagrams includes only those diagrams each of whose connected components end on at least one external line. The vacuum bubbles are the same whatever the external lines, and give an overall multiplicative factor. The denominator is the sum over all vacuum bubbles, and dividing gets rid of the second factor.

The vacuum bubbles then are only useful for determining Z itself, which from the definition of the path integral is equal to:

$$Z=\int e^{-S}D\phi=e^{-HT}=e^{-\rho V}$$

where ρ is the energy density in the vacuum. Each vacuum bubble contains a factor of $\delta(k)$ zeroing the total k at each vertex, and when there are no external lines, this contains a factor of $\delta(0)$, because the momentum conservation is over-enforced. In finite volume, this factor can be identified as the total volume of space time. Dividing by the volume, the remaining integral for the vacuum bubble has an interpretation: it is a contribution to the energy density of the vacuum.

Sources

Correlation functions are the sum of the connected Feynman diagrams, but the formalism treats the connected and disconnected diagrams differently. Internal lines end on vertices, while external lines go off to insertions. Introducing *sources* unifies the formalism, by making new vertices where one line can end.

Sources are external fields, fields that contribute to the action, but are not dynamical variables. A scalar field source is another scalar field h that contributes a term to the (Lorentz) Lagrangian:

$$\int h(x)\phi(x)d^d x = \int h(k)\phi(k)d^d k$$

In the Feynman expansion, this contributes H terms with one half-line ending on a vertex. Lines in a Feynman diagram can now end either on an X vertex, or on an H vertex, and only one line enters an H vertex. The Feynman rule for an H vertex is that a line from an H with momentum k gets a factor of $h(k)$.

The sum of the connected diagrams in the presence of sources includes a term for each connected diagram in the absence of sources, except now the diagrams can end on the source. Traditionally, a source is represented by a little "×" with one line extending out, exactly as an insertion.

$$\log\big(Z[h]\big) = \sum_{n,C} h(k_1)h(k_2)\cdots h(k_n)C(k_1,\cdots,k_n)$$

where $C(k_1,...,k_n)$ is the connected diagram with n external lines carrying momentum as indicated. The sum is over all connected diagrams, as before.

The field h is not dynamical, which means that there is no path integral over h: h is just a parameter in the Lagrangian, which varies from point to point. The path integral for the field is:

$$Z[h] = \int e^{iS+i\int h\phi}\, D\phi$$

and it is a function of the values of h at every point. One way to interpret this expression is that it is taking the Fourier transform in field space. If there is a probability density on $\mathbb{R}^n$, the Fourier transform of the probability density is:

$$\int \rho(y)e^{iky}\, d^n y = \big\langle e^{iky} \big\rangle = \Big\langle \prod_{i=1}^{n} e^{ih_i y_i} \Big\rangle$$

The Fourier transform is the expectation of an oscillatory exponential. The path integral in the presence of a source $h(x)$ is:

$$Z[h] = \int e^{iS} e^{i\int_x h(x)\phi(x)}\, D\phi = \big\langle e^{ih\phi} \big\rangle$$

which, on a lattice, is the product of an oscillatory exponential for each field value:

$$\left\langle \prod_x e^{ih_x\phi_x} \right\rangle$$

The fourier transform of a delta-function is a constant, which gives a formal expression for a delta function:

$$\delta(x-y) = \int e^{ik(x-y)}\,dk$$

This tells you what a field delta function looks like in a path-integral. For two scalar fields φ and η,

$$\delta(\phi-\eta) = \int e^{ih(x)\left(\phi(x)-\eta(x)\right)d^d x}\,Dh,$$

which integrates over the Fourier transform coordinate, over h. This expression is useful for formally changing field coordinates in the path integral, much as a delta function is used to change coordinates in an ordinary multi-dimensional integral.

The partition function is now a function of the field h, and the physical partition function is the value when h is the zero function:

The correlation functions are derivatives of the path integral with respect to the source:

$$\langle\phi(x)\rangle = \frac{1}{Z}\frac{\partial}{\partial h(x)}Z[h] = \frac{\partial}{\partial h(x)}\log\left(Z[h]\right).$$

In Euclidean space, source contributions to the action can still appear with a factor of i, so that they still do a Fourier transform.

Spin 1/2; "Photons" and "Ghosts"

Spin 1/2: Grassmann Integrals

The field path integral can be extended to the Fermi case, but only if the notion of integration is expanded. A Grassmann integral of a free Fermi field is a high-dimensional determinant or Pfaffian, which defines the new type of Gaussian integration appropriate for Fermi fields.

The two fundamental formulas of Grassmann integration are:

$$\int e^{M_{ij}\bar\psi^i\psi^j}\,D\bar\psi\,D\psi = \mathrm{Det}(M),$$

where M is an arbitrary matrix and ψ, ψ are independent Grassmann variables for each index i, and

$$\int e^{\frac{1}{2}A_{ij}\psi^i\psi^j}\,D\psi = \mathrm{Pfaff}(A),$$

where A is an antisymmetric matrix, ψ is a collection of Grassmann variables, and the 1/2 is to prevent double-counting (since $\psi^i\psi^j = -\psi^j\psi^i$).

In matrix notation, where ψ and η are Grassmann-valued row vectors, η and ψ are Grassmann-valued column vectors, and M is a real-valued matrix:

$$Z = \int e^{\bar{\psi}M\psi + \bar{\eta}\psi + \bar{\psi}\eta} \, D\bar{\psi}\,D\psi = \int e^{\left(\bar{\psi} + \bar{\eta}M^{-1}\right)M\left(\psi + M^{-1}\eta\right) - \bar{\eta}M^{-1}\eta} \, D\bar{\psi}\,D\psi = \mathrm{Det}(M)e^{-\bar{\eta}M^{-1}\eta},$$

where the last equality is a consequence of the translation invariance of the Grassmann integral. The Grassmann variables η are external sources for ψ, and differentiating with respect to η pulls down factors of ψ.

$$\langle \bar{\psi}\psi \rangle = \frac{1}{Z}\frac{\partial}{\partial\eta}\frac{\partial}{\partial\bar{\eta}}Z\big|_{\eta=\bar{\eta}=0} = M^{-1}$$

again, in a schematic matrix notation. The meaning of the formula above is that the derivative with respect to the appropriate component of η and η gives the matrix element of M^{-1}. This is exactly analogous to the bosonic path integration formula for a Gaussian integral of a complex bosonic field:

$$\int e^{\phi^* M\phi + h^*\phi + \phi^* h} \, D\phi^*\, D\phi = \frac{e^{h^* M^{-1} h}}{\mathrm{Det}(M)}$$

$$\langle \phi^*\phi \rangle = \frac{1}{Z}\frac{\partial}{\partial h}\frac{\partial}{\partial h^*}Z\big|_{h=h^*=0} = M^{-1}.$$

So that the propagator is the inverse of the matrix in the quadratic part of the action in both the Bose and Fermi case.

For real Grassmann fields, for Majorana fermions, the path integral a Pfaffian times a source quadratic form, and the formulas give the square root of the determinant, just as they do for real Bosonic fields. The propagator is still the inverse of the quadratic part.

The free Dirac Lagrangian:

$$\int \bar{\psi}\left(\gamma^\mu \partial_\mu - m\right)\psi$$

formally gives the equations of motion and the anticommutation relations of the Dirac field, just as the Klein Gordon Lagrangian in an ordinary path integral gives the equations of motion and commutation relations of the scalar field. By using the spatial Fourier transform of the Dirac field as a new basis for the Grassmann algebra, the quadratic part of the Dirac action becomes simple to invert:

$$S = \int_k \bar{\psi}\left(i\gamma^\mu k_\mu - m\right)\psi.$$

The propagator is the inverse of the matrix M linking $\psi(k)$ and $\psi(k)$, since different values of k do not mix together.

$$\left\langle \overline{\psi}(k')\psi(k) \right\rangle = \delta(k+k')\frac{1}{\gamma \cdot k - m} = \delta(k+k')\frac{\gamma \cdot k + m}{k^2 - m^2}$$

The analog of Wick's theorem matches ψ and ψ in pairs:

$$\left\langle \overline{\psi}(k_1)\overline{\psi}(k_2)\cdots\overline{\psi}(k_n)\psi(k_1')\cdots\psi(k_n') \right\rangle = \sum_{\text{pairings}} (-1)^S \prod_{\text{pairs } i,j} \delta\left(k_i - k_j\right)\frac{1}{\gamma \cdot k_i - m}$$

where S is the sign of the permutation that reorders the sequence of ψ and ψ to put the ones that are paired up to make the delta-functions next to each other, with the ψ coming right before the ψ. Since a ψ, ψ pair is a commuting element of the Grassmann algebra, it doesn't matter what order the pairs are in. If more than one ψ, ψ pair have the same k, the integral is zero, and it is easy to check that the sum over pairings gives zero in this case (there are always an even number of them). This is the Grassmann analog of the higher Gaussian moments that completed the Bosonic Wick's theorem earlier.

The rules for spin-1/2 Dirac particles are as follows: The propagator is the inverse of the Dirac operator, the lines have arrows just as for a complex scalar field, and the diagram acquires an overall factor of −1 for each closed Fermi loop. If there are an odd number of Fermi loops, the diagram changes sign. Historically, the −1 rule was very difficult for Feynman to discover. He discovered it after a long process of trial and error, since he lacked a proper theory of Grassmann integration.

The rule follows from the observation that the number of Fermi lines at a vertex is always even. Each term in the Lagrangian must always be Bosonic. A Fermi loop is counted by following Fermionic lines until one comes back to the starting point, then removing those lines from the diagram. Repeating this process eventually erases all the Fermionic lines: this is the Euler algorithm to 2-color a graph, which works whenever each vertex has even degree. Note that the number of steps in the Euler algorithm is only equal to the number of independent Fermionic homology cycles in the common special case that all terms in the Lagrangian are exactly quadratic in the Fermi fields, so that each vertex has exactly two Fermionic lines. When there are four-Fermi interactions (like in the Fermi effective theory of the weak nuclear interactions) there are more k-integrals than Fermi loops. In this case, the counting rule should apply the Euler algorithm by pairing up the Fermi lines at each vertex into pairs that together form a bosonic factor of the term in the Lagrangian, and when entering a vertex by one line, the algorithm should always leave with the partner line.

To clarify and prove the rule, consider a Feynman diagram formed from vertices, terms in the Lagrangian, with Fermion fields. The full term is Bosonic, it is a commuting element of the Grassmann algebra, so the order in which the vertices appear is not important. The Fermi lines are linked into loops, and when traversing the loop, one can reorder the vertex terms one after the other as one goes around without any sign cost. The exception is when you return to the starting point, and the final half-line must be joined with the unlinked first half-line. This requires one permutation to move the last ψ to go in front of the first ψ, and this gives the sign.

This rule is the only visible effect of the exclusion principle in internal lines. When there are external lines, the amplitudes are antisymmetric when two Fermi insertions for identical particles are interchanged. This is automatic in the source formalism, because the sources for Fermi fields are themselves Grassmann valued.

Spin 1: Photons

The naive propagator for photons is infinite, since the Lagrangian for the A-field is:

$$S = \int \tfrac{1}{4} F^{\mu\nu} F_{\mu\nu} = \int -\tfrac{1}{2} \left(\partial^\mu A_\nu \partial_\mu A^\nu - \partial^\mu A_\mu \partial_\nu A^\nu \right).$$

The quadratic form defining the propagator is non-invertible. The reason is the gauge invariance of the field; adding a gradient to A does not change the physics.

To fix this problem, one needs to fix a gauge. The most convenient way is to demand that the divergence of A is some function f, whose value is random from point to point. It does no harm to integrate over the values of f, since it only determines the choice of gauge. This procedure inserts the following factor into the path integral for A:

$$\int \delta \left(\partial_\mu A^\mu - f \right) e^{-\frac{f^2}{2}} \, Df.$$

The first factor, the delta function, fixes the gauge. The second factor sums over different values of f that are inequivalent gauge fixings. This is simply

$$e^{-\frac{\left(\partial_i A_i \right)^2}{2}}$$

The additional contribution from gauge-fixing cancels the second half of the free Lagrangian, giving the Feynman Lagrangian:

$$S = \int \partial^\mu A^\nu \partial_\mu A_\nu$$

which is just like four independent free scalar fields, one for each component of A. The Feynman propagator is:

$$\left\langle A_\mu(k) A_\nu(k') \right\rangle = \delta \left(k + k' \right) \frac{g_{\mu\nu}}{k^2}.$$

The one difference is that the sign of one propagator is wrong in the Lorentz case: the timelike component has an opposite sign propagator. This means that these particle states have negative norm—they are not physical states. In the case of photons, it is easy to show by diagram methods that these states are not physical—their contribution cancels with longitudinal photons to only leave two physical photon polarization contributions for any value of k.

If the averaging over f is done with a coefficient different from 1/2, the two terms don't cancel completely. This gives a covariant Lagrangian with a coefficient λ, which does not affect anything:

$$S = \int \tfrac{1}{2} \left(\partial^\mu A^\nu \partial_\mu A_\nu - \lambda \left(\partial_\mu A^\mu \right)^2 \right)$$

and the covariant propagator for QED is:

$$\langle A_\mu(k)A_\nu(k')\rangle = \delta(k+k')\frac{g_{\mu\nu} - \lambda\frac{k_\mu k_\nu}{k^2}}{k^2}.$$

Spin 1: Non-Abelian Ghosts

To find the Feynman rules for non-Abelian gauge fields, the procedure that performs the gauge fixing must be carefully corrected to account for a change of variables in the path-integral.

The gauge fixing factor has an extra determinant from popping the delta function:

$$\delta(\partial_\mu A_\mu - f)e^{-\frac{f^2}{2}}\det M$$

To find the form of the determinant, consider first a simple two-dimensional integral of a function f that depends only on r, not on the angle θ. Inserting an integral over θ:

$$\int f(r)dxdy = \int f(r)\int d\theta\delta(y)\left|\frac{dy}{d\theta}\right|dxdy$$

The derivative-factor ensures that popping the delta function in θ removes the integral. Exchanging the order of integration,

$$\int f(r)dxdy = \int d\theta\int f(r)\delta(y)\left|\frac{dy}{d\theta}\right|dxdy$$

but now the delta-function can be popped in y,

$$\int f(r)dxdy = \int d\theta_0\int f(x)\left|\frac{dy}{d\theta}\right|dx.$$

The integral over θ just gives an overall factor of 2π, while the rate of change of y with a change in θ is just x, so this exercise reproduces the standard formula for polar integration of a radial function:

$$\int f(r)dxdy = 2\pi\int f(x)xdx$$

In the path-integral for a nonabelian gauge field, the analogous manipulation is:

$$\int DA\int \delta(F(A))\det\left(\frac{\partial F}{\partial G}\right)DGe^{iS} = \int DG\int \delta(F(A))\det\left(\frac{\partial F}{\partial G}\right)e^{iS}$$

The factor in front is the volume of the gauge group, and it contributes a constant, which can be discarded. The remaining integral is over the gauge fixed action.

$$\int \det\left(\frac{\partial F}{\partial G}\right)e^{iS_{GF}}DA$$

To get a covariant gauge, the gauge fixing condition is the same as in the Abelian case:

$$\partial_\mu A^\mu = f,$$

Whose variation under an infinitesimal gauge transformation is given by:

$$\partial_\mu D_\mu \alpha,$$

where α is the adjoint valued element of the Lie algebra at every point that performs the infinitesimal gauge transformation. This adds the Faddeev Popov determinant to the action:

$$\det\left(\partial_\mu D_\mu\right)$$

which can be rewritten as a Grassman integral by introducing ghost fields:

$$\int e^{\overline{\eta}\partial_\mu D^\mu \eta}\, D\overline{\eta}\, D\eta$$

The determinant is independent of f, so the path-integral over f can give the Feynman propagator (or a covariant propagator) by choosing the measure for f as in the abelian case. The full gauge fixed action is then the Yang Mills action in Feynman gauge with an additional ghost action:

$$S = \int \mathrm{Tr}\,\partial_\mu A_\nu \partial^\mu A^\nu + f^i_{jk}\partial^\nu A^\mu_i A^j_\mu A^k_\nu + f^i_{jr}f^r_{kl}A_i A_j A^k A^l + \mathrm{Tr}\,\partial_\mu \overline{\eta}\partial^\mu \eta + \overline{\eta}\,A_j\eta$$

The diagrams are derived from this action. The propagator for the spin-1 fields has the usual Feynman form. There are vertices of degree 3 with momentum factors whose couplings are the structure constants, and vertices of degree 4 whose couplings are products of structure constants. There are additional ghost loops, which cancel out timelike and longitudinal states in A loops.

In the Abelian case, the determinant for covariant gauges does not depend on A, so the ghosts do not contribute to the connected diagrams.

Particle-Path Representation

Feynman diagrams were originally discovered by Feynman, by trial and error, as a way to represent the contribution to the S-matrix from different classes of particle trajectories.

Schwinger Representation

The Euclidean scalar propagator has a suggestive representation:

$$\frac{1}{p^2 + m^2} = \int_0^\infty e^{-\tau\left(p^2 + m^2\right)}\, d\tau$$

The meaning of this identity (which is an elementary integration) is made clearer by Fourier transforming to real space.

$$\Delta(x) = \int_0^\infty d\tau e^{-m^2\tau} \frac{1}{(4\pi\tau)^{d/2}} e^{\frac{-x^2}{4\tau}}$$

The contribution at any one value of τ to the propagator is a Gaussian of width $\sqrt{\tau}$. The total propagation function from 0 to x is a weighted sum over all proper times τ of a normalized Gaussian, the probability of ending up at x after a random walk of time τ.

The path-integral representation for the propagator is then:

$$\Delta(x) = \int_0^\infty d\tau \int DX e^{-\int_0^\tau \left(\frac{\dot{x}^2}{2} + m^2\right) d\tau'}$$

which is a path-integral rewrite of the Schwinger representation.

The Schwinger representation is both useful for making manifest the particle aspect of the propagator, and for symmetrizing denominators of loop diagrams.

Combining Denominators

The Schwinger representation has an immediate practical application to loop diagrams. For example, for the diagram in the φ^4 theory formed by joining two xs together in two half-lines, and making the remaining lines external, the integral over the internal propagators in the loop is:

$$\int_k \frac{1}{k^2 + m^2} \frac{1}{(k+p)^2 + m^2}.$$

Here one line carries momentum k and the other $k + p$. The asymmetry can be fixed by putting everything in the Schwinger representation.

$$\int_{t,t'} e^{-t(k^2+m^2)-t'\left((k+p)^2+m^2\right)} dt \, dt'.$$

Now the exponent mostly depends on $t + t'$,

$$\int_{t,t'} e^{-(t+t')(k^2+m^2)-t'2p\cdot k-t'p^2},$$

except for the asymmetrical little bit. Defining the variable $u = t + t'$ and $v = t'/u$, the variable u goes from 0 to ∞, while v goes from 0 to 1. The variable u is the total proper time for the loop, while v parametrizes the fraction of the proper time on the top of the loop versus the bottom.

The Jacobian for this transformation of variables is easy to work out from the identities:

$$d(uv) = dt' \quad du = dt + dt',$$

and "wedging" gives

$$u\,du \wedge dv = dt \wedge dt'.$$

This allows the u integral to be evaluated explicitly:

$$\int_{u,v} u e^{-u\left(k^2+m^2+v2\,p\cdot k+vp^2\right)} = \int \frac{1}{\left(k^2+m^2+v2\,p\cdot k-vp^2\right)^2}\,dv$$

leaving only the v-integral. This method, invented by Schwinger but usually attributed to Feynman, is called *combining denominator*. Abstractly, it is the elementary identity:

$$\frac{1}{AB} = \int_0^1 \frac{1}{\left(vA+(1-v)B\right)^2}\,dv$$

But this form does not provide the physical motivation for introducing v; v is the proportion of proper time on one of the legs of the loop.

Once the denominators are combined, a shift in k to $k' = k + vp$ symmetrizes everything:

$$\int_0^1\int \frac{1}{\left(k^2+m^2+2vp\cdot k+vp^2\right)^2}\,dk\,dv = \int_0^1\int \frac{1}{\left(k'^2+m^2+v(1-v)p^2\right)^2}\,dk'\,dv$$

This form shows that the moment that p^2 is more negative than four times the mass of the particle in the loop, which happens in a physical region of Lorentz space, the integral has a cut. This is exactly when the external momentum can create physical particles.

When the loop has more vertices, there are more denominators to combine:

$$\int dk \frac{1}{k^2+m^2} \frac{1}{(k+p_1)^2+m^2} \cdots \frac{1}{(k+p_n)^2+m^2}$$

The general rule follows from the Schwinger prescription for $n + 1$ denominators:

$$\frac{1}{D_0 D_1 \cdots D_n} = \int_0^\infty \cdots \int_0^\infty e^{-u_0 D_0 \cdots -u_n D_n}\,du_0 \cdots du_n.$$

The integral over the Schwinger parameters u_i can be split up as before into an integral over the total proper time $u = u_0 + u_1 \dots + u_n$ and an integral over the fraction of the proper time in all but the first segment of the loop $v_i = u_i/u$ for $i \in \{1,2,...,n\}$. The v_i are positive and add up to less than 1, so that the v integral is over an n-dimensional simplex.

The Jacobian for the coordinate transformation can be worked out as before:

$$du = du_0 + du_1 \cdots + du_n$$

$$d(uv_i) = du_i.$$

Wedging all these equations together, one obtains

$$u^n \, du \wedge dv_1 \wedge dv_2 \cdots \wedge dv_n = du_0 \wedge du_1 \cdots \wedge du_n.$$

This gives the integral:

$$\int_0^\infty \int_{\text{simplex}} u^n e^{-u\left(v_0 D_0 + v_1 D_1 + v_2 D_2 \cdots + v_n D_n\right)} \, dv_1 \cdots dv_n \, du,$$

where the simplex is the region defined by the conditions

$$v_i > 0 \quad \text{and} \quad \sum_{i=1}^n v_i < 1$$

as well as

$$v_0 = 1 - \sum_{i=1}^n v_i.$$

Performing the u integral gives the general prescription for combining denominators:

$$\frac{1}{D_0 \cdots D_n} = n! \int_{\text{simplex}} \frac{1}{\left(v_0 D_0 + v_1 D_1 \cdots + v_n D_n\right)^{n+1}} dv_1 \, dv_2 \cdots dv_n$$

Since the numerator of the integrand is not involved, the same prescription works for any loop, no matter what the spins are carried by the legs. The interpretation of the parameters v_i is that they are the fraction of the total proper time spent on each leg.

Scattering

The correlation functions of a quantum field theory describe the scattering of particles. The definition of "particle" in relativistic field theory is not self-evident, because if you try to determine the position so that the uncertainty is less than the compton wavelength, the uncertainty in energy is large enough to produce more particles and antiparticles of the same type from the vacuum. This means that the notion of a single-particle state is to some extent incompatible with the notion of an object localized in space.

In the 1930s, Wigner gave a mathematical definition for single-particle states: they are a collection of states that form an irreducible representation of the Poincaré group. Single particle states describe an object with a finite mass, a well defined momentum, and a spin. This definition is fine for protons and neutrons, electrons and photons, but it excludes quarks, which are permanently confined, so the modern point of view is more accommodating: a particle is anything whose interaction can be described in terms of Feynman diagrams, which have an interpretation as a sum over particle trajectories.

A field operator can act to produce a one-particle state from the vacuum, which means that the field operator $\varphi(x)$ produces a superposition of Wigner particle states. In the free field theory, the field produces one particle states only. But when there are interactions, the field operator can also produce 3-particle, 5-particle (if there is no +/− symmetry also 2, 4, 6 particle) states too. To compute the scattering amplitude for single particle states only requires a careful limit, sending the fields to infinity and integrating over space to get rid of the higher-order corrections.

The relation between scattering and correlation functions is the LSZ-theorem: The scattering amplitude for n particles to go to m particles in a scattering event is the given by the sum of the Feynman diagrams that go into the correlation function for $n + m$ field insertions, leaving out the propagators for the external legs.

For example, for the $\lambda\varphi^4$ interaction of the previous section, the order λ contribution to the (Lorentz) correlation function is:

$$\langle \phi(k_1)\phi(k_2)\phi(k_3)\phi(k_4) \rangle = \frac{i}{k_1^2} \frac{i}{k_2^2} \frac{i}{k_3^2} \frac{i}{k_4^2} i\lambda$$

Stripping off the external propagators, that is, removing the factors of i/k^2, gives the invariant scattering amplitude M:

$$M = i\lambda$$

which is a constant, independent of the incoming and outgoing momentum. The interpretation of the scattering amplitude is that the sum of $|M|^2$ over all possible final states is the probability for the scattering event. The normalization of the single-particle states must be chosen carefully, however, to ensure that M is a relativistic invariant.

Non-relativistic single particle states are labeled by the momentum k, and they are chosen to have the same norm at every value of k. This is because the nonrelativistic unit operator on single particle states is:

$$\int dk |k\rangle\langle k|.$$

In relativity, the integral over the k-states for a particle of mass m integrates over a hyperbola in E,k space defined by the energy–momentum relation:

$$E^2 - k^2 = m^2.$$

If the integral weighs each k point equally, the measure is not Lorentz-invariant. The invariant measure integrates over all values of k and E, restricting to the hyperbola with a Lorentz-invariant delta function:

$$\int \delta(E^2 - k^2 - m^2) |E,k\rangle\langle E,k| dE\, dk = \int \frac{dk}{2E} |k\rangle\langle k|.$$

So the normalized k-states are different from the relativistically normalized k-states by a factor of

$$\sqrt{E} = \left(k^2 - m^2\right)^{\frac{1}{4}}.$$

The invariant amplitude M is then the probability amplitude for relativistically normalized incoming states to become relativistically normalized outgoing states.

For nonrelativistic values of k, the relativistic normalization is the same as the nonrelativistic normalization (up to a constant factor $\sqrt{m}$). In this limit, the φ^4 invariant scattering amplitude is still constant. The particles created by the field φ scatter in all directions with equal amplitude.

The nonrelativistic potential, which scatters in all directions with an equal amplitude (in the Born approximation), is one whose Fourier transform is constant—a delta-function potential. The lowest order scattering of the theory reveals the non-relativistic interpretation of this theory—it describes a collection of particles with a delta-function repulsion. Two such particles have an aversion to occupying the same point at the same time.

Nonperturbative Effects

Thinking of Feynman diagrams as a perturbation series, nonperturbative effects like tunneling do not show up, because any effect that goes to zero faster than any polynomial does not affect the Taylor series. Even bound states are absent, since at any finite order particles are only exchanged a finite number of times, and to make a bound state, the binding force must last forever.

But this point of view is misleading, because the diagrams not only describe scattering, but they also are a representation of the short-distance field theory correlations. They encode not only asymptotic processes like particle scattering, they also describe the multiplication rules for fields, the operator product expansion. Nonperturbative tunneling processes involve field configurations that on average get big when the coupling constant gets small, but each configuration is a coherent superposition of particles whose local interactions are described by Feynman diagrams. When the coupling is small, these become collective processes that involve large numbers of particles, but where the interactions between each of the particles is simple.

This means that nonperturbative effects show up asymptotically in resummations of infinite classes of diagrams, and these diagrams can be locally simple. The graphs determine the local equations of motion, while the allowed large-scale configurations describe non-perturbative physics. But because Feynman propagators are nonlocal in time, translating a field process to a coherent particle language is not completely intuitive, and has only been explicitly worked out in certain special cases. In the case of nonrelativistic bound states, the Bethe–Salpeter equation describes the class of diagrams to include to describe a relativistic atom. For quantum chromodynamics, the Shifman Vainshtein Zakharov sum rules describe non-perturbatively excited long-wavelength field modes in particle language, but only in a phenomenological way.

The number of Feynman diagrams at high orders of perturbation theory is very large, because there are as many diagrams as there are graphs with a given number of nodes. Nonperturbative effects leave a signature on the way in which the number of diagrams and resummations diverge at high order. It is only because non-perturbative effects appear in hidden form in diagrams that it was possible to analyze nonperturbative effects in string theory, where in many cases a Feynman description is the only one available.

In Popular Culture

- The use of the above diagram of the virtual particle producing a quark–antiquark pair was featured in the television sit-com *The Big Bang Theory*, in the episode "The Bat Jar Conjecture".

- *PhD Comics* of January 11, 2012, shows Feynman diagrams that *visualize and describe quantum academic interactions*, i.e. the paths followed by Ph.D. students when interacting with their advisors.

References

- Dirac, Paul A. M. (1933). "The Lagrangian in Quantum Mechanics" (PDF). Physikalische Zeitschrift der Sowjetunion. 3: 64–72.

- For a simplified, step by step, derivation of the above relation see Path Integrals in Quantum Theories: A Pedagogic 1st Step

- Feynman, Richard P.; Hibbs, Albert R.; Styer, Daniel F. (2010). Quantum Mechanics and Path Integrals. Mineola, NY: Dover Publications. pp. 29–31. ISBN 0-486-47722-3.

- Gell-Mann, Murray. "Most of the Good Stuff". In Brown, Laurie M.; Rigden, John S. Memories Of Richard Feynman. American Institute of Physics.

- Dirac, P. A. M. (1958). The Principles of Quantum Mechanics, Fourth Edition. Oxford. ISBN 0-19-851208-2.

- Richard P. Feynman (1958). Feynman's Thesis: A New Approach to Quantum Theory. World Scientific. ISBN 981-256-366-0.

Symmetry in Quantum Mechanics

In physics, symmetry is the physical feature of the system that remains unchanged under some transformation. Spacetime symmetries, supersymmetry, Noether's theorem and parity are some of the aspects of symmetry. This section helps the reader in understanding the features of symmetry in quantum mechanics.

Symmetry (Physics)

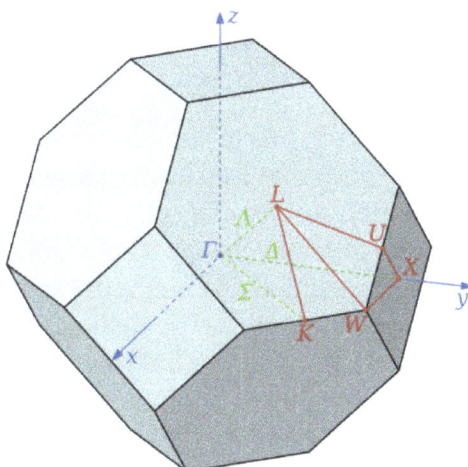

First Brillouin zone of FCC lattice showing symmetry labels

In physics, a symmetry of a physical system is a physical or mathematical feature of the system (observed or intrinsic) that is preserved or remains unchanged under some transformation.

A family of particular transformations may be *continuous* (such as rotation of a circle) or *discrete* (e.g., reflection of a bilaterally symmetric figure, or rotation of a regular polygon). Continuous and discrete transformations give rise to corresponding types of symmetries. Continuous symmetries can be described by Lie groups while discrete symmetries are described by finite groups.

These two concepts, Lie and finite groups, are the foundation for the fundamental theories of modern physics. Symmetries are frequently amenable to mathematical formulations such as group representations and can, in addition, be exploited to simplify many problems.

Arguably the most important example of a symmetry in physics is that the speed of light has the same value in all frames of reference, which is known in mathematical terms as Poincaré group, the symmetry group of special relativity. Another important example is the invariance of the form of physical laws under arbitrary differentiable coordinate transformations, which is an important idea in general relativity.

Symmetry as a Kind of Invariance

Invariance is specified mathematically by transformations that leave some property (e.g. quantity) unchanged. This idea can apply to basic real-world observations. For example, temperature may be homogeneous throughout a room. Since the temperature does not depend on the position of an observer within the room, we say that the temperature is *invariant* under a shift in an observer's position within the room.

Similarly, a uniform sphere rotated about its center will appear exactly as it did before the rotation. The sphere is said to exhibit spherical symmetry. A rotation about any axis of the sphere will preserve how the sphere "looks".

Invariance in Force

The above ideas lead to the useful idea of *invariance* when discussing observed physical symmetry; this can be applied to symmetries in forces as well.

For example, an electric field due to a wire is said to exhibit cylindrical symmetry, because the electric field strength at a given distance r from the electrically charged wire of infinite length will have the same magnitude at each point on the surface of a cylinder (whose axis is the wire) with radius r. Rotating the wire about its own axis does not change its position or charge density, hence it will preserve the field. The field strength at a rotated position is the same. Suppose some configuration of charges (may be non-stationary) produce an electric field in some direction, then rotating the configuration of the charges (without disturbing the internal dynamics that produces the particular field) will lead to a net rotation of the direction of the electric field. These two properties are interconnected through the more general property that rotating *any* system of charges causes a corresponding rotation of the electric field.

In Newton's theory of mechanics, given two bodies, each with mass m, starting from rest at the origin and moving along the x-axis in opposite directions, one with speed v_1 and the other with speed v_2 the total kinetic energy of the system (as calculated from an observer at the origin) is $\frac{1}{2}m(v_1^2 + v_2^2)$ and remains the same if the velocities are interchanged. The total kinetic energy is preserved under a reflection in the y-axis.

The last example above illustrates another way of expressing symmetries, namely through the equations that describe some aspect of the physical system. The above example shows that the total kinetic energy will be the same if v_1 and v_2 are interchanged.

Local and Global Symmetries

Symmetries may be broadly classified as *global* or *local*. A *global symmetry* is one that holds at all points of spacetime, whereas a *local symmetry* is one that has a different symmetry transformation at different points of spacetime; specifically a local symmetry transformation is parameterised by the spacetime co-ordinates. Local symmetries play an important role in physics as they form the basis for gauge theories.

Continuous Symmetries

The two examples of rotational symmetry described above - spherical and cylindrical - are each instances of continuous symmetry. These are characterised by invariance following a continuous change in the geometry of the system. For example, the wire may be rotated through any angle about its axis and the field strength will be the same on a given cylinder. Mathematically, continuous symmetries are described by continuous or smooth functions. An important subclass of continuous symmetries in physics are spacetime symmetries.

Spacetime Symmetries

Continuous *spacetime symmetries* are symmetries involving transformations of space and time. These may be further classified as *spatial symmetries*, involving only the spatial geometry associated with a physical system; *temporal symmetries*, involving only changes in time; or *spatio-temporal symmetries*, involving changes in both space and time.

- *Time translation*: A physical system may have the same features over a certain interval of time δt; this is expressed mathematically as invariance under the transformation for any real numbers t and a in the interval. For example, in classical mechanics, a particle solely acted upon by gravity will have gravitational potential energy $t \to t + a$ when suspended from a height mgh above the Earth's surface. Assuming no change in the height of the particle, this will be the total gravitational potential energy of the particle at all times. In other words, by considering the state of the particle at some time (in seconds) t_0 and also at $t_0 + 3,$, say, the particle's total gravitational potential energy will be preserved.

- *Spatial translation*: These spatial symmetries are represented by transformations of the form $\vec{r} \to \vec{r} + \vec{a}$ and describe those situations where a property of the system does not change with a continuous change in location. For example, the temperature in a room may be independent of where the thermometer is located in the room.

- *Spatial rotation*: These spatial symmetries are classified as proper rotations and improper rotations. The former are just the 'ordinary' rotations; mathematically, they are represented by square matrices with unit determinant. The latter are represented by square matrices with determinant −1 and consist of a proper rotation combined with a spatial reflection (inversion). For example, a sphere has proper rotational symmetry.

- *Poincaré transformations*: These are spatio-temporal symmetries which preserve distances in Minkowski spacetime, i.e. they are isometries of Minkowski space. They are studied primarily in special relativity. Those isometries that leave the origin fixed are called Lorentz transformations and give rise to the symmetry known as Lorentz covariance.

- *Projective symmetries*: These are spatio-temporal symmetries which preserve the geodesic structure of spacetime. They may be defined on any smooth manifold, but find many applications in the study of exact solutions in general relativity.

- *Inversion transformations*: These are spatio-temporal symmetries which generalise Poincaré transformations to include other conformal one-to-one transformations on the spacetime coordinates. Lengths are not invariant under inversion transformations but there is a cross-ratio on four points that is invariant.

Mathematically, spacetime symmetries are usually described by smooth vector fields on a smooth manifold. The underlying local diffeomorphisms associated with the vector fields correspond more directly to the physical symmetries, but the vector fields themselves are more often used when classifying the symmetries of the physical system.

Some of the most important vector fields are Killing vector fields which are those spacetime symmetries that preserve the underlying metric structure of a manifold. In rough terms, Killing vector fields preserve the distance between any two points of the manifold and often go by the name of isometries.

Discrete Symmetries

A discrete symmetry is a symmetry that describes non-continuous changes in a system. For example, a square possesses discrete rotational symmetry, as only rotations by multiples of right angles will preserve the square's original appearance. Discrete symmetries sometimes involve some type of 'swapping', these swaps usually being called *reflections* or *interchanges*.

- *Time reversal*: Many laws of physics describe real phenomena when the direction of time is reversed. Mathematically, this is represented by the transformation, $t \rightarrow -t.$. For example, Newton's second law of motion still holds if, in the equation $F = m\ddot{r}$, t is replaced by $-t$. This may be illustrated by recording the motion of an object thrown up vertically (neglecting air resistance) and then playing it back. The object will follow the same parabolic trajectory through the air, whether the recording is played normally or in reverse. Thus, position is symmetric with respect to the instant that the object is at its maximum height.

- *Spatial inversion*: These are represented by transformations of the form $\vec{r} \rightarrow -\vec{r}$ and indicate an invariance property of a system when the coordinates are 'inverted'. Said another way, these are symmetries between a certain object and its mirror image.

- *Glide reflection*: These are represented by a composition of a translation and a reflection. These symmetries occur in some crystals and in some planar symmetries, known as wallpaper symmetries.

C, P, and T Symmetries

The Standard model of particle physics has three related natural near-symmetries. These state that the actual universe about us is indistinguishable from one where:

- Every particle is replaced with its antiparticle. This is C-symmetry (charge symmetry);

- Everything appears as if reflected in a mirror. This is P-symmetry (parity symmetry);

- The direction of time is reversed. This is T-symmetry (time symmetry).

T-symmetry is counterintuitive (surely the future and the past are not symmetrical) but explained by the fact that the Standard model describes local properties, not global ones like entropy. To properly reverse the direction of time, one would have to put the big bang and the resulting low-entropy state in the "future." Since we perceive the "past" ("future") as having lower (higher) entropy than the present, the inhabitants of this hypothetical time-reversed universe would perceive the future in the same way as we perceive the past.

These symmetries are near-symmetries because each is broken in the present-day universe. However, the Standard Model predicts that the combination of the three (that is, the simultaneous application of all three transformations) must be a symmetry, called CPT symmetry. In the 4 dimensional matrix description of P,T is through a diagonal matrix, the negative identity, as well as C. Hence CPT is the identity operator. CP violation, the violation of the combination of C- and P-symmetry, is necessary for the presence of significant amounts of baryonic matter in the universe. CP violation is a fruitful area of current research in particle physics.

Supersymmetry

A type of symmetry known as supersymmetry has been used to try to make theoretical advances in the standard model. Supersymmetry is based on the idea that there is another physical symmetry beyond those already developed in the standard model, specifically a symmetry between bosons and fermions. Supersymmetry asserts that each type of boson has, as a supersymmetric partner, a fermion, called a superpartner, and vice versa. Supersymmetry has not yet been experimentally verified: no known particle has the correct properties to be a superpartner of any other known particle. If superpartners exist they must have masses greater than current particle accelerators can generate.

Mathematics of Physical Symmetry

The transformations describing physical symmetries typically form a mathematical group. Group theory is an important area of mathematics for physicists.

Continuous symmetries are specified mathematically by *continuous groups* (called Lie groups). Many physical symmetries are isometries and are specified by symmetry groups. Sometimes this term is used for more general types of symmetries. The set of all proper rotations (about any angle) through any axis of a sphere form a Lie group called the special orthogonal group $SO(3)$.. (The 3 refers to the three-dimensional space of an ordinary sphere.) Thus, the symmetry group of the sphere with proper rotations is $SO(3)$. Any rotation preserves distances on the surface of the ball. The set of all Lorentz transformations form a group called the Lorentz group (this may be generalised to the Poincaré group).

Discrete symmetries are described by discrete groups. For example, the symmetries of an equilateral triangle are described by the symmetric group S_3.

An important type of physical theory based on *local* symmetries is called a *gauge* theory and the symmetries natural to such a theory are called gauge symmetries. Gauge symmetries in the Standard model, used to describe three of the fundamental interactions, are based on the SU(3) × SU(2) × U(1) group. (Roughly speaking, the symmetries of the SU(3) group describe the strong force, the SU(2) group describes the weak interaction and the U(1) group describes the electromagnetic force.)

Also, the reduction by symmetry of the energy functional under the action by a group and spontaneous symmetry breaking of transformations of symmetric groups appear to elucidate topics in particle physics (for example, the unification of electromagnetism and the weak force in physical cosmology).

Conservation Laws and Symmetry

The symmetry properties of a physical system are intimately related to the conservation laws characterizing that system. Noether's theorem gives a precise description of this relation. The theorem states that each continuous symmetry of a physical system implies that some physical property of that system is conserved. Conversely, each conserved quantity has a corresponding symmetry. For example, the isometry of space gives rise to conservation of (linear) momentum, and isometry of time gives rise to conservation of energy.

The following table summarizes some fundamental symmetries and the associated conserved quantity.

Class	Invariance	Conserved quantity
Proper orthochronous Lorentz symmetry	translation in time (homogeneity)	energy
	translation in space (homogeneity)	linear momentum
	rotation in space (isotropy)	angular momentum
Discrete symmetry	P, coordinate inversion	spatial parity
	C, charge conjugation	charge parity
	T, time reversal	time parity
	CPT	product of parities
Internal symmetry (independent of spacetime coordinates)	U(1) gauge transformation	electric charge
	U(1) gauge transformation	lepton generation number
	U(1) gauge transformation	hypercharge
	U(1)$_Y$ gauge transformation	weak hypercharge
	U(2) [U(1) × SU(2)]	electroweak force
	SU(2) gauge transformation	isospin
	SU(2)$_L$ gauge transformation	weak isospin
	P × SU(2)	G-parity
	SU(3) "winding number"	baryon number
	SU(3) gauge transformation	quark color
	SU(3) (approximate)	quark flavor
	S(U(2) × U(3)) [U(1) × SU(2) × SU(3)]	Standard Model

Mathematics

Continuous symmetries in physics preserve transformations. One can specify a symmetry by showing how a very small transformation affects various particle fields. The commutator of two of these infinitesimal transformations are equivalent to a third infinitesimal transformation of the same kind hence they form a Lie algebra.

A general coordinate transformation (also known as a diffeomorphism) has the infinitesimal effect on a scalar, spinor and vector field for example:

$$\delta\phi(x) = h^\mu(x)\partial_\mu\phi(x)$$

$$\delta\psi^\alpha(x) = h^\mu(x)\partial_\mu\psi^\alpha(x) + \partial_\mu h_\nu(x)\sigma_{\mu\nu}^{\alpha\beta}\psi^\beta(x)$$

$$\delta A_\mu(x) = h^\nu(x)\partial_\nu A_\mu(x) + A_\nu(x)\partial_\mu h^\nu(x)$$

for a general field, $h(x)$. Without gravity only the Poincaré symmetries are preserved which restricts $h(x)$ to be of the form:

$$h^\mu(x) = M^{\mu\nu}x_\nu + P^\mu$$

where M is an antisymmetric matrix (giving the Lorentz and rotational symmetries) and P is a general vector (giving the translational symmetries). Other symmetries affect multiple fields simultaneously. For example, local gauge transformations apply to both a vector and spinor field:

$$\delta\psi^\alpha(x) = \lambda(x).\tau^{\alpha\beta}\psi^\beta(x)$$

$$\delta A_\mu(x) = \partial_\mu\lambda(x)$$

where τ are generators of a particular Lie group. So far the transformations on the right have only included fields of the same type. Supersymmetries are defined according to how the mix fields of *different* types.

Another symmetry which is part of some theories of physics and not in others is scale invariance which involve Weyl transformations of the following kind:

$$\delta\phi(x) = \Omega(x)\phi(x)$$

If the fields have this symmetry then it can be shown that the field theory is almost certainly conformally invariant also. This means that in the absence of gravity h(x) would restricted to the form:

$$h^\mu(x) = M^{\mu\nu}x_\nu + P^\mu + Dx_\mu + K^\mu|x|^2 - 2K^\nu x_\nu x_\mu$$

with D generating scale transformations and K generating special conformal transformations. For example, N=4 super-Yang-Mills theory has this symmetry while General Relativity doesn't although other theories of gravity such as conformal gravity do. The 'action' of a field theory is an invariant under all the symmetries of the theory. Much of modern theoretical physics is to do with speculating on the various symmetries the Universe may have and finding the invariants to construct field theories as models.

In string theories, since a string can be decomposed into an infinite number of particle fields, the symmetries on the string world sheet is equivalent to special transformations which mix an infinite number of fields.

Spacetime Symmetries

Spacetime symmetries are features of spacetime that can be described as exhibiting some form of symmetry. The role of symmetry in physics is important in simplifying solutions to many problems. Spacetime symmetries are used in the study of exact solutions of Einstein's field equations of general relativity.

Physical Motivation

Physical problems are often investigated and solved by noticing features which have some form of symmetry. For example, in the Schwarzschild solution, the role of spherical symmetry is important in deriving the Schwarzschild solution and deducing the physical consequences of this symmetry (such as the non-existence of gravitational radiation in a spherically pulsating star). In cosmological problems, symmetry finds a role to play in the cosmological principle which restricts the type of universes that are consistent with large-scale observations (e.g. the Friedmann-Lemaître-Robertson-Walker (FLRW) metric). Symmetries usually require some form of preserving property, the most important of which in general relativity include the following:

- preserving geodesics of the spacetime

- preserving the metric tensor

- preserving the curvature tensor

These and other symmetries will be discussed below in more detail. This preservation[which?] feature can be used to motivate a useful definition of symmetries.

Mathematical Definition

A rigorous definition of symmetries in general relativity has been given by Hall (2004). In this approach, the idea is to use (smooth) vector fields whose local flow diffeomorphisms preserve some property of the spacetime. (Note that one should emphasize in one's thinking this is a diffeomorphism--a transformation on a differential element. The implication is that the behavior of objects with extent may not be as manifestly symmetric.) This preserving property of the diffeomorphisms is made precise as follows. A smooth vector field X on a spacetime M is said to *preserve* a smooth tensor T on M (or T is invariant under X) if, for each smooth local flow diffeomorphism ϕ_t associated with X, the tensors T and $\phi_t{}^*(T)$ are equal on the domain of ϕ_t. This statement is equivalent to the more usable condition that the Lie derivative of the tensor under the vector field vanishes:

$$\mathcal{L}_X T = 0$$

on M. This has the consequence that, given any two points p and q on M, the coordinates of T in a coordinate system around p are equal to the coordinates of T in a coordinate system around q. A *symmetry on the spacetime* is a smooth vector field whose local flow diffeomorphisms preserve some (usually geometrical) feature of the spacetime. The (geometrical) feature may refer to specific tensors (such as the metric, or the energy-momentum tensor) or to other aspects of the spacetime such as its geodesic structure. The vector fields are sometimes referred to as *collineations*, *symmetry vector fields* or just *symmetries*. The set of all symmetry vector fields on M forms a Lie algebra under the Lie bracket operation as can be seen from the identity:

$$\mathcal{L}_{[X,Y]} T = \mathcal{L}_X(\mathcal{L}_Y T) - \mathcal{L}_Y(\mathcal{L}_X T)$$

the term on the right usually being written, with an abuse of notation, as $[\mathcal{L}_X, \mathcal{L}_Y] T$.

Killing Symmetry

A Killing vector field is one of the most important types of symmetries and is defined to be a smooth vector field that preserves the metric tensor:

$$\mathcal{L}_X g_{ab} = 0$$

This is usually written in the expanded form as:

$$X_{a;b} + X_{b;a} = 0$$

Killing vector fields find extensive applications (including in classical mechanics) and are related to conservation laws.

Homothetic Symmetry

A homothetic vector field is one which satisfies:

$$\mathcal{L}_X g_{ab} = 2c g_{ab}$$

where c is a real constant. Homothetic vector fields find application in the study of singularities in general relativity.

Affine Symmetry

An affine vector field is one that satisfies:

$$(\mathcal{L}_X g_{ab})_{;c} = 0$$

An affine vector field preserves geodesics and preserves the affine parameter.

The above three vector field types are special cases of projective vector fields which preserve geodesics without necessarily preserving the affine parameter.

Conformal Symmetry

A conformal vector field is one which satisfies:

$$\mathcal{L}_X g_{ab} = \phi g_{ab}$$

where ϕ is a smooth real-valued function on M.

Curvature Symmetry

A curvature collineation is a vector field which preserves the Riemann tensor:

$$\mathcal{L}_X R^a{}_{bcd} = 0$$

where $R^a{}_{bcd}$ are the components of the Riemann tensor. The set of all smooth curvature collineations forms a Lie algebra under the Lie bracket operation (if the smoothness condition is dropped, the set of all curvature collineations need not form a Lie algebra). The Lie algebra is denoted by $CC(M)$ and may be infinite-dimensional. Every affine vector field is a curvature collineation.

Matter Symmetry

A less well-known form of symmetry concerns vector fields that preserve the energy-momentum tensor. These are variously referred to as matter collineations or matter symmetries and are defined by:

$$\mathcal{L}_X T_{ab} = 0$$

where T_{ab} are the energy-momentum tensor components. The intimate relation between geometry and physics may be highlighted here, as the vector field X is regarded as preserving certain physical quantities along the flow lines of X, this being true for any two observers. In connection with this, it may be shown that *every Killing vector field is a matter collineation* (by the Einstein field equations, with or without cosmological constant). Thus, given a solution of the EFE, *a vector field that preserves the metric necessarily preserves the corresponding energy-momentum tensor.* When the energy-momentum tensor represents a perfect fluid, every Killing vector field preserves the energy density, pressure and the fluid flow vector field. When the energy-momentum tensor represents an electromagnetic field, a Killing vector field does *not necessarily* preserve the electric and magnetic fields.

Local and Global Symmetries

Applications

As mentioned at the start of this article, the main application of these symmetries occur in general relativity, where solutions of Einstein's equations may be classified by imposing some certain symmetries on the spacetime.

Spacetime Classifications

Classifying solutions of the EFE constitutes a large part of general relativity research. Various approaches to classifying spacetimes, including using the Segre classification of the energy-momentum tensor or the Petrov classification of the Weyl tensor have been studied extensively by many researchers, most notably Stephani et al. (2003). They also classify spacetimes using symmetry vector fields (especially Killing and homothetic symmetries). For example, Killing vector fields may be used to classify spacetimes, as there is a limit to the number of global, smooth Killing vector fields that a spacetime may possess (the maximum being 10 for 4-dimensional spacetimes). Generally speaking, the higher the dimension of the algebra of symmetry vector fields on a spacetime, the more symmetry the spacetime admits. For example, the Schwarzschild solution has a Killing algebra of dimension 4 (3 spatial rotational vector fields and a time translation), whereas the Friedmann-Lemaître-Robertson-Walker (FLRW) metric (excluding the Einstein static subcase) has a Killing algebra of dimension 6 (3 translations and 3 rotations). The Einstein static metric has a Killing algebra of dimension 7 (the previous 6 plus a time translation).

The assumption of a spacetime admitting a certain symmetry vector field can place restrictions on the spacetime.

Supersymmetry

In particle physics, supersymmetry (SUSY) is a proposed type of spacetime symmetry that relates two basic classes of elementary particles: bosons, which have an integer-valued spin, and fermions, which have a half-integer spin. Each particle from one group is associated with a particle from the other, known as its superpartner, the spin of which differs by a half-integer. In a theory with perfectly "unbroken" supersymmetry, each pair of superpartners would share the same mass and internal quantum numbers besides spin. For example, there would be a "selectron" (superpartner electron), a bosonic version of the electron with the same mass as the electron, that would be easy to find in a laboratory. Thus, since no superpartners have been observed, if supersymmetry exists it must be a spontaneously broken symmetry so that superpartners may differ in mass. Spontaneously-broken supersymmetry could solve many mysterious problems in particle physics including the hierarchy problem. The simplest realization of spontaneously-broken supersymmetry, the so-called Minimal Supersymmetric Standard Model, is one of the best studied candidates for physics beyond the Standard Model.

There is only indirect evidence and motivation for the existence of supersymmetry. Direct confirmation would entail production of superpartners in collider experiments, such as the Large Hadron Collider (LHC). The first run of the LHC found no evidence for supersymmetry (all results were consistent with the Standard Model), and thus set limits on superpartner masses in supersymmetric theories. While some remain enthusiastic about supersymmetry, this first run at the LHC led some physicists to explore other ideas. The LHC resumed its search for supersymmetry and other new physics in its second run.

Motivations

There are numerous phenomenological motivations for supersymmetry close to the electroweak scale, as well as technical motivations for supersymmetry at any scale.

The Hierarchy Problem

Supersymmetry close to the electroweak scale ameliorates the hierarchy problem that afflicts the Standard Model. In the Standard Model, the electroweak scale receives enormous Planck-scale quantum corrections. The observed hierarchy between the electroweak scale and the Planck scale must be achieved with extraordinary fine tuning. In a supersymmetric theory, on the other hand, Planck-scale quantum corrections cancel between partners and superpartners (owing to a minus sign associated with fermionic loops). The hierarchy between the electroweak scale and the Planck scale is achieved in a natural manner, without miraculous fine-tuning.

Gauge Coupling Unification

The idea that the gauge symmetry groups unify at high-energy is called Grand unification theory. In the Standard Model, however, the weak, strong and electromagnetic couplings fail to unify at high energy. In a supersymmetry theory, the running of the gauge couplings are modified, and precise high-energy unification of the gauge couplings is achieved. The modified running also provides a natural mechanism for radiative electroweak symmetry breaking.

Dark Matter

TeV-scale supersymmetry (augmented with a discrete symmetry) typically provides a candidate dark matter particle at a mass scale consistent with thermal relic abundance calculations.

Other Technical Motivations

Supersymmetry is also motivated by solutions to several theoretical problems, for generally providing many desirable mathematical properties, and for ensuring sensible behavior at high energies. Supersymmetric quantum field theory is often much easier to analyze, as many more problems become exactly solvable. When supersymmetry is imposed as a *local* symmetry, Einstein's theory of general relativity is included automatically, and the result is said to be a theory of supergravity. It is also a necessary feature of the most popular candidate for a theory of everything, superstring theory, and a SUSY theory could explain the issue of cosmological inflation.

Another theoretically appealing property of supersymmetry is that it offers the only "loophole" to the Coleman–Mandula theorem, which prohibits spacetime and internal symmetries from being combined in any nontrivial way, for quantum field theories like the Standard Model with very general assumptions. The Haag-Lopuszanski-Sohnius theorem demonstrates that supersymmetry is the only way spacetime and internal symmetries can be combined consistently.

History

A supersymmetry relating mesons and baryons was first proposed, in the context of hadronic physics, by Hironari Miyazawa during 1966. This supersymmetry did not involve spacetime, that is, it concerned internal symmetry, and was broken badly. Miyazawa's work was largely ignored at the time.

J. L. Gervais and B. Sakita (during 1971), Yu. A. Golfand and E. P. Likhtman (also during 1971), and D.V. Volkov and V.P. Akulov (1972), independently rediscovered supersymmetry in the context of quantum field theory, a radically new type of symmetry of spacetime and fundamental fields, which establishes a relationship between elementary particles of different quantum nature, bosons and fermions, and unifies spacetime and internal symmetries of microscopic phenomena. Supersymmetry with a consistent Lie-algebraic graded structure on which the Gervais–Sakita rediscovery was based directly first arose during 1971 in the context of an early version of string theory by Pierre Ramond, John H. Schwarz and André Neveu.

Finally, Julius Wess and Bruno Zumino (during 1974) identified the characteristic renormalization features of four-dimensional supersymmetric field theories, which identified them as remarkable QFTs, and they and Abdus Salam and their fellow researchers introduced early particle physics applications. The mathematical structure of supersymmetry (Graded Lie superalgebras) has subsequently been applied successfully to other topics of physics, ranging from nuclear physics, critical phenomena, quantum mechanics to statistical physics. It remains a vital part of many proposed theories of physics.

The first realistic supersymmetric version of the Standard Model was proposed during 1977 by Pierre Fayet and is known as the Minimal Supersymmetric Standard Model or MSSM for short. It was proposed to solve, amongst other things, the hierarchy problem.

Applications

Extension of Possible Symmetry Groups

One reason that physicists explored supersymmetry is because it offers an extension to the more familiar symmetries of quantum field theory. These symmetries are grouped into the Poincaré group and internal symmetries and the Coleman–Mandula theorem showed that under certain assumptions, the symmetries of the S-matrix must be a direct product of the Poincaré group with a compact internal symmetry group or if there is not any mass gap, the conformal group with a compact internal symmetry group. During 1971 Golfand and Likhtman were the first to show that the Poincaré algebra can be extended through introduction of four anticommuting spinor generators (in four dimensions), which later became known as supercharges. During 1975 the Haag-Lopuszanski-Sohnius theorem analyzed all possible superalgebras in the general form, including those with an extended number of the supergenerators and central charges. This extended super-Poincaré algebra paved the way for obtaining a very large and important class of supersymmetric field theories.

The Supersymmetry Algebra

Traditional symmetries of physics are generated by objects that transform by the tensor representations of the Poincaré group and internal symmetries. Supersymmetries, however, are generated by objects that transform by the spinor representations. According to the spin-statistics theorem, bosonic fields commute while fermionic fields anticommute. Combining the two kinds of fields into a single algebra requires the introduction of a Z_2-grading under which the bosons are the even elements and the fermions are the odd elements. Such an algebra is called a Lie superalgebra.

The simplest supersymmetric extension of the Poincaré algebra is the Super-Poincaré algebra. Expressed in terms of two Weyl spinors, has the following anti-commutation relation:

$$\{Q_\alpha, \overline{Q_{\dot\beta}}\} = 2(\sigma^\mu)_{\alpha\dot\beta} P_\mu$$

and all other anti-commutation relations between the Qs and commutation relations between the Qs and Ps vanish. In the above expression $P_\mu = -i\partial_\mu$ are the generators of translation and σ^μ are the Pauli matrices.

There are representations of a Lie superalgebra that are analogous to representations of a Lie algebra. Each Lie algebra has an associated Lie group and a Lie superalgebra can sometimes be extended into representations of a Lie supergroup.

The Supersymmetric Standard Model

Incorporating supersymmetry into the Standard Model requires doubling the number of particles since there is no way that any of the particles in the Standard Model can be superpartners of each other. With the addition of new particles, there are many possible new interactions. The simplest possible supersymmetric model consistent with the Standard Model is the Minimal Supersymmetric Standard Model (MSSM) which can include the necessary additional new particles that are able to be superpartners of those in the Standard Model.

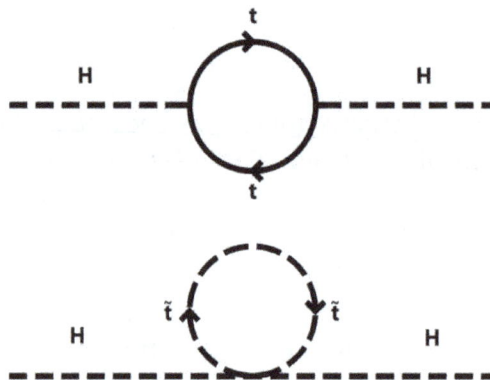

Cancellation of the Higgs boson quadratic mass renormalization between fermionic top quark loop and scalar stop squark tadpole Feynman diagrams in a supersymmetric extension of the Standard Model

One of the main motivations for SUSY comes from the quadratically divergent contributions to the Higgs mass squared. The quantum mechanical interactions of the Higgs boson causes a large renormalization of the Higgs mass and unless there is an accidental cancellation, the natural size of the Higgs mass is the greatest scale possible. This problem is known as the hierarchy problem. Supersymmetry reduces the size of the quantum corrections by having automatic cancellations between fermionic and bosonic Higgs interactions. If supersymmetry is restored at the weak scale, then the Higgs mass is related to supersymmetry breaking which can be induced from small non-perturbative effects explaining the vastly different scales in the weak interactions and gravitational interactions.

In many supersymmetric Standard Models there is a heavy stable particle (such as neutralino) which could serve as a weakly interacting massive particle (WIMP) dark matter candidate. The existence of a supersymmetric dark matter candidate is related closely to R-parity.

The standard paradigm for incorporating supersymmetry into a realistic theory is to have the underlying dynamics of the theory be supersymmetric, but the ground state of the theory does not respect the symmetry and supersymmetry is broken spontaneously. The supersymmetry break can not be done permanently by the particles of the MSSM as they currently appear. This means that there is a new sector of the theory that is responsible for the breaking. The only constraint on this new sector is that it must break supersymmetry permanently and must give superparticles TeV scale masses. There are many models that can do this and most of their details do not matter. In order to parameterize the relevant features of supersymmetry breaking, arbitrary soft SUSY breaking terms are added to the theory which temporarily break SUSY explicitly but could never arise from a complete theory of supersymmetry breaking.

Gauge-Coupling Unification

One piece of evidence for supersymmetry existing is gauge coupling unification. The renormalization group evolution of the three gauge coupling constants of the Standard Model is somewhat sensitive to the present particle content of the theory. These coupling constants do not quite meet together at a common energy scale if we run the renormalization group using the Standard Model. With the addition of minimal SUSY joint convergence of the coupling constants is projected at approximately 10^{16} GeV.

Supersymmetric Quantum Mechanics

Supersymmetric quantum mechanics adds the SUSY superalgebra to quantum mechanics as opposed to quantum field theory. Supersymmetric quantum mechanics often becomes relevant when studying the dynamics of supersymmetric solitons, and due to the simplified nature of having fields which are only functions of time (rather than space-time), a great deal of progress has been made in this subject and it is now studied in its own right.

SUSY quantum mechanics involves pairs of Hamiltonians which share a particular mathematical relationship, which are called *partner Hamiltonians*. (The potential energy terms which occur in the Hamiltonians are then known as *partner potentials*.) An introductory theorem shows that for every eigenstate of one Hamiltonian, its partner Hamiltonian has a corresponding eigenstate with the same energy. This fact can be exploited to deduce many properties of the eigenstate spectrum. It is analogous to the original description of SUSY, which referred to bosons and fermions. We can imagine a "bosonic Hamiltonian", whose eigenstates are the various bosons of our theory. The SUSY partner of this Hamiltonian would be "fermionic", and its eigenstates would be the theory's fermions. Each boson would have a fermionic partner of equal energy.

Supersymmetry: Applications to Condensed Matter Physics

SUSY concepts have provided useful extensions to the WKB approximation. Additionally, SUSY has been applied to disorder averaged systems both quantum and non-quantum (through statistical mechanics), the Fokker-Planck equation being an example of a non-quantum theory. The 'supersymmetry' in all these systems arises from the fact that one is modelling one particle and as such the 'statistics' don't matter. The use of the supersymmetry method provides a mathematical rigorous alternative to the replica trick, but only in non-interacting systems, which attempts to address the so-called 'problem of the denominator' under disorder averaging.

Supersymmetry in Optics

Integrated optics was recently found to provide a fertile ground on which certain ramifications of SUSY can be explored in readily-accessible laboratory settings. Making use of the analogous mathematical structure of the quantum-mechanical Schrödinger equation and the wave equation governing the evolution of light in one-dimensional settings, one may interpret the refractive index distribution of a structure as a potential landscape in which optical wave packets propagate. In this manner, a new class of functional optical structures with possible applications in phase matching, mode conversion and space-division multiplexing becomes possible. SUSY transformations have been also proposed as a way to address inverse scattering problems in optics and as a one-dimensional transformation optics

Mathematics

SUSY is also sometimes studied mathematically for its intrinsic properties. This is because it describes complex fields satisfying a property known as holomorphy, which allows holomorphic quantities to be exactly computed. This makes supersymmetric models useful "toy models" of more realistic theories. A prime example of this has been the demonstration of S-duality in four-dimensional gauge theories that interchanges particles and monopoles.

The proof of the Atiyah-Singer index theorem is much simplified by the use of supersymmetric quantum mechanics.

General Supersymmetry

Supersymmetry appears in many related contexts of theoretical physics. It is possible to have multiple supersymmetries and also have supersymmetric extra dimensions.

Extended Supersymmetry

It is possible to have more than one kind of supersymmetry transformation. Theories with more than one supersymmetry transformation are known as extended supersymmetric theories. The more supersymmetry a theory has, the more constrained are the field content and interactions. Typically the number of copies of a supersymmetry is a power of 2, i.e. 1, 2, 4, 8. In four dimensions, a spinor has four degrees of freedom and thus the minimal number of supersymmetry generators is four in four dimensions and having eight copies of supersymmetry means that there are 32 supersymmetry generators.

The maximal number of supersymmetry generators possible is 32. Theories with more than 32 supersymmetry generators automatically have massless fields with spin greater than 2. It is not known how to make massless fields with spin greater than two interact, so the maximal number of supersymmetry generators considered is 32. This is due to the Weinberg-Witten theorem. This corresponds to an $N = 8$ supersymmetry theory. Theories with 32 supersymmetries automatically have a graviton.

For four dimensions there are the following theories, with the corresponding multiplets(CPT adds a copy, whenever they are not invariant under such symmetry)

- $N = 1$

Chiral multiplet: $(0,\tfrac{1}{2})$ Vector multiplet: $(\tfrac{1}{2},1)$ Gravitino multiplet: $(1,\tfrac{3}{2})$ Graviton multiplet: $(\tfrac{3}{2},2)$

- $N = 2$

hypermultiplet: $(-\tfrac{1}{2},0^2,\tfrac{1}{2})$ vector multiplet: $(0,\tfrac{1}{2}^2,1)$ supergravity multiplet: $(1,\tfrac{3}{2}^2,2)$

- $N = 4$

Vector multiplet: $(-1,-\tfrac{1}{2}^4,0^6,\tfrac{1}{2}^4,1)$ Supergravity multiplet: $(0,\tfrac{1}{2}^4,1^6,\tfrac{3}{2}^4,2)$

- $N = 8$

Supergravity multiplet: $(-2,-\tfrac{3}{2}^8,-1^{28},-\tfrac{1}{2}^{56},0^{70},\tfrac{1}{2}^{56},1^{28},\tfrac{3}{2}^8,2)$

Supersymmetry in Alternate Numbers of Dimensions

It is possible to have supersymmetry in dimensions other than four. Because the properties of spinors change drastically between different dimensions, each dimension has its characteristic. In d dimensions, the size of spinors is approximately $2^{d/2}$ or $2^{(d-1)/2}$. Since the maximum number of supersymmetries is 32, the greatest number of dimensions in which a supersymmetric theory can exist is eleven.

Supersymmetry in Quantum Gravity

Supersymmetry is part of a larger enterprise of theoretical physics to unify everything we know about the universe into a single consistent set of physical principles, known as the quest for a Theory of Everything (TOE). A significant part of this larger enterprise is the quest for a theory of quantum gravity, which would unify the classical theory of general relativity and the Standard Model, which explains the other three basic forces in physics (electromagnetism, the strong interaction, and the weak interaction), and provides a palette of fundamental particles upon which all four forces act. Two of the most active methods of forming a theory of quantum gravity are string theory and loop quantum gravity (LQG), although in theory, supersymmetry could be a component of other theories as well.

For string theory to be consistent, supersymmetry seems to be required at some level (although it may be a strongly broken symmetry). In particle theory, supersymmetry is recognized as a way to stabilize the hierarchy between the unification scale and the electroweak scale (or the Higgs boson mass), and can also provide a natural dark matter candidate. String theory also requires extra spatial dimensions which have to be compactified as in Kaluza–Klein theory.

Loop quantum gravity (LQG) predicts no additional spatial dimensions, nor anything else about particle physics. These theories can be formulated in three spatial dimensions and one dimension of time, although in some LQG theories dimensionality is an emergent property of the theory, rather than a fundamental assumption of the theory. Also, LQG is a theory of quantum gravity which does not require supersymmetry. Lee Smolin, one of the originators of LQG, has proposed that a loop quantum gravity theory incorporating either supersymmetry or extra dimensions, or both, be called "loop quantum gravity II".

If experimental evidence confirms supersymmetry in the form of supersymmetric particles such as the neutralino that is often believed to be the lightest superpartner, some people believe this would be a major boost to string theory. Since supersymmetry is a required component of string theory, any discovered supersymmetry would be consistent with string theory. If the Large Hadron Collider and other major particle physics experiments fail to detect supersymmetric partners or evidence of extra dimensions, many versions of string theory which had predicted certain low mass superpartners to existing particles may need to be significantly revised. The failure of experiments to discover either supersymmetric partners or extra spatial dimensions, as of 2013[update], has encouraged loop quantum gravity researchers.

Current Status

Supersymmetric models are constrained by a variety of experiments, including measurements of low-energy observables – for example, the anomalous magnetic moment of the muon at Brookhaven; the WMAP dark matter density measurement and direct detection experiments – for example, XENON-100 and LUX; and by particle collider experiments, including B-physics, Higgs phenomenology and direct searches for superpartners (sparticles), at the Large Electron–Positron Collider, Tevatron and the LHC.

Historically, the tightest limits were from direct production at colliders. The first mass limits for squarks and gluinos were made at CERN by the UA1 experiment and the UA2 experiment at the Super Proton Synchrotron. LEP later set very strong limits., which in 2006 were extended by the D0 experiment at the Tevatron. From 2003-2015, WMAP's and Planck's dark matter density measurements have strongly constrained supersymmetry models, which, if they explain dark matter, have to be tuned to invoke a particular mechanism to sufficiently reduce the neutralino density.

Prior to the beginning of the LHC, in 2009 fits of available data to CMSSM and NUHM1 indicated that squarks and gluinos were most likely to have masses in the 500 to 800 GeV range, though values as high as 2.5 TeV were allowed with low probabilities. Neutralinos and sleptons were expected to be quite light, with the lightest neutralino and the lightest stau most likely to be found between 100 and 150 GeV.

The first run of the LHC found no evidence for supersymmetry, and, as a result, surpassed existing experimental limits from the Large Electron–Positron Collider and Tevatron and partially excluded the aforementioned expected ranges.

During 2011 and 2012, the LHC discovered a Higgs boson with a mass of about 125 GeV, and with couplings to fermions and bosons which are consistent with the Standard Model. The MSSM predicts that the mass of the lightest Higgs boson should not be much higher than the mass of the Z boson, and, in the absence of fine tuning (with the supersymmetry breaking scale on the order of 1 TeV), should not exceed 130 GeV. Furthermore, for values of the MSSM parameter $tan\ \beta \le 3$, it predicts a Higgs mass below 114 GeV over most of the parameter space. This region of Higgs mass was excluded by LEP by 2000. The LHC result is somewhat problematic for the minimal supersymmetric model, as the value of 125 GeV is relatively large for the model and can only be achieved with large radiative loop corrections from top squarks, which many theorists consider to be "unnatural" . On the other hand, the lightest Higgs boson in the MSSM is Standard Model-like, which is consistent with measurements of the Higgs boson couplings at the LHC.

Noether's Theorem

Emmy Noether was an influential mathematician known for her groundbreaking contributions to abstract algebra and theoretical physics.

Noether's (first) theorem states that every differentiable symmetry of the action of a physical system has a corresponding conservation law. The theorem was proven by mathematician Emmy Noether in 1915 and published in 1918. The action of a physical system is the integral over time of a Lagrangian function (which may or may not be an integral over space of a Lagrangian density function), from which the system's behavior can be determined by the principle of least action.

Noether's theorem is used in theoretical physics and the calculus of variations. A generalization of the formulations on constants of motion in Lagrangian and Hamiltonian mechanics (developed in 1788 and 1833, respectively), it does not apply to systems that cannot be modeled with a Lagrangian alone (e.g. systems with a Rayleigh dissipation function). In particular, dissipative systems with continuous symmetries need not have a corresponding conservation law.

Basic Illustrations and Background

As an illustration, if a physical system behaves the same regardless of how it is oriented in space, its Lagrangian is rotationally symmetric: from this symmetry, Noether's theorem dictates that the angular momentum of the system be conserved, as a consequence of its laws of motion. The physical system itself need not be symmetric; a jagged asteroid tumbling in space conserves angular momentum despite its asymmetry. It is the laws of its motion that are symmetric.

As another example, if a physical process exhibits the same outcomes regardless of place or time, then its Lagrangian is symmetric under continuous translations in space and time: by Noether's theorem, these symmetries account for the conservation laws of linear momentum and energy within this system, respectively.

Noether's theorem is important, both because of the insight it gives into conservation laws, and also as a practical calculational tool. It allows investigators to determine the conserved quantities (invariants) from the observed symmetries of a physical system. Conversely, it allows researchers to consider whole classes of hypothetical Lagrangians with given invariants, to describe a physical system. As an illustration, suppose that a physical theory is proposed which conserves a quantity X. A researcher can calculate the types of Lagrangians that conserve X through a continuous symmetry. Due to Noether's theorem, the properties of these Lagrangians provide further criteria to understand the implications and judge the fitness of the new theory.

There are numerous versions of Noether's theorem, with varying degrees of generality. The original version applied only to ordinary differential equations (particles) and not partial differential equations (fields). The original versions also assume that the Lagrangian depends only upon the first derivative, while later versions generalize the theorem to Lagrangians depending on the nth derivative. There are natural quantum counterparts of this theorem, expressed in the Ward–Takahashi identities. Generalizations of Noether's theorem to superspaces are also available.

Informal Statement of the Theorem

All fine technical points aside, Noether's theorem can be stated informally

If a system has a continuous symmetry property, then there are corresponding quantities whose values are conserved in time.

A more sophisticated version of the theorem involving fields states that:

To every differentiable symmetry generated by local actions, there corresponds a conserved current.

The word "symmetry" in the above statement refers more precisely to the covariance of the form that a physical law takes with respect to a one-dimensional Lie group of transformations satisfying certain technical criteria. The conservation law of a physical quantity is usually expressed as a continuity equation.

The formal proof of the theorem utilizes the condition of invariance to derive an expression for a current associated with a conserved physical quantity. In modern (since ca. 1980) terminology, the conserved quantity is called the *Noether charge*, while the flow carrying that charge is called the *Noether current*. The Noether current is defined up to a solenoidal (divergenceless) vector field.

In the context of gravitation, Felix Klein's statement of Noether's theorem for action I stipulates for the invariants:

If an integral I is invariant under a continuous group G_ρ with ρ parameters, then ρ linearly independent combinations of the Lagrangian expressions are divergences.

Historical Context

A conservation law states that some quantity X in the mathematical description of a system's evolution remains constant throughout its motion — it is an invariant. Mathematically, the rate of change of X (its derivative with respect to time) vanishes,

$$\frac{dX}{dt} = 0 \ .$$

Such quantities are said to be conserved; they are often called constants of motion (although motion *per se* need not be involved, just evolution in time). For example, if the energy of a system is conserved, its energy is invariant at all times, which imposes a constraint on the system's motion and may help in solving for it. Aside from insights that such constants of motion give into the nature of a system, they are a useful calculational tool; for example, an approximate solution can be corrected by finding the nearest state that satisfies the suitable conservation laws.

The earliest constants of motion discovered were momentum and energy, which were proposed in the 17th century by René Descartes and Gottfried Leibniz on the basis of collision experiments, and refined by subsequent researchers. Isaac Newton was the first to enunciate the conservation of momentum in its modern form, and showed that it was a consequence of Newton's third law. According to general relativity, the conservation laws of linear momentum, energy and angular momentum are only exactly true globally when expressed in terms of the sum of the stress–energy tensor (non-gravitational stress–energy) and the Landau–Lifshitz stress–energy–momentum pseudotensor (gravitational stress–energy). The local conservation of non-gravitational linear momentum and energy in a free-falling reference frame is expressed by the vanishing of the covariant divergence of the stress–energy tensor. Another important conserved quantity, discovered in studies of the celestial mechanics of astronomical bodies, is the Laplace–Runge–Lenz vector.

In the late 18th and early 19th centuries, physicists developed more systematic methods for discovering invariants. A major advance came in 1788 with the development of Lagrangian mechanics, which is related to the principle of least action. In this approach, the state of the system can be described by any type of generalized coordinates q; the laws of motion need not be expressed in a Cartesian coordinate system, as was customary in Newtonian mechanics. The action is defined as the time integral I of a function known as the Lagrangian L

$$I = \int L(\mathbf{q},\dot{\mathbf{q}},t)dt \ ,$$

where the dot over q signifies the rate of change of the coordinates q,

$$\dot{\mathbf{q}} = \frac{d\mathbf{q}}{dt} \ .$$

Hamilton's principle states that the physical path q(t)—the one actually taken by the system—is a path for which infinitesimal variations in that path cause no change in I, at least up to first order. This principle results in the Euler–Lagrange equations,

$$\frac{d}{dt}\left(\frac{\partial L}{\partial \dot{\mathbf{q}}}\right) = \frac{\partial L}{\partial \mathbf{q}} \cdot$$

Thus, if one of the coordinates, say q_k, does not appear in the Lagrangian, the right-hand side of the equation is zero, and the left-hand side requires that

$$\frac{d}{dt}\left(\frac{\partial L}{\partial \dot{q}_k}\right) = \frac{dp_k}{dt} = 0 ,$$

where the momentum

$$p_k = \frac{\partial L}{\partial \dot{q}_k}$$

is conserved throughout the motion (on the physical path).

Thus, the absence of the ignorable coordinate q_k from the Lagrangian implies that the Lagrangian is unaffected by changes or transformations of q_k; the Lagrangian is invariant, and is said to exhibit a symmetry under such transformations. This is the seed idea generalized in Noether's theorem.

Several alternative methods for finding conserved quantities were developed in the 19th century, especially by William Rowan Hamilton. For example, he developed a theory of canonical transformations which allowed changing coordinates so that some coordinates disappeared from the Lagrangian, as above, resulting in conserved canonical momenta. Another approach, and perhaps the most efficient for finding conserved quantities, is the Hamilton–Jacobi equation.

Mathematical Expression

Simple form using Perturbations

The essence of Noether's theorem is generalizing the ignorable coordinates outlined.

One can assume that the action L defined above is invariant under small perturbations (warpings) of the time variable t and the generalized coordinates q. One may write

$$t \rightarrow t' = t + \delta t$$

$$\mathbf{q} \rightarrow \mathbf{q}' = \mathbf{q} + \delta \mathbf{q} ,$$

where the perturbations δt and δq are both small, but variable. For generality, assume there are (say) N such symmetry transformations of the action, i.e. transformations leaving the action unchanged; labelled by an index $r = 1, 2, 3, ..., N$.

Then the resultant perturbation can be written as a linear sum of the individual types of perturbations,

$$\delta t = \sum_r \varepsilon_r T_r$$

$$\delta \mathbf{q} = \sum_r \varepsilon_r \mathbf{Q}_r \ ,$$

where ε_r are infinitesimal parameter coefficients corresponding to each:

- generator T_r of time evolution, and
- generator Q_r of the generalized coordinates.

For translations, Q_r is a constant with units of length; for rotations, it is an expression linear in the components of q, and the parameters make up an angle.

Using these definitions, Noether showed that the N quantities

$$\left(\frac{\partial L}{\partial \dot{\mathbf{q}}} \cdot \dot{\mathbf{q}} - L \right) T_r - \frac{\partial L}{\partial \dot{\mathbf{q}}} \cdot \mathbf{Q}_r$$

(which have the dimensions of [energy]·[time] + [momentum]·[length] = [action]) are conserved (constants of motion).

Examples

Time invariance

For illustration, consider a Lagrangian that does not depend on time, i.e., that is invariant (symmetric) under changes $t \to t + \delta t$, without any change in the coordinates q. In this case, $N = 1$, $T = 1$ and $Q = 0$; the corresponding conserved quantity is the total energy H

$$H = \frac{\partial L}{\partial \dot{\mathbf{q}}} \cdot \dot{\mathbf{q}} - L.$$

Translational invariance

Consider a Lagrangian which does not depend on an ("ignorable", as above) coordinate q_k; so it is invariant (symmetric) under changes $q_k \to q_k + \delta q_k$. In that case, $N = 1$, $T = 0$, and $Q_k = 1$; the conserved quantity is the corresponding momentum p_k

$$p_k = \frac{\partial L}{\partial \dot{q}_k}.$$

In special and general relativity, these apparently separate conservation laws are aspects of a single conservation law, that of the stress–energy tensor, that is derived in the next section.

Rotational invariance

The conservation of the angular momentum L = r × p is analogous to its linear momentum counterpart. It is assumed that the symmetry of the Lagrangian is rotational, i.e., that the Lagrangian does not depend on the absolute orientation of the physical system in space. For concreteness, assume that the Lagrangian does not change under small rotations of an angle $\delta\theta$ about an axis n; such a rotation transforms the Cartesian coordinates by the equation

$$\mathbf{r} \rightarrow \mathbf{r} + \delta\theta\,\mathbf{n} \times \mathbf{r}.$$

Since time is not being transformed, $T=0$. Taking $\delta\theta$ as the ε parameter and the Cartesian coordinates r as the generalized coordinates q, the corresponding Q variables are given by

$$\mathbf{Q} = \mathbf{n} \times \mathbf{r}.$$

Then Noether's theorem states that the following quantity is conserved,

$$\frac{\partial L}{\partial \dot{\mathbf{q}}} \cdot \mathbf{Q}_r = \mathbf{p} \cdot (\mathbf{n} \times \mathbf{r}) = \mathbf{n} \cdot (\mathbf{r} \times \mathbf{p}) = \mathbf{n} \cdot \mathbf{L}.$$

In other words, the component of the angular momentum L along the n axis is conserved.

If n is arbitrary, i.e., if the system is insensitive to any rotation, then every component of L is conserved; in short, angular momentum is conserved.

Field Theory Version

Although useful in its own right, the version of Noether's theorem just given is a special case of the general version derived in 1915. To give the flavor of the general theorem, a version of the Noether theorem for continuous fields in four-dimensional space–time is now given. Since field theory problems are more common in modern physics than mechanics problems, this field theory version is the most commonly used version (or most often implemented) of Noether's theorem.

Let there be a set of differentiable fields ϕ defined over all space and time; for example, the temperature $T(\mathbf{x}, t)$ would be representative of such a field, being a number defined at every place and time. The principle of least action can be applied to such fields, but the action is now an integral over space and time

$$S = \int \mathcal{L}\left(\phi, \partial_\mu \phi, x^\mu\right) d^4 x$$

(the theorem can actually be further generalized to the case where the Lagrangian depends on up to the nth derivative using jet bundles)

A continuous transformation of the fields ϕ can be written infinitesimally as

$$\phi \mapsto \phi + \epsilon \Psi,$$

where Ψ is in general a function that may depend on both x^μ and ϕ. The condition for Ψ to generate a physical symmetry is that the action S is left invariant. This will certainly be true if the Lagrangian density $\mathcal{L}$ is left invariant, but it will also be true if the Lagrangian changes by a divergence,

$$\mathcal{L} \mapsto \mathcal{L} + \epsilon \partial_\mu \Lambda^\mu,$$

since the integral of a divergence becomes a boundary term according to the divergence theorem. A system described by a given action might have multiple independent symmetries of this type, indexed by $r = 1, 2, \ldots N,$ so the most general symmetry transformation would be written as

$$\phi \mapsto \phi + \epsilon_r \Psi_r$$

with the consequence

$$\mathcal{L} \mapsto \mathcal{L} + \epsilon_r \partial_\mu \Lambda_r^\mu$$

For such systems, Noether's theorem states that there are N conserved current densities

$$j_r^\nu = \Lambda_r^\nu - \frac{\partial \mathcal{L}}{\partial \phi_{,\nu}} \cdot \Psi_r$$

(where the dot product is understood to contract the *field* indices, not the ν index or r index)

In such cases, the conservation law is expressed in a four-dimensional way

$$\partial_\nu j^\nu = 0$$

which expresses the idea that the amount of a conserved quantity within a sphere cannot change unless some of it flows out of the sphere. For example, electric charge is conserved; the amount of charge within a sphere cannot change unless some of the charge leaves the sphere.

For illustration, consider a physical system of fields that behaves the same under translations in time and space, as considered above; in other words, $L\left(\phi, \partial_\mu \phi, x^\mu\right)$ is constant in its third argument. In that case, $N = 4$, one for each dimension of space and time. An infinitesimal translation in space, $x^\mu \mapsto x^\mu + \epsilon_r \delta_r^\mu$ (with δ denoting the Kronecker delta), affects the fields as $\phi(x^\mu) \mapsto \phi(x^\mu - \epsilon_r \delta_r^\mu)$: that is, relabelling the coordinates is equivalent to leaving the coordinates in place while translating the field itself, which in turn is equivalent to transforming the field by replacing its value at each point x^μ with the value at the point $x^\mu - \epsilon X^\mu$ "behind" it which would be mapped onto x^μ by the infinitesimal displacement under consideration. Since this is infinitesimal, we may write this transformation as

$$\Psi_r = -\delta_r^\mu \partial_\mu \phi$$

The Lagrangian density transforms in the same way, $\mathcal{L}(x^\mu) \mapsto \mathcal{L}(x^\mu - \epsilon_r \delta_r^\mu)$, so

$$\Lambda_r^\mu = -\delta_r^\mu \mathcal{L}$$

and thus Noether's theorem corresponds to the conservation law for the stress–energy tensor T_μ^ν, where we have used μ in place of r. To wit, by using the expression given earlier, and collecting the four conserved currents (one for each μ) into a tensor T, Noether's theorem gives

$$T_\mu^{\ \nu} = -\delta_\mu^\nu \mathcal{L} + \delta_\mu^\sigma \partial_\sigma \phi \frac{\partial \mathcal{L}}{\partial \phi_{,\nu}} = \left(\frac{\partial \mathcal{L}}{\partial \phi_{,\nu}} \right) \cdot \phi_{,\mu} - \delta_\mu^\nu \mathcal{L}$$

with

$$T_\mu^{\ \nu}{}_{,\nu} = 0$$

(note that we relabelled μ as σ at an intermediate step to avoid conflict). (However, note that the T obtained in this way may differ from the symmetric tensor used as the source term in general relativity.)

The conservation of electric charge, by contrast, can be derived by considering Ψ linear in the fields φ rather than in the derivatives. In quantum mechanics, the probability amplitude $\psi(x)$ of finding a particle at a point x is a complex field φ, because it ascribes a complex number to every point in space and time. The probability amplitude itself is physically unmeasurable; only the probability $p = |\psi|^2$ can be inferred from a set of measurements. Therefore, the system is invariant under transformations of the ψ field and its complex conjugate field ψ^* that leave $|\psi|^2$ unchanged, such as

$$\psi \to e^{i\theta} \psi \ , \ \psi^* \to e^{-i\theta} \psi^* \ ,$$

a complex rotation. In the limit when the phase θ becomes infinitesimally small, $\delta\theta$, it may be taken as the parameter ε, while the Ψ are equal to $i\psi$ and $-i\psi^*$, respectively. A specific example is the Klein–Gordon equation, the relativistically correct version of the Schrödinger equation for spinless particles, which has the Lagrangian density

$$L = \psi_{,\nu} \psi^*_{,\mu} \eta^{\nu\mu} + m^2 \psi \psi^*.$$

In this case, Noether's theorem states that the conserved ($\partial \cdot j = 0$) current equals

$$j^\nu = i \left(\frac{\partial \psi}{\partial x^\mu} \psi^* - \frac{\partial \psi^*}{\partial x^\mu} \psi \right) \eta^{\nu\mu} \ ,$$

which, when multiplied by the charge on that species of particle, equals the electric current density due to that type of particle. This "gauge invariance" was first noted by Hermann Weyl, and is one of the prototype gauge symmetries of physics.

Derivations

One Independent Variable

Consider the simplest case, a system with one independent variable, time. Suppose the dependent variables q are such that the action integral

$$I = \int_{t_1}^{t_2} L[\mathbf{q}[t], \dot{\mathbf{q}}[t], t] dt$$

is invariant under brief infinitesimal variations in the dependent variables. In other words, they satisfy the Euler–Lagrange equations

$$\frac{d}{dt}\frac{\partial L}{\partial \dot{\mathbf{q}}}[t] = \frac{\partial L}{\partial \mathbf{q}}[t].$$

And suppose that the integral is invariant under a continuous symmetry. Mathematically such a symmetry is represented as a flow, φ, which acts on the variables as follows

$$t \to t' = t + \varepsilon T$$

$$\mathbf{q}[t] \to \mathbf{q}'[t'] = \phi[\mathbf{q}[t], \varepsilon] = \phi[\mathbf{q}[t' - \varepsilon T], \varepsilon]$$

where ε is a real variable indicating the amount of flow, and T is a real constant (which could be zero) indicating how much the flow shifts time.

$$\dot{\mathbf{q}}[t] \to \dot{\mathbf{q}}'[t'] = \frac{d}{dt}\phi[\mathbf{q}[t], \varepsilon] = \frac{\partial \phi}{\partial \mathbf{q}}[\mathbf{q}[t' - \varepsilon T], \varepsilon]\dot{\mathbf{q}}[t' - \varepsilon T].$$

The action integral flows to

$$I'[\varepsilon] = \int_{t_1 + \varepsilon T}^{t_2 + \varepsilon T} L[\mathbf{q}'[t'], \dot{\mathbf{q}}'[t'], t']dt'$$

$$= \int_{t_1 + \varepsilon T}^{t_2 + \varepsilon T} L[\phi[\mathbf{q}[t' - \varepsilon T], \varepsilon], \frac{\partial \phi}{\partial \mathbf{q}}[\mathbf{q}[t' - \varepsilon T], \varepsilon]\dot{\mathbf{q}}[t' - \varepsilon T], t']dt'$$

which may be regarded as a function of ε. Calculating the derivative at $\varepsilon = 0$ and using Leibniz's rule, we get

$$0 = \frac{dI'}{d\varepsilon}[0] = L[\mathbf{q}[t_2], \dot{\mathbf{q}}[t_2], t_2]T - L[\mathbf{q}[t_1], \dot{\mathbf{q}}[t_1], t_1]T$$

$$+ \int_{t_1}^{t_2} \frac{\partial L}{\partial \mathbf{q}}\left(-\frac{\partial \phi}{\partial \mathbf{q}}\dot{\mathbf{q}}T + \frac{\partial \phi}{\partial \varepsilon}\right) + \frac{\partial L}{\partial \dot{\mathbf{q}}}\left(-\frac{\partial^2 \phi}{(\partial \mathbf{q})^2}\dot{\mathbf{q}}^2 T + \frac{\partial^2 \phi}{\partial \varepsilon \partial \mathbf{q}}\dot{\mathbf{q}} - \frac{\partial \phi}{\partial \mathbf{q}}\ddot{\mathbf{q}}T\right)dt.$$

Notice that the Euler–Lagrange equations imply

$$\frac{d}{dt}\left(\frac{\partial L}{\partial \dot{\mathbf{q}}}\frac{\partial \phi}{\partial \mathbf{q}}\dot{\mathbf{q}}T\right) = \left(\frac{d}{dt}\frac{\partial L}{\partial \dot{\mathbf{q}}}\right)\frac{\partial \phi}{\partial \mathbf{q}}\dot{\mathbf{q}}T + \frac{\partial L}{\partial \dot{\mathbf{q}}}\left(\frac{d}{dt}\frac{\partial \phi}{\partial \mathbf{q}}\right)\dot{\mathbf{q}}T + \frac{\partial L}{\partial \dot{\mathbf{q}}}\frac{\partial \phi}{\partial \mathbf{q}}\ddot{\mathbf{q}}T$$

$$= \frac{\partial L}{\partial \mathbf{q}}\frac{\partial \phi}{\partial \mathbf{q}}\dot{\mathbf{q}}T + \frac{\partial L}{\partial \dot{\mathbf{q}}}\left(\frac{\partial^2 \phi}{(\partial \mathbf{q})^2}\dot{\mathbf{q}}\right)\dot{\mathbf{q}}T + \frac{\partial L}{\partial \dot{\mathbf{q}}}\frac{\partial \phi}{\partial \mathbf{q}}\ddot{\mathbf{q}}T.$$

Substituting this into the previous equation, one gets

$$0 = \frac{dI'}{d\varepsilon}[0] = L[\mathbf{q}[t_2], \dot{\mathbf{q}}[t_2], t_2]T - L[\mathbf{q}[t_1], \dot{\mathbf{q}}[t_1], t_1]T - \frac{\partial L}{\partial \dot{\mathbf{q}}} \frac{\partial \phi}{\partial \mathbf{q}} \dot{\mathbf{q}}[t_2]T$$

$$+ \frac{\partial L}{\partial \dot{\mathbf{q}}} \frac{\partial \phi}{\partial \mathbf{q}} \dot{\mathbf{q}}[t_1]T + \int_{t_1}^{t_2} \frac{\partial L}{\partial \mathbf{q}} \frac{\partial \phi}{\partial \varepsilon} + \frac{\partial L}{\partial \dot{\mathbf{q}}} \frac{\partial^2 \phi}{\partial \varepsilon \partial \mathbf{q}} \dot{\mathbf{q}} \, dt$$

Again using the Euler–Lagrange equations we get

$$\frac{d}{dt}\left(\frac{\partial L}{\partial \dot{\mathbf{q}}} \frac{\partial \phi}{\partial \varepsilon} \right) = \left(\frac{d}{dt} \frac{\partial L}{\partial \dot{\mathbf{q}}} \right) \frac{\partial \phi}{\partial \varepsilon} + \frac{\partial L}{\partial \dot{\mathbf{q}}} \frac{\partial^2 \phi}{\partial \varepsilon \partial \mathbf{q}} \dot{\mathbf{q}} = \frac{\partial L}{\partial \mathbf{q}} \frac{\partial \phi}{\partial \varepsilon} + \frac{\partial L}{\partial \dot{\mathbf{q}}} \frac{\partial^2 \phi}{\partial \varepsilon \partial \mathbf{q}} \dot{\mathbf{q}}.$$

Substituting this into the previous equation, one gets

$$0 = L[\mathbf{q}[t_2], \dot{\mathbf{q}}[t_2], t_2]T - L[\mathbf{q}[t_1], \dot{\mathbf{q}}[t_1], t_1]T - \frac{\partial L}{\partial \dot{\mathbf{q}}} \frac{\partial \phi}{\partial \mathbf{q}} \dot{\mathbf{q}}[t_2]T + \frac{\partial L}{\partial \dot{\mathbf{q}}} \frac{\partial \phi}{\partial \mathbf{q}} \dot{\mathbf{q}}[t_1]T$$

$$+ \frac{\partial L}{\partial \dot{\mathbf{q}}} \frac{\partial \phi}{\partial \varepsilon}[t_2] - \frac{\partial L}{\partial \dot{\mathbf{q}}} \frac{\partial \phi}{\partial \varepsilon}[t_1].$$

From which one can see that

$$\left(\frac{\partial L}{\partial \dot{\mathbf{q}}} \frac{\partial \phi}{\partial \mathbf{q}} \dot{\mathbf{q}} - L \right) T - \frac{\partial L}{\partial \dot{\mathbf{q}}} \frac{\partial \phi}{\partial \varepsilon}$$

is a constant of the motion, i.e., it is a conserved quantity. Since $\varphi[q, 0] = q$, we get $\dfrac{\partial \phi}{\partial \mathbf{q}} = 1$ and so the conserved quantity simplifies to

$$\left(\frac{\partial L}{\partial \dot{\mathbf{q}}} \dot{\mathbf{q}} - L \right) T - \frac{\partial L}{\partial \dot{\mathbf{q}}} \frac{\partial \phi}{\partial \varepsilon}.$$

To avoid excessive complication of the formulas, this derivation assumed that the flow does not change as time passes. The same result can be obtained in the more general case.

Field-Theoretic Derivation

Noether's theorem may also be derived for tensor fields φ^A where the index A ranges over the various components of the various tensor fields. These field quantities are functions defined over a four-dimensional space whose points are labeled by coordinates x^μ where the index μ ranges over time ($\mu = 0$) and three spatial dimensions ($\mu = 1, 2, 3$). These four coordinates are the independent variables; and the values of the fields at each event are the dependent variables. Under an infinitesimal transformation, the variation in the coordinates is written

$$x^\mu \rightarrow \xi^\mu = x^\mu + \delta x^\mu$$

whereas the transformation of the field variables is expressed as

$$\phi^A \rightarrow \alpha^A(\xi^\mu) = \phi^A(x^\mu) + \delta\phi^A(x^\mu).$$

By this definition, the field variations $\delta\varphi^A$ result from two factors: intrinsic changes in the field themselves and changes in coordinates, since the transformed field α^A depends on the transformed coordinates ξ^μ. To isolate the intrinsic changes, the field variation at a single point x^μ may be defined

$$\alpha^A(x^\mu) = \phi^A(x^\mu) + \overline{\delta}\phi^A(x^\mu).$$

If the coordinates are changed, the boundary of the region of space–time over which the Lagrangian is being integrated also changes; the original boundary and its transformed version are denoted as Ω and Ω', respectively.

Noether's theorem begins with the assumption that a specific transformation of the coordinates and field variables does not change the action, which is defined as the integral of the Lagrangian density over the given region of spacetime. Expressed mathematically, this assumption may be written as

$$\int_{\Omega'} L\left(\alpha^A, \alpha^A{}_{,v}, \xi^\mu\right) d^4\xi - \int_{\Omega} L\left(\phi^A, \phi^A{}_{,v}, x^\mu\right) d^4x = 0$$

where the comma subscript indicates a partial derivative with respect to the coordinate(s) that follows the comma, e.g.

$$\phi^A{}_{,\sigma} = \frac{\partial\phi^A}{\partial x^\sigma}.$$

Since ξ is a dummy variable of integration, and since the change in the boundary Ω is infinitesimal by assumption, the two integrals may be combined using the four-dimensional version of the divergence theorem into the following form

$$\int_{\Omega} \left\{\left[L\left(\alpha^A, \alpha^A{}_{,v}, x^\mu\right) - L\left(\phi^A, \phi^A{}_{,v}, x^\mu\right)\right] + \frac{\partial}{\partial x^\sigma}\left[L\left(\phi^A, \phi^A{}_{,v}, x^\mu\right)\delta x^\sigma\right]\right\} d^4x = 0.$$

The difference in Lagrangians can be written to first-order in the infinitesimal variations as

$$\left[L\left(\alpha^A, \alpha^A{}_{,v}, x^\mu\right) - L\left(\phi^A, \phi^A{}_{,v}, x^\mu\right)\right] = \frac{\partial L}{\partial\phi^A}\overline{\delta}\phi^A + \frac{\partial L}{\partial\phi^A{}_{,\sigma}}\overline{\delta}\phi^A{}_{,\sigma}.$$

However, because the variations are defined at the same point as described above, the variation and the derivative can be done in reverse order; they commute

$$\bar{\delta}\phi^A{}_{,\sigma} = \bar{\delta}\frac{\partial \phi^A}{\partial x^\sigma} = \frac{\partial}{\partial x^\sigma}(\bar{\delta}\phi^A).$$

Using the Euler–Lagrange field equations

$$\frac{\partial}{\partial x^\sigma}\left(\frac{\partial L}{\partial \phi^A{}_{,\sigma}}\right) = \frac{\partial L}{\partial \phi^A}$$

the difference in Lagrangians can be written neatly as

$$\left[L\left(\alpha^A,\alpha^A{}_{,\nu},x^\mu\right)-L\left(\phi^A,\phi^A{}_{,\nu},x^\mu\right)\right] = \frac{\partial}{\partial x^\sigma}\left(\frac{\partial L}{\partial \phi^A{}_{,\sigma}}\right)\bar{\delta}\phi^A + \frac{\partial L}{\partial \phi^A{}_{,\sigma}}\bar{\delta}\phi^A{}_{,\sigma} = \frac{\partial}{\partial x^\sigma}\left(\frac{\partial L}{\partial \phi^A{}_{,\sigma}}\bar{\delta}\phi^A\right).$$

Thus, the change in the action can be written as

$$\int_\Omega \frac{\partial}{\partial x^\sigma}\left\{\frac{\partial L}{\partial \phi^A{}_{,\sigma}}\bar{\delta}\phi^A + L\left(\phi^A,\phi^A{}_{,\nu},x^\mu\right)\delta x^\sigma\right\}d^4x = 0.$$

Since this holds for any region Ω, the integrand must be zero

$$\frac{\partial}{\partial x^\sigma}\left\{\frac{\partial L}{\partial \phi^A{}_{,\sigma}}\bar{\delta}\phi^A + L\left(\phi^A,\phi^A{}_{,\nu},x^\mu\right)\delta x^\sigma\right\} = 0.$$

For any combination of the various symmetry transformations, the perturbation can be written

$$\delta x^\mu = \varepsilon X^\mu$$

$$\delta \phi^A = \varepsilon \Psi^A = \bar{\delta}\phi^A + \varepsilon \mathcal{L}_X \phi^A$$

where $\mathcal{L}_X \phi^A$ is the Lie derivative of φ^A in the X^μ direction. When φ^A is a scalar or $X^\mu{}_{,\nu} = 0$,

$$\mathcal{L}_X \phi^A = \frac{\partial \phi^A}{\partial x^\mu}X^\mu.$$

These equations imply that the field variation taken at one point equals

$$\bar{\delta}\phi^A = \varepsilon \Psi^A - \varepsilon \mathcal{L}_X \phi^A.$$

Differentiating the above divergence with respect to ε at $\varepsilon=0$ and changing the sign yields the conservation law

$$\frac{\partial}{\partial x^\sigma}j^\sigma = 0$$

where the conserved current equals

$$j^{\sigma} = \left[\frac{\partial L}{\partial \phi^{A}_{,\sigma}} \mathcal{L}_{X} \phi^{A} - L X^{\sigma} \right] - \left(\frac{\partial L}{\partial \phi^{A}_{,\sigma}} \right) \varnothing^{A}.$$

Manifold/Fiber Bundle Derivation

Suppose we have an n-dimensional oriented Riemannian manifold, M and a target manifold T. Let be the configuration space of smooth functions from M to T. (More generally, we can have smooth sections of a fiber bundle over M.)

Examples of this M in physics include:

- In classical mechanics, in the Hamiltonian formulation, M is the one-dimensional manifold R, representing time and the target space is the cotangent bundle of space of generalized positions.

- In field theory, M is the spacetime manifold and the target space is the set of values the fields can take at any given point. For example, if there are m real-valued scalar fields, , then the target manifold is R^m. If the field is a real vector field, then the target manifold is isomorphic to R^3.

Now suppose there is a functional

$$S : \mathcal{C} \rightarrow \mathbf{R},$$

called the action. (Note that it takes values into R, rather than C; this is for physical reasons, and doesn't really matter for this proof.)

To get to the usual version of Noether's theorem, we need additional restrictions on the action. We assume is the integral over M of a function

$$\mathcal{L}(\phi, \partial_{\mu}\phi, x)$$

called the Lagrangian density, depending on φ, its derivative and the position. In other words, for φ in

$$S[\phi] = \int_{M} \mathcal{L}[\phi(x), \partial_{\mu}\phi(x), x] \mathrm{d}^{n}x.$$

Suppose we are given boundary conditions, i.e., a specification of the value of φ at the boundary if M is compact, or some limit on φ as x approaches ∞. Then the subspace of consisting of functions φ such that all functional derivatives of at φ are zero, that is:

$$\frac{\delta S[\phi]}{\delta \phi(x)} \approx 0$$

and that φ satisfies the given boundary conditions, is the subspace of on shell solutions.

Now, suppose we have an infinitesimal transformation on $\mathcal{C}$, generated by a functional derivation, Q such that

$$Q\left[\int_N \mathcal{L}\mathrm{d}^n x\right] \approx \int_{\partial N} f^\mu[\phi(x), \partial\phi, \partial\partial\phi, \ldots]\mathrm{d}s_\mu$$

for all compact submanifolds N or in other words,

$$Q[\mathcal{L}(x)] \approx \partial_\mu f^\mu(x)$$

for all x, where we set

$$\mathcal{L}(x) = \mathcal{L}[\phi(x), \partial_\mu\phi(x), x].$$

If this holds on shell and off shell, we say Q generates an off-shell symmetry. If this only holds on shell, we say Q generates an on-shell symmetry. Then, we say Q is a generator of a one parameter symmetry Lie group.

Now, for any N, because of the Euler–Lagrange theorem, on shell (and only on-shell), we have

$$Q\left[\int_N \mathcal{L}\mathrm{d}^n x\right] \quad = \int_N \left[\frac{\partial\mathcal{L}}{\partial\phi} - \partial_\mu \frac{\partial\mathcal{L}}{\partial(\partial_\mu\phi)}\right]Q[\phi]\mathrm{d}^n x + \int_{\partial N} \frac{\partial\mathcal{L}}{\partial(\partial_\mu\phi)}Q[\phi]\mathrm{d}s_\mu$$

$$\approx \int_{\partial N} f^\mu \mathrm{d}s_\mu.$$

Since this is true for any N, we have

$$\partial_\mu\left[\frac{\partial\mathcal{L}}{\partial(\partial_\mu\phi)}Q[\phi] - f^\mu\right] \approx 0.$$

But this is the continuity equation for the current defined by:

$$J^\mu = \frac{\partial\mathcal{L}}{\partial(\partial_\mu\phi)}Q[\phi] - f^\mu,$$

which is called the Noether current associated with the symmetry. The continuity equation tells us that if we integrate this current over a space-like slice, we get a conserved quantity called the Noether charge (provided, of course, if M is noncompact, the currents fall off sufficiently fast at infinity).

Comments

Noether's theorem is an on shell theorem: it relies on use of the equations of motion—the classical path. It reflects the relation between the boundary conditions and the variational principle. Assuming no boundary terms in the action, Noether's theorem implies that

$$\int_{\partial N} J^\mu ds_\mu \approx 0 \ .$$

The quantum analogs of Noether's theorem involving expectation values, e.g. $\langle \int d^4x \ \partial \cdot J \rangle = 0$, probing off shell quantities as well are the Ward–Takahashi identities.

Generalization to Lie Algebras

Suppose say we have two symmetry derivations Q_1 and Q_2. Then, $[Q_1, Q_2]$ is also a symmetry derivation. Let's see this explicitly. Let's say

$$Q_1[\mathcal{L}] \approx \partial_\mu f_1^\mu$$

and

$$Q_2[\mathcal{L}] \approx \partial_\mu f_2^\mu$$

Then,

$$[Q_1, Q_2][\mathcal{L}] = Q_1[Q_2[\mathcal{L}]] - Q_2[Q_1[\mathcal{L}]] \approx \partial_\mu f_{12}^\mu$$

where $f_{12} = Q_1[f_2^\mu] - Q_2[f_1^\mu]$. So,

$$j_{12}^\mu = \left(\frac{\partial}{\partial(\partial_\mu \phi)} \mathcal{L} \right) (Q_1[Q_2[\phi]] - Q_2[Q_1[\phi]]) - f_{12}^\mu .$$

This shows we can extend Noether's theorem to larger Lie algebras in a natural way.

Generalization of the Proof

This applies to *any* local symmetry derivation Q satisfying $QS \approx 0$, and also to more general local functional differentiable actions, including ones where the Lagrangian depends on higher derivatives of the fields. Let ε be any arbitrary smooth function of the spacetime (or time) manifold such that the closure of its support is disjoint from the boundary. ε is a test function. Then, because of the variational principle (which does *not* apply to the boundary, by the way), the derivation distribution q generated by $q[\varepsilon][\Phi(x)] = \varepsilon(x)Q[\Phi(x)]$ satisfies $q[\varepsilon][S] \approx 0$ for every ε, or more compactly, $q(x)[S] \approx 0$ for all x not on the boundary (but remember that $q(x)$ is a shorthand for a derivation *distribution*, not a derivation parametrized by x in general). This is the generalization of Noether's theorem.

To see how the generalization is related to the version given above, assume that the action is the spacetime integral of a Lagrangian that only depends on φ and its first derivatives. Also, assume

$$Q[\mathcal{L}] \approx \partial_\mu f^\mu$$

Then,

$$q[\varepsilon][\mathcal{S}] \quad \int q[\varepsilon][\mathcal{L}]\mathrm{d}\ x$$

$$= \int \left\{ \left(\frac{\partial}{\partial \phi}\mathcal{L}\right)\varepsilon Q[\phi] + \left[\frac{\partial}{\partial(\partial\ \phi)}\mathcal{L}\right]\partial\ (\varepsilon Q[\phi]) \right\}\mathrm{d}\ x$$

$$= \int \left\{ \varepsilon Q[\mathcal{L}] + \partial\ \varepsilon \left[\frac{\partial}{\partial(\partial\ )}\mathcal{L}\right]Q[\phi] \right\}\mathrm{d}\ x$$

$$\approx \int \varepsilon\partial\ \left\{ f\ -\left[\frac{\partial}{\partial(\partial\ )}\ \right]Q[\phi] \right\}\mathrm{d}\ x$$

for all ε.

More generally, if the Lagrangian depends on higher derivatives, then

$$\partial_\mu \left[f^\mu - \left[\frac{\partial}{\partial(\partial_\mu\phi)}\mathcal{L}\right]Q[\phi] - 2\left[\frac{\partial}{\partial(\partial_\mu\partial_\nu\phi)}\mathcal{L}\right]\partial_\nu Q[\phi] + \partial_\nu\left[\left[\frac{\partial}{\partial(\partial_\mu\partial_\nu\phi)}\mathcal{L}\right]Q[\phi]\right] - \cdots \right] \approx 0.$$

Examples

Example 1: Conservation of Energy

Looking at the specific case of a Newtonian particle of mass m, coordinate x, moving under the influence of a potential V, coordinatized by time t. The action, S, is:

$$S[x] = \int L[x(t), \dot{x}(t)]dt$$

$$= \int \left(\frac{m}{2}\sum_{i=1}^{3}\dot{x}_i^2 - V(x(t))\right)dt$$

The first term in the brackets is the kinetic energy of the particle, whilst the second is its potential energy. Consider the generator of time translations $Q = \partial/\partial t$. In other words, . Note that x has an explicit dependence on time, whilst V does not; consequently:

$$Q[L] = m\sum_i \dot{x}_i\ddot{x}_i - \sum_i \frac{\partial V(x)}{\partial x_i}\dot{x}_i = \frac{d}{dt}\left[\frac{m}{2}\sum_i \dot{x}_i^2 - V(x)\right]$$

so we can set

$$L = \frac{m}{2}\sum_i \dot{x}_i^2 - V(x).$$

Then,

$$j = \sum_{i=1}^{3} \frac{\partial L}{\partial \dot{x}_i} Q[x_i] - L$$

$$= m \sum_i \dot{x}_i^2 - \left[\frac{m}{2} \sum_i \dot{x}_i^2 - V(x) \right]$$

$$= \frac{m}{2} \sum_i \dot{x}_i^2 + V(x).$$

The right hand side is the energy, and Noether's theorem states that (i.e. the principle of conservation of energy is a consequence of invariance under time translations).

More generally, if the Lagrangian does not depend explicitly on time, the quantity

$$\sum_{i=1}^{3} \frac{\partial L}{\partial \dot{x}_i} \dot{x}_i - L$$

(called the Hamiltonian) is conserved.

Example 2: Conservation of Center of Momentum

Still considering 1-dimensional time, let

$$S[\vec{x}] = \int \mathcal{L}[\vec{x}(t), \dot{\vec{x}}(t)] dt$$

$$= \int \left[\sum_{\alpha=1}^{N} \frac{m_\alpha}{2} (\dot{\vec{x}}_\alpha)^2 - \sum_{\alpha<\beta} V_{\alpha\beta}(\vec{x}_\beta - \vec{x}_\alpha) \right] dt$$

i.e. N Newtonian particles where the potential only depends pairwise upon the relative displacement.

For , let's consider the generator of Galilean transformations (i.e. a change in the frame of reference). In other words,

$$Q_i[x_\alpha^j(t)] = t\delta_i^j.$$

Note that

$$Q_i[\mathcal{L}] = \sum_\alpha m_\alpha \dot{x}_\alpha^i - \sum_{\alpha<\beta} \partial_i V_{\alpha\beta}(\vec{x}_\beta - \vec{x}_\alpha)(t-t)$$

$$= \sum_\alpha m_\alpha \dot{x}_\alpha^i.$$

This has the form of so we can set

$$\vec{f} = \sum_\alpha m_\alpha \vec{x}_\alpha.$$

Then,

$$\vec{j} = \sum_\alpha \left(\frac{\partial}{\partial \dot{\vec{x}}_\alpha} \mathcal{L} \right) \cdot \vec{Q}[\vec{x}_\alpha] - \vec{f}$$

$$= \sum_\alpha (m_\alpha \dot{\vec{x}}_\alpha t - m_\alpha \vec{x}_\alpha)$$

$$= \vec{P}t - M\vec{x}_{CM}$$

where $\vec{P}$ is the total momentum, M is the total mass and $\vec{x}_{CM}$ is the center of mass. Noether's theorem states:

$$\dot{\vec{j}} = 0 \Rightarrow \vec{P} - M\dot{\vec{x}}_{CM} = 0.$$

Example 3: Conformal Transformation

Both examples 1 and 2 are over a 1-dimensional manifold (time). An example involving spacetime is a conformal transformation of a massless real scalar field with a quartic potential in $(3 + 1)$-Minkowski spacetime.

$$\mathcal{S}[\phi] = \int \mathcal{L}[\phi(x), \partial_\mu \phi(x)] \mathrm{d}^4 x$$

$$= \int \left(\frac{1}{2} \partial^\mu \phi \partial_\mu \phi - \lambda \phi^4 \right) \mathrm{d}^4 x$$

For Q, consider the generator of a spacetime rescaling. In other words,

$$Q[\phi(x)] = x^\mu \partial_\mu \phi(x) + \phi(x).$$

The second term on the right hand side is due to the "conformal weight" of ϕ. Note that

$$Q[\mathcal{L}] = \partial^\mu \phi \left(\partial_\mu \phi + x^\nu \partial_\mu \partial_\nu \phi + \partial_\mu \phi \right) - 4\lambda \phi^3 \left(x^\mu \partial_\mu \phi + \phi \right).$$

This has the form of

$$\partial_\mu \left[\frac{1}{2} x^\mu \partial^\nu \phi \partial_\nu \phi - \lambda x^\mu \phi^4 \right] = \partial_\mu \left(x^\mu \mathcal{L} \right)$$

(where we have performed a change of dummy indices) so set

$$f^\mu = x^\mu \mathcal{L}.$$

Then,

$$j^{\mu} = \left[\frac{\partial}{\partial(\partial_{\mu}\phi)} \mathcal{L} \right] Q[\phi] - f^{\mu}$$

$$= \partial^{\mu}\phi \left(x^{\nu}\partial_{\nu}\phi + \phi \right) - x^{\mu} \left(\frac{1}{2}\partial^{\nu}\phi\partial_{\nu}\phi - \lambda\phi^{4} \right).$$

Noether's theorem states that $\partial_{\mu}j^{\mu} = 0$ (as one may explicitly check by substituting the Euler–Lagrange equations into the left hand side).

Note that if one tries to find the Ward–Takahashi analog of this equation, one runs into a problem because of anomalies.

Applications

Application of Noether's theorem allows physicists to gain powerful insights into any general theory in physics, by just analyzing the various transformations that would make the form of the laws involved invariant. For example:

- the invariance of physical systems with respect to spatial translation (in other words, that the laws of physics do not vary with locations in space) gives the law of conservation of linear momentum;

- invariance with respect to rotation gives the law of conservation of angular momentum;

- invariance with respect to time translation gives the well-known law of conservation of energy

In quantum field theory, the analog to Noether's theorem, the Ward–Takahashi identity, yields further conservation laws, such as the conservation of electric charge from the invariance with respect to a change in the phase factor of the complex field of the charged particle and the associated gauge of the electric potential and vector potential.

The Noether charge is also used in calculating the entropy of stationary black holes.

Parity (Physics)

In quantum mechanics, a parity transformation (also called parity inversion) is the flip in the sign of *one* spatial coordinate. In three dimensions, it is also often described by the simultaneous flip in the sign of all three spatial coordinates (a point reflection):

$$\mathbf{P}: \begin{pmatrix} x \\ y \\ z \end{pmatrix} \mapsto \begin{pmatrix} -x \\ -y \\ -z \end{pmatrix}.$$

It can also be thought of as a test for chirality of a physical phenomenon, in that a parity inversion transforms a phenomenon into its mirror image. A parity transformation on something achiral, on the other hand, can be viewed as an identity transformation. All fundamental interactions of elementary particles, with the exception of the weak interaction, are symmetric under parity. The weak interaction is chiral and thus provides a means for probing chirality in physics. In interactions that are symmetric under parity, such as electromagnetism in atomic and molecular physics, parity serves as a powerful controlling principle underlying quantum transitions.

A matrix representation of P (in any number of dimensions) has determinant equal to −1, and hence is distinct from a rotation, which has a determinant equal to 1. In a two-dimensional plane, a simultaneous flip of all coordinates in sign is *not* a parity transformation; it is the same as a 180°-rotation.

Simple Symmetry Relations

Under rotations, classical geometrical objects can be classified into scalars, vectors, and tensors of higher rank. In classical physics, physical configurations need to transform under representations of every symmetry group.

Quantum theory predicts that states in a Hilbert space do not need to transform under representations of the group of rotations, but only under projective representations. The word *projective* refers to the fact that if one projects out the phase of each state, where we recall that the overall phase of a quantum state is not an observable, then a projective representation reduces to an ordinary representation. All representations are also projective representations, but the converse is not true, therefore the projective representation condition on quantum states is weaker than the representation condition on classical states.

The projective representations of any group are isomorphic to the ordinary representations of a central extension of the group. For example, projective representations of the 3-dimensional rotation group, which is the special orthogonal group SO(3), are ordinary representations of the special unitary group SU(2) . Projective representations of the rotation group that are not representations are called spinors, and so quantum states may transform not only as tensors but also as spinors.

If one adds to this a classification by parity, these can be extended, for example, into notions of

- *scalars* ($P = 1$) and *pseudoscalars* ($P = -1$) which are rotationally invariant.

- *vectors* ($P = -1$) and *axial vectors* (also called *pseudovectors*) ($P = 1$) which both transform as vectors under rotation.

One can define reflections such as

$$V_x : \begin{pmatrix} x \\ y \\ z \end{pmatrix} \mapsto \begin{pmatrix} -x \\ y \\ z \end{pmatrix},$$

which also have negative determinant and form a valid parity transformation. Then, combining them with rotations (or successively performing x-, y-, and z-reflections) one can recover the particular parity transformation defined earlier. The first parity transformation given does not work in an even number of dimensions, though, because it results in a positive determinant. In odd number of dimensions only the latter example of a parity transformation (or any reflection of an odd number of coordinates) can be used.

Parity forms the abelian group Z_2 due to the relation $P^2 = 1$. All Abelian groups have only one-dimensional irreducible representations. For Z_2, there are two irreducible representations: one is even under parity ($P\varphi = \varphi$), the other is odd ($P\varphi = -\varphi$). These are useful in quantum mechanics. However, as is elaborated below, in quantum mechanics states need not transform under actual representations of parity but only under projective representations and so in principle a parity transformation may rotate a state by any phase.

Classical Mechanics

Newton's equation of motion $F = ma$ (if the mass is constant) equates two vectors, and hence is invariant under parity. The law of gravity also involves only vectors and is also, therefore, invariant under parity.

However, angular momentum L is an axial vector,

$$L = r \times p,$$

$$P(L) = (-r) \times (-p) = L.$$

In classical electrodynamics, the charge density ρ is a scalar, the electric field, E, and current j are vectors, but the magnetic field, H is an axial vector. However, Maxwell's equations are invariant under parity because the curl of an axial vector is a vector.

Effect of Spatial Inversion on Some Variables of Classical Physics

Even

Classical variables, predominantly scalar quantities, which do not change upon spatial inversion include:

$t,$, the time when an event occurs

m , the mass of a particle

E , the energy of the particle

P , power (rate of work done)

ρ, the electric charge density

V , the electric potential (voltage)

ρ , energy density of the electromagnetic field

L , the angular momentum of a particle (both orbital and spin) (axial vector)

B the magnetic field (axial vector)

H, the auxiliary magnetic field

M, the magnetization

T_{ij} Maxwell stress tensor.

All masses, charges, coupling constants, and other physical constants, except those associated with the weak force

Odd

Classical variables, predominantly vector quantities, which have their sign flipped by spatial inversion include:

h, the helicity

Φ, the magnetic flux

x, the position of a particle in three-space

v the velocity of a particle

a, the acceleration of the particle

p, the linear momentum of a particle

F, the force exerted on a particle

J, the electric current density

E, the electric field

D, the electric displacement field

P, the electric polarization

A, the electromagnetic vector potential

S, Poynting vector.

Quantum Mechanics

Possible Eigenvalues

In quantum mechanics, spacetime transformations act on quantum states. The parity transformation, P, is a unitary operator, in general acting on a state ψ as follows: $P\psi(r) = e^{i\varphi/2}\psi(-r)$.

One must then have $P^2\psi(r) = e^{i\varphi}\psi(r)$, since an overall phase is unobservable. The operator P^2, which reverses the parity of a state twice, leaves the spacetime invariant, and so is an internal symmetry which rotates its eigenstates by phases $e^{i\varphi}$. If P^2 is an element e^{iQ} of a continuous U(1) symmetry group of phase rotations, then $e^{-iQ/2}$ is part of this U(1) and so is also a symmetry. In particular, we can define $P' = Pe^{-iQ/2}$, which is also a symmetry, and so we can choose to call P' our parity operator, instead of P. Note that $P'^2 = 1$ and so P' has eigenvalues ± 1. However, when no such symmetry group exists, it may be that all parity transformations have some eigenvalues which are phases other than ± 1.

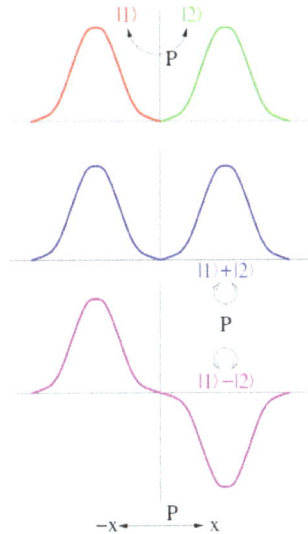

Two dimensional representations of parity are given by a pair of quantum states which go into each other under parity. However, this representation can always be reduced to linear combinations of states, each of which is either even or odd under parity. One says that all irreducible representations of parity are one-dimensional.

For electronic wavefunctions, even states are usually indicated by a subscript g for *gerade* (German: even) and odd states by a subscript u for *ungerade* (German: odd). For example, the lowest energy level of the hydrogen molecule ion (H_2^+) is labelled $1\sigma_g$ and the next-closest (higher) energy level is labelled $1\sigma_u$.

Consequences of Parity Symmetry

When parity generates the Abelian group Z_2, one can always take linear combinations of quantum states such that they are either even or odd under parity. Thus the parity of such states is ± 1. The parity of a multiparticle state is the product of the parities of each state; in other words parity is a multiplicative quantum number

In quantum mechanics, Hamiltonians are invariant (symmetric) under a parity transformation if P commutes with the Hamiltonian. In non-relativistic quantum mechanics, this happens for any potential which is scalar, i.e., $V = V(r)$, hence the potential is spherically symmetric. The following facts can be easily proven:

- If $|A\rangle$ and $|B\rangle$ have the same parity, then $\langle A| X |B\rangle = 0$ where X is the position operator.

- For a state $|L, L_z\rangle$ of orbital angular momentum L with z-axis projection L_z, $P|L, L_z\rangle = (-1)^L|L, L_z\rangle$.

- If $[H, P] = 0$, then atomic dipole transitions only occur between states of opposite parity.

- If $[H, P] = 0$, then a non-degenerate eigenstate of H is also an eigenstate of the parity operator; i.e., a non-degenerate eigenfunction of H is either invariant to P or is changed in sign by P.

Some of the non-degenerate eigenfunctions of H are unaffected (invariant) by parity P and the others will be merely reversed in sign when the Hamiltonian operator and the parity operator commute:

$$P \Psi = c \Psi,$$

where c is a constant, the eigenvalue of P,

$$P^2 \Psi = cP \Psi.$$

Quantum Field Theory

If we can show that the vacuum state is invariant under parity $(P|0\rangle = |0\rangle)$, the Hamiltonian is parity invariant $([H, P] = 0)$ and the quantization conditions remain unchanged under parity, then it follows that every state has good parity, and this parity is conserved in any reaction.

To show that quantum electrodynamics is invariant under parity, we have to prove that the action is invariant and the quantization is also invariant. For simplicity we will assume that canonical quantization is used; the vacuum state is then invariant under parity by construction. The invariance of the action follows from the classical invariance of Maxwell's equations. The invariance of the canonical quantization procedure can be worked out, and turns out to depend on the transformation of the annihilation operator:

$$Pa(p, \pm)P^+ = -a(-p, \pm)$$

where p denotes the momentum of a photon and ± refers to its polarization state. This is equivalent to the statement that the photon has odd intrinsic parity. Similarly all vector bosons can be shown to have odd intrinsic parity, and all axial-vectors to have even intrinsic parity.

There is a straightforward extension of these arguments to scalar field theories which shows that scalars have even parity, since

$$Pa(p)P^+ = a(-p).$$

This is true even for a complex scalar field.

With fermions, there is a slight complication because there is more than one spin group.

Parity in the Standard Model

Fixing the Global Symmetries

In the Standard Model of fundamental interactions there are precisely three global internal U(1) symmetry groups available, with charges equal to the baryon number B, the lepton number L and the electric charge Q. The product of the parity operator with any combination of these rotations is another parity operator. It is conventional to choose one specific combination of these rotations to define a standard parity operator, and other parity operators are related to the standard one by internal rotations. One way to fix a standard parity operator is to assign the parities of three particles with linearly independent charges B, L and Q. In general one assigns the parity of the most common massive particles, the proton, the neutron and the electron, to be +1.

Steven Weinberg has shown that if $P^2 = (-1)^F$, where F is the fermion number operator, then, since the fermion number is the sum of the lepton number plus the baryon number, $F = B + L$, for all particles in the Standard Model and since lepton number and baryon number are charges Q of continuous symmetries e^{iQ}, it is possible to redefine the parity operator so that $P^2 = 1$. However, if there exist Majorana neutrinos, which experimentalists today believe is possible, their fermion number is equal to one because they are neutrinos while their baryon and lepton numbers are zero because they are Majorana, and so $(-1)^F$ would not be embedded in a continuous symmetry group. Thus Majorana neutrinos would have parity $\pm i$.

Parity of the Pion

In 1954, a paper by William Chinowsky and Jack Steinberger demonstrated that the pion has negative parity. They studied the decay of an "atom" made from a deuteron ($^2_1$H+) and a negatively charged pion ($\pi-$) in a state with zero orbital angular momentum $L = 0$ into two neutrons (n).

Neutrons are fermions and so obey Fermi–Dirac statistics, which implies that the final state is antisymmetric. Using the fact that the deuteron has spin one and the pion spin zero together with the antisymmetry of the final state they concluded that the two neutrons must have orbital angular momentum $L = 1$. The total parity is the product of the intrinsic parities of the particles and the extrinsic parity of the spherical harmonic function $(-1)^L$. Since the orbital momentum changes from zero to one in this process, if the process is to conserve the total parity then the products of the intrinsic parities of the initial and final particles must have opposite sign. A deuteron nucleus is made from a proton and a neutron, and so using the aforementioned convention that protons and neutrons have intrinsic parities equal to +1 they argued that the parity of the pion is equal to minus the product of the parities of the two neutrons divided by that of the proton and neutron in the deuteron, $(-1)(1)^2/(1)^2$, which is equal to minus one. Thus they concluded that the pion is a pseudoscalar particle.

Parity Violation

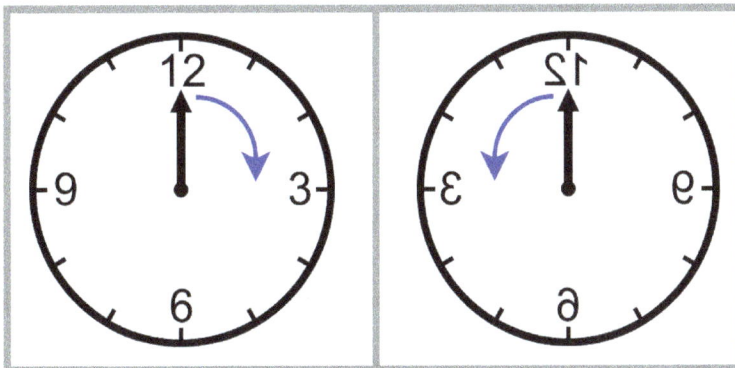

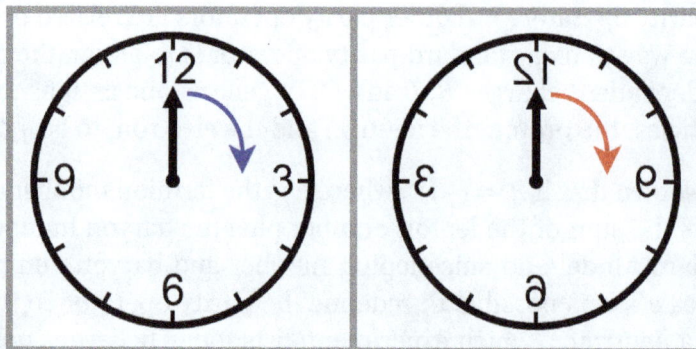

Top: P-symmetry: A clock built like its mirrored image will behave like the mirrored image of the original clock.
Bottom: P-asymmetry: A clock built like its mirrored image will *not* behave like the mirrored image of the original clock.

Although parity is conserved in electromagnetism, strong interactions and gravity, it turns out to be violated in weak interactions. The Standard Model incorporates parity violation by expressing the weak interaction as a chiral gauge interaction. Only the left-handed components of particles and right-handed components of antiparticles participate in weak interactions in the Standard Model. This implies that parity is not a symmetry of our universe, unless a hidden mirror sector exists in which parity is violated in the opposite way.

By the mid-20th Century, it had been suggested by several scientists that parity might not be conserved (in different contexts), but without solid evidence these suggestions were not considered important. Then, in 1956, a careful review and analysis by theoretical physicists Tsung Dao Lee and Chen Ning Yang went further, showing that while parity conservation had been verified in decays by the strong or electromagnetic interactions, it was untested in the weak interaction. They proposed several possible direct experimental tests. They were mostly ignored, but Lee was able to convince his Columbia colleague Chien-Shiung Wu to try it. She needed special cryogenic facilities and expertise, so the experiment was done at the National Bureau of Standards.

In 1957 C. S. Wu, E. Ambler, R. W. Hayward, D. D. Hoppes, and R. P. Hudson found a clear violation of parity conservation in the beta decay of cobalt-60. As the experiment was winding down, with double-checking in progress, Wu informed Lee and Yang of their positive results, and saying the results need further examination, she asked them not to publicize the results first. However, Lee revealed the results to his Columbia colleagues on 4 January 1957 at a "Friday Lunch" gathering of the Physics Department of Columbia. Three of them, R. L. Garwin, Leon Lederman, and R. Weinrich modified an existing cyclotron experiment, and they immediately verified the parity violation. They delayed publication of their results until after Wu's group was ready, and the two papers appeared back to back in the same physics journal.

After the fact, it was noted that an obscure 1928 experiment had in effect reported parity violation in weak decays, but since the appropriate concepts had not yet been developed, those results had no impact. The discovery of parity violation immediately explained the outstanding $\tau-\theta$ puzzle in the physics of kaons.

In 2010, it was reported that physicists working with the Relativistic Heavy Ion Collider (RHIC) had created a short-lived parity symmetry-breaking bubble in quark-gluon plasmas. An experiment conducted by several physicists including Yale's Jack Sandweiss as part of the STAR collaboration, suggested that parity may also be violated in the strong interaction.

Intrinsic Parity of Hadrons

To every particle one can assign an intrinsic parity as long as nature preserves parity. Although weak interactions do not, one can still assign a parity to any hadron by examining the strong interaction reaction that produces it, or through decays not involving the weak interaction, such as rho meson decay to pions.

References

- David, Curtin (August 2011). "MODEL BUILDING AND COLLIDER PHYSICS ABOVE THE WEAK SCAE" (PDF) (COLLIDER PHYSICS ABOVE THE WEAK SCALE).

- R. Haag, J. T. Lopuszanski and M. Sohnius, "All Possible Generators Of Supersymmetries Of The S Matrix", Nucl. Phys. B 88 (1975) 257

- H. Miyazawa (1966). "Baryon Number Changing Currents". Prog. Theor. Phys. 36 (6): 1266–1276. Bibcode:1966PThPh..36.1266M. doi:10.1143/PTP.36.1266.

- H. Miyazawa (1968). "Spinor Currents and Symmetries of Baryons and Mesons". Phys. Rev. 170 (5): 1586–1590. Bibcode:1968PhRv..170.1586M. doi:10.1103/PhysRev.170.1586.

- Gervais, J. -L.; Sakita, B. (1971). "Field theory interpretation of supergauges in dual models". Nuclear Physics B. 34 (2): 632–639. Bibcode:1971NuPhB..34..632G. doi:10.1016/0550-3213(71)90351-8.

- D.V. Volkov, V.P. Akulov, Pisma Zh.Eksp.Teor.Fiz. 16 (1972) 621; Phys.Lett. B46 (1973) 109; V.P. Akulov, D.V. Volkov, Teor.Mat.Fiz. 18 (1974) 39

- Ramond, P. (1971). "Dual Theory for Free Fermions". Physical Review D. 3 (10): 2415–2418. Bibcode:1971PhRvD...3.2415R. doi:10.1103/PhysRevD.3.2415.

- Wess, J.; Zumino, B. (1974). "Supergauge transformations in four dimensions". Nuclear Physics B. 70: 39–50. Bibcode:1974NuPhB..70...39W. doi:10.1016/0550-3213(74)90355-1.

- Iachello, F. (1980). "Dynamical Supersymmetries in Nuclei". Physical Review Letters. 44 (12): 772–775. Bibcode:1980PhRvL..44..772I. doi:10.1103/PhysRevLett.44.772.

- Gordon L. Kane, The Dawn of Physics Beyond the Standard Model, Scientific American, June 2003, page 60 and The frontiers of physics, special edition, Vol 15, #3, page 8

Quantum Field Theories

Quantum field theories include theories such as common integrals in quantum field theory, first quantization and second quantization. Quantum field theory is the theoretical framework; this framework is used in constructing quantum mechanical models. The topics discussed in the chapter are of great importance to broaden the existing knowledge on quantum field theories.

Quantum Field Theory

In theoretical physics, quantum field theory (QFT) is the theoretical framework for constructing quantum mechanical models of subatomic particles in particle physics and quasiparticles in condensed matter physics. QFT treats particles as excited states of the underlying physical field, so these are called field quanta.

In quantum field theory, quantum mechanical interactions among particles are described by interaction terms among the corresponding underlying quantum fields. These interactions are conveniently visualized by Feynman diagrams, which are a formal tool of relativistically covariant perturbation theory, serving to evaluate particle processes.

History

Even though QFT is an unavoidable consequence of the reconciliation of quantum mechanics with special relativity (Weinberg (2005)), historically, it emerged in the 1920s with the quantization of the electromagnetic field (the quantization being based on an analogy of the eigenmode expansion of a vibrating string with fixed endpoints).

Early Development

Max Born (1882–1970), one of the founders of quantum field theory.

He is also known for the Born rule that introduced the probabilistic interpretation in quantum mechanics. He received the 1954 Nobel Prize in Physics together with Walther Bothe.

The first achievement of quantum field theory, namely quantum electrodynamics (QED), is "still the paradigmatic example of a successful quantum field theory" (Weinberg (2005)). Ordinarily, quantum mechanics (QM) cannot give an account of photons which constitute the prime case of relativistic 'particles'. Since photons have rest mass zero, and correspondingly travel in the vacuum at the speed c, a non-relativistic theory such as ordinary QM cannot give even an approximate description. Photons are implicit in the emission and absorption processes which have to be postulated; for instance, when one of an atom's electrons makes a transition between energy levels. The formalism of QFT is needed for an explicit description of photons. In fact most topics in the early development of quantum theory (the so-called old quantum theory, 1900–25) were related to the interaction of radiation and matter and thus should be treated by quantum field theoretical methods. However, quantum mechanics as formulated by Dirac, Heisenberg, and Schrödinger in 1926–27 started from atomic spectra and did not focus much on problems of radiation.

As soon as the conceptual framework of quantum mechanics was developed, a small group of theoreticians tried to extend quantum methods to electromagnetic fields. A good example is the famous paper by Born, Jordan & Heisenberg (1926). (P. Jordan was especially acquainted with the literature on light quanta and made seminal contributions to QFT.) The basic idea was that in QFT the electromagnetic field should be represented by matrices in the same way that position and momentum were represented in QM by matrices (matrix mechanics oscillator operators). The ideas of QM were thus extended to systems having an infinite number of degrees of freedom, so an infinite array of quantum oscillators.

The inception of QFT is usually considered to be Dirac's famous 1927 paper on "The quantum theory of the emission and absorption of radiation". Here Dirac coined the name "quantum electrodynamics" (QED) for the part of QFT that was developed first. Dirac supplied a systematic procedure for transferring the characteristic quantum phenomenon of discreteness of physical quantities from the quantum-mechanical treatment of particles to a corresponding treatment of fields. Employing the theory of the quantum harmonic oscillator, Dirac gave a theoretical description of how photons appear in the quantization of the electromagnetic radiation field. Later, Dirac's procedure became a model for the quantization of other fields as well. These first approaches to QFT were further developed during the following three years. P. Jordan introduced creation and annihilation operators for fields obeying Fermi–Dirac statistics. These differ from the corresponding operators for Bose–Einstein statistics in that the former satisfy *anti-commutation relations* while the latter satisfy commutation relations.

The methods of QFT could be applied to derive equations resulting from the quantum-mechanical (field-like) treatment of particles, e.g. the Dirac equation, the Klein–Gordon equation and the Maxwell equations. Schweber points out that the idea and procedure of second quantization goes back to Jordan, in a number of papers from 1927, while the expression itself was coined by Dirac. Some difficult problems concerning commutation relations, statistics, and Lorentz invariance were eventually solved. The first comprehensive account of a general theory of quantum fields, in particular, the method of canonical quantization, was presented by Heisenberg & Pauli in 1929. Whereas Jordan's second quantization procedure applied to the coefficients of the normal modes of the field, Heisenberg & Pauli started with the fields themselves and subjected them to the canonical procedure. Heisenberg and Pauli thus established the basic structure of QFT as presented in modern introductions to QFT. Fermi and Dirac, as well as Fock and Podolsky, presented different formulations which played a heuristic role in the following years.

Quantum electrodynamics rests on two pillars, see e.g., the short and lucid "Historical Introduction" of Scharf (2014). The first pillar is the quantization of the electromagnetic field, i.e., it is about photons as the quantized excitations or 'quanta' of the electromagnetic field. This procedure will be described in some more detail in the section on the particle interpretation. As Weinberg points out the "photon is the only particle that was known as a field before it was detected as a particle" so that it is natural that QED began with the analysis of the radiation field. The second pillar of QED consists of the relativistic theory of the electron, centered on the Dirac equation.

The Problem of Infinities

The Emergence of Infinities

Pascual Jordan (1902–1980), doctoral student of Max Born, was a pioneer in quantum field theory, coauthoring a number of seminal papers with Born and Heisenberg.

Jordan algebras were introduced by him to formalize the notion of an algebra of observables in quantum mechanics. He was awarded the Max Planck medal 1954.

Quantum field theory started with a theoretical framework that was built in analogy to quantum mechanics. Although there was no unique and fully developed theory, quantum field theoretical tools could be applied to concrete processes. Examples are the scattering of radiation by free electrons, Compton scattering, the collision between relativistic electrons or the production of electron-positron pairs by photons. Calculations to the first order of approximation were quite successful, but most people working in the field thought that QFT still had to undergo a major change. On the one side, some calculations of effects for cosmic rays clearly differed from measurements. On the other side and, from a theoretical point of view more threatening, calculations of higher orders of the perturbation series led to infinite results. The self-energy of the electron as well as vacuum fluctuations of the electromagnetic field seemed to be infinite. The perturbation expansions did not converge to a finite sum and even most individual terms were divergent.

The various forms of infinities suggested that the divergences were more than failures of specific calculations. Many physicists tried to avoid the divergences by formal tricks (truncating the integrals at some value of momentum, or even ignoring infinite terms) but such rules were not reliable, violated the requirements of relativity and were not considered as satisfactory. Others came up with the first ideas for coping with infinities by a redefinition of the parameters of the theory and using a measured finite value, for example of the charge of the electron, instead of the infinite 'bare' value. This process is called renormalization.

From the point of view of the philosophy of science, it is remarkable that these divergences did not give enough reason to discard the theory. The years from 1930 to the beginning of World War II were characterized by a variety of attitudes towards QFT. Some physicists tried to circumvent the infinities by more-or-less arbitrary prescriptions, others worked on transformations and improvements of the theoretical framework. Most of the theoreticians believed that QED would break down at high energies. There was also a considerable number of proposals in favor of alternative approaches. These proposals included changes in the basic concepts e.g. negative probabilities and interactions at a distance instead of a field theoretical approach, and a methodological change to phenomenological methods that focusses on relations between observable quantities without an analysis of the microphysical details of the interaction, the so-called S-matrix theory where the basic elements are amplitudes for various scattering processes.

Despite the feeling that QFT was imperfect and lacking rigor, its methods were extended to new areas of applications. In 1933 Fermi's theory of the beta decay started with conceptions describing the emission and absorption of photons, transferred them to beta radiation and analyzed the creation and annihilation of electrons and neutrinos described by the weak interaction. Further applications of QFT outside of quantum electrodynamics succeeded in nuclear physics with the strong interaction. In 1934 Pauli & Weisskopf showed that a new type of field (the scalar field), described by the Klein–Gordon equation, could be quantized. This is another example of second quantization. This new theory for matter fields could be applied a decade later when new particles, pions, were detected.

The Taming of Infinities

Werner Heisenberg (1901–1976), doctoral student of Arnold Sommerfeld, was one of the founding fathers of quantum mechanics and QFT.

In particular, he introduced the version of quantum mechanics known as matrix mechanics, but is now more known for the Heisenberg uncertainty relations. He was awarded the Nobel prize in physics 1932 together with Erwin Schrödinger and Paul Dirac.

After the end of World War II more reliable and effective methods for dealing with infinities in QFT were developed, namely coherent and systematic rules for performing relativistic field theoretical calculations, and a general renormalization theory. On three famous conferences, the Shelter Island Conference 1947, the Pocono Conference 1948, and the 1949 Oldstone Conference, developments in theoretical physics were confronted with relevant new experimental results. In the late forties, there were two different ways to address the problem of divergences. One of these was discovered by Richard Feynman, the other one (based on an operator formalism) by Julian Schwinger and, independently, by Sin-Itiro Tomonaga.

In 1949, Freeman Dyson showed that the two approaches are in fact equivalent and fit into an elegant field-theoretic framework. Thus, Freeman Dyson, Feynman, Schwinger, and Tomonaga became the inventors of renormalization theory. The most spectacular successes of renormalization theory were the calculations of the anomalous magnetic moment of the electron and the Lamb shift in the spectrum of hydrogen. These successes were so outstanding because the theoretical results were in better agreement with high-precision experiments than anything in physics encountered before. Nevertheless, mathematical problems lingered on and prompted a search for rigorous formulations (discussed below).

The rationale behind renormalization is to avoid divergences that appear in physical predictions by shifting them into a part of the theory where they do not influence empirical statements. Dyson could show that a rescaling of charge and mass ('renormalization') is sufficient to remove all divergences in QED consistently, to all orders of perturbation theory. A QFT is called renormalizable if all infinities can be absorbed into a redefinition of a *finite number* of coupling constants and masses. A consequence is that the physical charge and mass of the electron must be measured and cannot be computed from first principles.

Perturbation theory yields well-defined predictions only in renormalizable quantum field theories; luckily, QED, the first fully developed QFT, belonged to this class of renormalizable theories. There are various technical procedures to renormalize a theory. One way is to cut off the integrals in the calculations at a certain value Λ of the momentum which is large but finite. This cut-off procedure is successful if, after taking the limit $\Lambda \to \infty$, the resulting quantities are independent of Λ.

His 1945 PhD thesis developed the path integral formulation of ordinary quantum mechanics. This was later generalized to field theory.

Feynman's formulation of QED is of special interest from a philosophical point of view. His so-called space-time approach is visualized by the celebrated Feynman diagrams that look like depicting paths of particles. Feynman's method of calculating scattering amplitudes is based on the functional integral formulation of field theory. A set of graphical rules can be derived so that the probability of a specific scattering process can be calculated by drawing a diagram of that process and then using that diagram to write down the precise mathematical expressions for calculating its amplitude in relativistically covariant perturbation theory.

Richard Feynman (1918–1988)

The diagrams provide an effective way to organize and visualize the various terms in the perturbation series, and they naturally account for the flow of electrons and photons during the scattering process. External lines in the diagrams represent incoming and outgoing particles, internal lines are connected with virtual particles and vertices with interactions. Each of these graphical elements is associated with mathematical expressions that contribute to the amplitude of the respective process. The diagrams are part of Feynman's very efficient and elegant algorithm for computing the probability of scattering processes.

The idea of particles traveling from one point to another was heuristically useful in constructing the theory. This heuristics, based on Huygen's principle, is useful for concrete calculations and actually give the correct particle propagators as derived more rigorously. Nevertheless, an analysis of the theoretical justification of the space-time approach shows that its success does not imply that particle paths need be taken seriously. General arguments against a particle interpretation of QFT clearly exclude that the diagrams represent actual paths of particles in the interaction area. Feynman himself was not particularly interested in ontological questions.

The Golden Age: Gauge Theory and the Standard Model

Chen-Ning Yang (b.1922), co-inventor of nonabelian gauge field theories.

Murray Gell-Mann (b. 1929) articulator and pioneer of group symmetry in QFT

In 1933, Enrico Fermi had already established that the creation, annihilation and transmutation of particles in the weak interaction beta decay could best be described in QFT, specifically his quartic fermion interaction. As a result, field theory had become a prospective tool for other particle interactions. In the beginning of the 1950s, QED had become a reliable theory which no longer counted as preliminary. However, it took two decades from writing down the first equations until QFT could be applied successfully to important physical problems in a systematic way.

The theories explored relied on—indeed, were virtually fully specified by—a rich variety of symmetries pioneered and articulated by Murray Gell-Mann. The new developments made it possible to apply QFT to new particles and new interactions and fully explain their structure.

In the following decades, QFT was extended to well-describe not only the electromagnetic force but also weak and strong interaction so that new Lagrangians were found which contain new classes of particles or quantum fields. The search still continues for a more comprehensive theory of matter and energy, a *unified theory of all interactions*.

Yoichiro Nambu (1921–2015), co-discoverer of field theoretic spontaneous symmetry breaking.

The new focus on symmetry led to the triumph of non-Abelian gauge theories (the development of such theories was pioneered in 1954 with the work of Yang and Mills) and spontaneous symmetry breaking (by Yoichiro Nambu). Today, there are reliable theories of the strong, weak, and electromagnetic interactions of elementary particles which have an analogous structure to QED: They are the dominant framework of particle physics.

A combined renormalizable theory associated with the gauge group SU(3) × SU(2) × U(1) is dubbed the *standard model of elementary particle physics* (even though it is a full theory, and not just a model) and was assembled by Sheldon Glashow, Steven Weinberg and Abdul Salam in 1968, and Frank Wilczek, David Gross and David Politzer in 1973, on the basis of conceptual breakthroughs by Peter Higgs, François Englert, Robert Brout, Martin Veltman, and Gerard 't Hooft.

Gerard 't Hooft (b.1946) proved gauge field theories are renormalizable.

According to the standard model, there are, on the one hand, six types of leptons (e.g. the electron and its neutrino) and six types of quarks, where the members of both groups are all fermions with spin 1/2. On the other hand, there are spin 1 particles (thus bosons) that mediate the interaction between elementary particles and the fundamental forces, namely the photon for electromagnetic interaction, two W and one Z-boson for weak interaction, and the gluons for strong interaction. The linchpin of the symmetry breaking mechanism of the theory is the spin 0 Higgs boson, discovered 40 years after its prediction.

Renormalization Group

Ken Wilson (1936–2013), Nobel laureate. He constructed the over-arching picture of the renormalization group which underlies the workings of all QFTs across all scales.

Parallel breakthroughs in the understanding of phase transitions in condensed matter physics led to novel insights based on the renormalization group. They involved the work of Leo Kadanoff (1966) and Michael Fisher (1973), which led to the seminal reformulation of quantum field theory by Ken Wilson in 1975. This reformulation provided insights into the evolution of effective field theories with scale, which classified all field theories, renormalizable or not (cf. subsequent section). The remarkable conclusion is that, in general, most observables are *"irrelevant"*, i.e., the macroscopic physics is *dominated by only a few observables* in most systems.

During the same period, Kadanoff (1969) introduced an operator algebra formalism for the two-dimensional Ising model, a widely studied mathematical model of ferromagnetism in statistical physics. This development suggested that quantum field theory describes its scaling limit. Later, there developed the idea that a finite number of generating operators could represent all the correlation functions of the Ising model.

Conformal Field Theory

The existence of a much stronger symmetry for the scaling limit of two-dimensional critical systems was suggested by Alexander Belavin, Alexander Polyakov and Alexander Zamolodchikov in 1984, which eventually led to the development of conformal field theory, a special case of quantum field theory, which is presently utilized in different areas of particle physics and condensed matter physics.

Historiography

The first chapter in Weinberg (2005) is a very good short description of the earlier history of QFT. Detailed accounts of the historical development of QFT can be found, e.g., in Darrigol 1986, Schweber (1994) and Cao 1997a. Various historical and conceptual studies of the standard model are gathered in Hoddeson et al. 1997 and of renormalization theory in Brown 1993.

Varieties of Approaches

There is currently no complete quantum theory of the remaining fundamental force, gravity. Many of the proposed theories to describe gravity as a QFT postulate the existence of a graviton particle that mediates the gravitational force. Presumably, the as yet unknown correct quantum field-theoretic treatment of the gravitational field will behave like Einstein's general theory of relativity in the low-energy limit. Quantum field theory of the fundamental forces itself has been postulated to be the low-energy effective field theory limit of a more fundamental theory such as superstring theory.

Most theories in standard particle physics are formulated as *relativistic quantum field theories*, such as QED, QCD, and the Standard Model. QED, the quantum field-theoretic description of the electromagnetic field, approximately reproduces Maxwell's theory of electrodynamics in the low-energy limit, with small non-linear corrections to the Maxwell equations required due to virtual electron–positron pairs.

In the perturbative approach to quantum field theory, the full field interaction terms are approximated as a perturbative expansion in the number of particles involved. Each term in the expansion can be thought of as forces between particles being mediated by other particles. In QED, the electromagnetic force between two electrons is caused by an exchange of photons. Similarly, intermediate vector bosons mediate the weak force and gluons mediate the strong force in QCD. The notion of a force-mediating particle comes from perturbation theory, and does not make sense in the context of non-perturbative approaches to QFT, such as with bound states.

Definition

Quantum electrodynamics (QED) has one electron field and one photon field; quantum chromodynamics (QCD) has one field for each type of quark; and, in condensed matter, there is an atomic displacement field that gives rise to phonon particles. Edward Witten describes QFT as "by far" the most difficult theory in modern physics – "so difficult that nobody fully believed it for 25 years."

Dynamics

Ordinary quantum mechanical systems have a fixed number of particles, with each particle having a finite number of degrees of freedom. In contrast, the excited states of a quantum field can represent any number of particles. This makes quantum field theories especially useful for describing systems where the particle count/number may change over time, a crucial feature of relativistic dynamics. A QFT is thus an organized infinite array of oscillators.

States

QFT interaction terms are similar in spirit to those between charges with electric and magnetic fields in Maxwell's equations. However, unlike the classical fields of Maxwell's theory, fields in QFT generally exist in quantum superpositions of states and are subject to the laws of quantum mechanics.

Because the fields are continuous quantities over space, there exist excited states with arbitrarily large numbers of particles in them, providing QFT systems with effectively an infinite number of degrees of freedom. Infinite degrees of freedom can easily lead to divergences of calculated quantities (e.g., the quantities become infinite). Techniques such as renormalization of QFT parameters or discretization of spacetime, as in lattice QCD, are often used to avoid such infinities so as to yield physically plausible results.

Fields and Radiation

The gravitational field and the electromagnetic field are the only two fundamental fields in nature that have infinite range and a corresponding classical low-energy limit, which greatly diminishes and hides their "particle-like" excitations. Albert Einstein in 1905, attributed "particle-like" and discrete exchanges of momenta and energy, characteristic of "field quanta", to the electromagnetic field. Originally, his principal motivation was to explain the thermodynamics of radiation. Although the photoelectric effect and Compton scattering strongly suggest the existence of the photon, it might alternatively be explained by a mere quantization of emission; more definitive evidence of the quantum nature of radiation is now taken up into modern quantum optics as in the antibunching effect.

Principles

Classical and Quantum Fields

A classical field is a function defined over some region of space and time. Two physical phenomena which are described by classical fields are Newtonian gravitation, described by Newtonian gravitational field g(x, t), and classical electromagnetism, described by the electric and magnetic fields E(x, t) and B(x, t). Because such fields can in principle take on distinct values at each point in space, they are said to have infinite degrees of freedom.

Classical field theory does not, however, account for the quantum-mechanical aspects of such physical phenomena. For instance, it is known from quantum mechanics that certain aspects of electromagnetism involve discrete particles—photons—rather than continuous fields. The business of *quantum field theory* is to write down a field that is, like a classical field, a function defined over space and time, but which also accommodates the observations of quantum mechanics. This is a *quantum field*.

To write down such a quantum field, one promotes the infinity of classical oscillators representing the modes of the classical fields to quantum harmonic oscillators. They thus become operator-valued functions (actually, distributions). (In its most general formulation, quantum mechanics is a theory of abstract operators (observables) acting on an abstract state space (Hilbert space), where the observables represent physically observable quantities and the state space represents the possible states of the system under study. For instance, the fundamental observables associated with the motion of a single quantum mechanical particle are the position and momentum operators and . Field theory, by sharp contrast, treats x as a label, an index of the field rather than as an operator.)

There are two common ways of handling a quantum field: canonical quantization and the path integral formalism. The latter of these is pursued in this article.

Lagrangian Formalism

Quantum field theory relies on the Lagrangian formalism from classical field theory. This formalism is analogous to the Lagrangian formalism used in classical mechanics to solve for the motion of a particle under the influence of a field. In classical field theory, one writes down a Lagrangian density, , involving a field, $\varphi(x,t)$, and possibly its first derivatives ($\partial\varphi/\partial t$ and $\nabla\varphi$), and then applies a field-theoretic form of the Euler–Lagrange equation. Writing coordinates $(t, x) = (x^0, x^1, x^2, x^3) = x^\mu$, this form of the Euler–Lagrange equation is

$$\frac{\partial}{\partial x^\mu}\left[\frac{\partial\mathcal{L}}{\partial(\partial\varphi/\partial x^\mu)}\right] - \frac{\partial\mathcal{L}}{\partial\varphi} = 0,$$

where a sum over μ is performed according to the rules of Einstein notation.

By solving this equation, one arrives at the "equations of motion" of the field. For example, if one begins with the Lagrangian density

$$\mathcal{L}(\varphi, \nabla\varphi) = -\rho(t, \mathbf{x})\varphi(t, \mathbf{x}) - \frac{1}{8\pi G}|\nabla\varphi|^2,$$

and then applies the Euler–Lagrange equation, one obtains the equation of motion

$$4\pi G\rho(t,\mathbf{x}) = \nabla^2\varphi.$$

This equation is Newton's law of universal gravitation, expressed in differential form in terms of the gravitational potential $\varphi(t, \mathbf{x})$ and the mass density $\rho(t, \mathbf{x})$. Despite the nomenclature, the "field" under study is the gravitational potential, φ, rather than the gravitational field, g. Similarly, when classical field theory is used to study electromagnetism, the "field" of interest is the electromagnetic four-potential $(V/c, \mathbf{A})$, rather than the electric and magnetic fields E and B.

Quantum field theory uses this same Lagrangian procedure to determine the equations of motion for quantum fields. These equations of motion are then supplemented by commutation relations derived from the canonical quantization procedure described below, thereby incorporating quantum mechanical effects into the behavior of the field.

Single- and Many-Particle Quantum Mechanics

In non-relativistic quantum mechanics, a particle (such as an electron or proton) is described by a complex wavefunction, $\psi(x, t)$, whose time-evolution is governed by the Schrödinger equation:

$$-\frac{\hbar^2}{2m}\frac{\partial^2}{\partial x^2}\psi(x,t) + V(x)\psi(x,t) = i\hbar\frac{\partial}{\partial t}\psi(x,t).$$

Here m is the particle's mass and $V(x)$ is the applied potential. Physical information about the behavior of the particle is extracted from the wavefunction by constructing expected values for various quantities; for example, the expected value of the particle's position is given by integrating $\psi^*(x)\,x\,\psi(x)$ over all space, and the expected value of the particle's momentum is found by integrating $-i\hbar\psi^*(x)\mathrm{d}\psi/\mathrm{d}x$. The quantity $\psi^*(x)\psi(x)$ is itself in the Copenhagen interpretation of quantum mechanics interpreted as a probability density function. This treatment of quantum mechanics, where a particle's wavefunction evolves against a classical background potential $V(x)$, is sometimes called *first quantization*.

This description of quantum mechanics can be extended to describe the behavior of multiple particles, so long as the number and the type of particles remain fixed. The particles are described by a wavefunction $\psi(x_1, x_2, ..., x_N, t)$, which is governed by an extended version of the Schrödinger equation.

Often one is interested in the case where N particles are all of the same type (for example, the 18 electrons orbiting a neutral argon nucleus). As described in the article on identical particles, this implies that the state of the entire system must be either symmetric (bosons) or antisymmetric (fermions) when the coordinates of its constituent particles are exchanged. This is achieved by using a Slater determinant as the wavefunction of a fermionic system (and a Slater permanent for a bosonic system), which is equivalent to an element of the symmetric or antisymmetric subspace of a tensor product.

For example, the general quantum state of a system of N bosons is written as

$$|\phi_1 \cdots \phi_N\rangle = \sqrt{\frac{\prod_j N_j!}{N!}} \sum_{p \in S_N} |\phi_{p(1)}\rangle \otimes \cdots \otimes |\phi_{p(N)}\rangle,$$

where $|\phi_i\rangle$ are the single-particle states, N_j is the number of particles occupying state j, and the sum is taken over all possible permutations p acting on N elements. In general, this is a sum of $N!$

(N factorial) distinct terms. $\sqrt{\dfrac{\prod_j N_j!}{N!}}$ is a normalizing factor.

There are several shortcomings to the above description of quantum mechanics, which are addressed by quantum field theory. First, it is unclear how to extend quantum mechanics to include the effects of special relativity. Attempted replacements for the Schrödinger equation, such as the Klein–Gordon equation or the Dirac equation, have many unsatisfactory qualities; for instance, they possess energy eigenvalues that extend to $-\infty$, so that there seems to be no easy definition of a ground state. It turns out that such inconsistencies arise from relativistic wavefunctions not having a well-defined probabilistic interpretation in position space, as probability conservation is not a relativistically covariant concept. The second shortcoming, related to the first, is that in quantum mechanics there is no mechanism to describe particle creation and annihilation; this is crucial for describing phenomena such as pair production, which result from the conversion between mass and energy according to the relativistic relation $E = mc^2$.

Second Quantization

In this section, we will describe a method for constructing a quantum field theory called second quantization. This basically involves choosing a way to index the quantum mechanical degrees of freedom in the space of multiple identical-particle states. It is based on the Hamiltonian formulation of quantum mechanics.

Several other approaches exist, such as the Feynman path integral, which uses a Lagrangian formulation.

Bosons

For simplicity, we will first discuss second quantization for bosons, which form perfectly symmetric quantum states. Let us denote the mutually orthogonal single-particle states which are possible in the system by $|\phi_1\rangle, |\phi_2\rangle, |\phi_3\rangle$, and so on. For example, the 3-particle state with one particle in state $|\phi_1\rangle$ and two in state $|\phi_2\rangle$ is

$$\frac{1}{\sqrt{3}}\left[|\phi_1\rangle|\phi_2\rangle|\phi_2\rangle + |\phi_2\rangle|\phi_1\rangle|\phi_2\rangle + |\phi_2\rangle|\phi_2\rangle|\phi_1\rangle\right].$$

The first step in second quantization is to express such quantum states in terms of occupation numbers, by listing the number of particles occupying each of the single-particle states $|\phi_1\rangle, |\phi_2\rangle$, etc. This is simply another way of labelling the states. For instance, the above 3-particle state is denoted as

$$|1,2,0,0,0,\ldots\rangle.$$

An N-particle state belongs to a space of states describing systems of N particles. The next step is to combine the individual N-particle state spaces into an extended state space, known as Fock space, which can describe systems of any number of particles. This is composed of the state space of a system with no particles (the so-called vacuum state, written as $|0\rangle$), plus the state space of a 1-particle system, plus the state space of a 2-particle system, and so forth. States describing a definite number of particles are known as Fock states: a general element of Fock space will be a linear combination of Fock states. There is a one-to-one correspondence between the occupation number representation and valid boson states in the Fock space.

At this point, the quantum mechanical system has become a quantum field in the sense we described above. The field's elementary degrees of freedom are the occupation numbers, and each occupation number is indexed by a number j indicating which of the single-particle states $|\phi_1\rangle, |\phi_2\rangle, \ldots, |\phi_j\rangle, \ldots$ it refers to:

$$| N_1, N_2, N_3, \ldots, N_j, \ldots \rangle.$$

The properties of this quantum field can be explored by defining creation and annihilation operators, which add and subtract particles. They are analogous to ladder operators in the quantum harmonic oscillator problem, which added and subtracted energy quanta. However, these operators literally create and annihilate particles of a given quantum state. The bosonic annihilation operator a_2 and creation operator $a_2^\dagger$ are easily defined in the occupation number representation as having the following effects:

$$a_2 \,| N_1, N_2, N_3, \ldots \rangle = \sqrt{N_2}\, | N_1, (N_2 - 1), N_3, \ldots \rangle,$$

$$a_2^\dagger \,| N_1, N_2, N_3, \ldots \rangle = \sqrt{N_2 + 1}\, | N_1, (N_2 + 1), N_3, \ldots \rangle.$$

It can be shown that these are operators in the usual quantum mechanical sense, i.e. linear operators acting on the Fock space. Furthermore, they are indeed Hermitian conjugates, which justifies the way we have written them. They can be shown to obey the commutation relation

$$\left[a_i, a_j \right] = 0 \quad , \quad \left[a_i^\dagger, a_j^\dagger \right] = 0 \quad , \quad \left[a_i, a_j^\dagger \right] = \delta_{ij},$$

where δ stands for the Kronecker delta. These are precisely the relations obeyed by the ladder operators for an infinite set of independent quantum harmonic oscillators, one for each single-particle state. Adding or removing bosons from each state is, therefore, analogous to exciting or de-exciting a quantum of energy in a harmonic oscillator.

Applying an annihilation operator a_k followed by its corresponding creation operator $a_k^\dagger$ returns the number N_k of particles in the kth single-particle eigenstate:

$$a_k^\dagger a_k \,| \ldots, N_k, \ldots \rangle = N_k \,| \ldots, N_k, \ldots \rangle.$$

The combination of operators $a_k^\dagger a_k$ is known as the number operator for the kth eigenstate.

The Hamiltonian operator of the quantum field (which, through the Schrödinger equation, determines its dynamics) can be written in terms of creation and annihilation operators. For instance, for a field of free (non-interacting) bosons, the total energy of the field is found by summing the energies of the bosons in each energy eigenstate. If the kth single-particle energy eigenstate has energy E_k and there are N_k bosons in this state, then the total energy of these bosons is $E_k N_k$. The energy in the *entire* field is then a sum over k:

$$E_{\text{tot}} = \sum_k E_k N_k$$

This can be turned into the Hamiltonian operator of the field by replacing N_k with the corresponding number operator, $a_k^\dagger a_k$. This yields

$$H = \sum_k E_k a_k^\dagger a_k.$$

Fermions

It turns out that a different definition of creation and annihilation must be used for describing fermions. According to the Pauli exclusion principle, fermions cannot share quantum states, so their occupation numbers N_i can only take on the value 0 or 1. The fermionic annihilation operators c and creation operators $c^\dagger$ are defined by their actions on a Fock state thus

$$c_j \,|\, N_1, N_2, \ldots, N_j = 0, \ldots \rangle = 0$$

$$c_j \,|\, N_1, N_2, \ldots, N_j = 1, \ldots \rangle = (-1)^{(N_1 + \cdots + N_{j-1})} \,|\, N_1, N_2, \ldots, N_j = 0, \ldots \rangle$$

$$c_j^\dagger \,|\, N_1, N_2, \ldots, N_j = 0, \ldots \rangle = (-1)^{(N_1 + \cdots + N_{j-1})} \,|\, N_1, N_2, \ldots, N_j = 1, \ldots \rangle$$

$$c_j^\dagger \,|\, N_1, N_2, \ldots, N_j = 1, \ldots \rangle = 0.$$

These obey an anticommutation relation:

$$\{c_i, c_j\} = 0 \quad , \quad \{c_i^\dagger, c_j^\dagger\} = 0 \quad , \quad \{c_i, c_j^\dagger\} = \delta_{ij}.$$

One may notice from this that applying a fermionic creation operator twice gives zero, so it is impossible for the particles to share single-particle states, in accordance with the exclusion principle.

Field Operators

We have previously mentioned that there can be more than one way of indexing the degrees of freedom in a quantum field. Second quantization indexes the field by enumerating the single-particle quantum states. However, as we have discussed, it is more natural to think about a "field", such as the electromagnetic field, as a set of degrees of freedom indexed by position.

To this end, we can define *field operators* that create or destroy a particle at a particular point in space. In particle physics, these operators turn out to be more convenient to work with, because they make it easier to formulate theories that satisfy the demands of relativity.

Single-particle states are usually enumerated in terms of their momenta (as in the particle in a box problem.) We can construct field operators by applying the Fourier transform to the creation and annihilation operators for these states. For example, the bosonic field annihilation operator $\phi(\mathbf{r})$ is

$$\phi(\mathbf{r}) \overset{\text{def}}{=} \sum_j e^{i\mathbf{k}_j \cdot \mathbf{r}} a_j.$$

The bosonic field operators obey the commutation relation

$$\left[\phi(\mathbf{r}), \phi(\mathbf{r}')\right] = 0 \quad , \quad \left[\phi^\dagger(\mathbf{r}), \phi^\dagger(\mathbf{r}')\right] = 0 \quad , \quad \left[\phi(\mathbf{r}), \phi^\dagger(\mathbf{r}')\right] = \delta^3(\mathbf{r} - \mathbf{r}')$$

where $\delta(\ )$ stands for the Dirac delta function. As before, the fermionic relations are the same, with the commutators replaced by anticommutators.

The field operator is not the same thing as a single-particle wavefunction. The former is an operator acting on the Fock space, and the latter is a quantum-mechanical amplitude for finding a particle in some position. However, they are closely related and are indeed commonly denoted with the same symbol. If we have a Hamiltonian with a space representation, say

$$H = -\frac{\hbar^2}{2m} \sum_i \nabla_i^2 + \sum_{i<j} U(|\mathbf{r}_i - \mathbf{r}_j|)$$

where the indices i and j run over all particles, then the field theory Hamiltonian (in the non-relativistic limit and for negligible self-interactions) is

$$H = -\frac{\hbar^2}{2m} \int d^3r \, \phi^\dagger(\mathbf{r}) \nabla^2 \phi(\mathbf{r}) + \frac{1}{2} \int d^3r \int d^3r' \, \phi^\dagger(\mathbf{r}) \phi^\dagger(\mathbf{r}') U(|\mathbf{r} - \mathbf{r}'|) \phi(\mathbf{r}') \phi(\mathbf{r}).$$

This looks remarkably like an expression for the expectation value of the energy, with ϕ playing the role of the wavefunction. This relationship between the field operators and wave functions makes it very easy to formulate field theories starting from space projected Hamiltonians.

Dynamics

Once the Hamiltonian operator is obtained as part of the canonical quantization process, the time dependence of the state is described with the Schrödinger equation, just as with other quantum theories. Alternatively, the Heisenberg picture can be used where the time dependence is in the operators rather than in the states.

Probability amplitudes of observables in such systems are quite hard to evaluate, an enterprise which has absorbed considerable ingenuity in the last three quarters of a century. In practice, most often, expectation values of operators are computed systematically through *covariant perturba-*

tion theory, formulated through Feynman diagrams, but path integral computer simulations have also produced important results. Contemporary particle physics relies on extraordinarily accurate predictions of such techniques.

Implications

Unification of Fields and Particles

The "second quantization" procedure outlined in the previous section takes a set of single-particle quantum states as a starting point. Sometimes, it is impossible to define such single-particle states, and one must proceed directly to quantum field theory. For example, a quantum theory of the electromagnetic field *must* be a quantum field theory, because it is impossible (for various reasons) to define a wavefunction for a single photon. In such situations, the quantum field theory can be constructed by examining the mechanical properties of the classical field and guessing the corresponding quantum theory. For free (non-interacting) quantum fields, the quantum field theories obtained in this way have the same properties as those obtained using second quantization, such as well-defined creation and annihilation operators obeying commutation or anticommutation relations.

Quantum field theory thus provides a unified framework for describing "field-like" objects (such as the electromagnetic field, whose excitations are photons) and "particle-like" objects (such as electrons, which are treated as excitations of an underlying electron field), so long as one can treat interactions as "perturbations" of free fields.

Physical Meaning of Particle Indistinguishability

The second quantization procedure relies crucially on the particles being identical. We would not have been able to construct a quantum field theory from a distinguishable many-particle system, because there would have been no way of separating and indexing the degrees of freedom.

Many physicists prefer to take the converse interpretation, which is that *quantum field theory explains what identical particles are*. In ordinary quantum mechanics, there is not much theoretical motivation for using symmetric (bosonic) or antisymmetric (fermionic) states, and the need for such states is simply regarded as an empirical fact. From the point of view of quantum field theory, particles are identical if and only if they are excitations of the same underlying quantum field. Thus, the question "why are all electrons identical?" arises from mistakenly regarding individual electrons as fundamental objects, when in fact it is only the electron field that is fundamental.

Particle Conservation and Non-Conservation

During second quantization, we started with a Hamiltonian and state space describing a fixed number of particles (N), and ended with a Hamiltonian and state space for an arbitrary number of particles. Of course, in many common situations N is an important and perfectly well-defined quantity, e.g. if we are describing a gas of atoms sealed in a box. From the point of view of quantum field theory, such situations are described by quantum states that are eigenstates of the number operator $\hat{N}$, which measures the total number of particles present. As with any quantum mechanical observable, $\hat{N}$ is conserved if it commutes with the Hamiltonian. In that case, the quantum state is trapped in the N-particle subspace of the total Fock space, and the situation could equally well be described by ordinary N-particle quantum mechanics. (Strictly speaking, this is only true in the noninteracting case or in the low energy density limit of renormalized quantum field theories)

For example, we can see that the free boson Hamiltonian described above conserves particle number. Whenever the Hamiltonian operates on a state, each particle destroyed by an annihilation operator a_k is immediately put back by the creation operator .

On the other hand, it is possible, and indeed common, to encounter quantum states that are *not* eigenstates of $\hat{N}$, which do not have well-defined particle numbers. Such states are difficult or impossible to handle using ordinary quantum mechanics, but they can be easily described in quantum field theory as quantum superpositions of states having different values of N. For example, suppose we have a bosonic field whose particles can be created or destroyed by interactions with a fermionic field. The Hamiltonian of the combined system would be given by the Hamiltonians of the free boson and free fermion fields, plus a "potential energy" term such as

$$H_I = \sum_{k,q} V_q (a_q + a_{-q}^\dagger) c_{k+q}^\dagger c_k,$$

where $a_k^\dagger$ and a_k denotes the bosonic creation and annihilation operators, $c_k^\dagger$ and c_k denotes the fermionic creation and annihilation operators, and V_q is a parameter that describes the strength of the interaction. This "interaction term" describes processes in which a fermion in state k either absorbs or emits a boson, thereby being kicked into a different eigenstate $k+q$. (In fact, this type of Hamiltonian is used to describe the interaction between conduction electrons and phonons in metals. The interaction between electrons and photons is treated in a similar way, but is a little more complicated because the role of spin must be taken into account.) One thing to notice here is that even if we start out with a fixed number of bosons, we will typically end up with a superposition of states with different numbers of bosons at later times. The number of fermions, however, is conserved in this case.

In condensed matter physics, states with ill-defined particle numbers are particularly important for describing the various superfluids. Many of the defining characteristics of a superfluid arise from the notion that its quantum state is a superposition of states with different particle numbers. In addition, the concept of a coherent state (used to model the laser and the BCS ground state) refers to a state with an ill-defined particle number but a well-defined phase.

Associated Phenomena

Beyond the most general features of quantum field theories, special aspects such as renormalizability, gauge symmetry, and supersymmetry are outlined below.

Renormalization

Early in the history of quantum field theory, as detailed above, it was found that many seemingly innocuous calculations, such as the perturbative shift in the energy of an electron due to the presence of the electromagnetic field, yield infinite results. The reason is that the perturbation theory for the shift in an energy involves a sum over *all other energy levels, and there are infinitely many levels* at short distances, so that each gives a finite contribution which results in a divergent series.

Many of these problems are related to failures in classical electrodynamics that were identified but unsolved in the 19th century, and they basically stem from the fact that many of the supposedly "intrinsic" properties of an electron are tied to the electromagnetic field that it carries around with it. The energy carried by a single electron—its self-energy—is not simply the bare value, but also includes the energy contained in its electromagnetic field, its attendant cloud of photons. The energy in a field of a spherical source diverges in both classical and quantum mechanics, but as discovered by Weisskopf with help from Furry, in quantum mechanics *the divergence is much milder*, going *only as the logarithm* of the radius of the sphere.

The solution to the problem, presciently suggested by Stueckelberg, independently by Bethe after the crucial experiment by Lamb, implemented at one loop by Schwinger, and systematically extended to all loops by Feynman and Dyson, with converging work by Tomonaga in isolated postwar Japan, comes from recognizing that all the infinities in the interactions of photons and electrons can be isolated into redefining a finite number of quantities in the equations by replacing them with the observed values: specifically the electron's mass and charge: this is called renormalization. The technique of renormalization recognizes that the problem is tractable and essentially purely mathematical; and that, physically, extremely short distances are at fault.

In order to define a theory on a continuum, one may first place a cutoff on the fields, by postulating that quanta cannot have energies above some extremely high value. This has the effect of replacing continuous space by a structure where very short wavelengths do not exist, as on a lattice. Lattices break rotational symmetry, and one of the crucial contributions made by Feynman, Pauli and Villars, and modernized by 't Hooft and Veltman, is a symmetry-preserving cutoff for perturbation theory (this process is called regularization). There is no known symmetrical cutoff outside of perturbation theory, so for rigorous or numerical work people often use an actual lattice.

On a lattice, every quantity is finite but depends on the spacing. When taking the limit to zero spacing, one makes sure that the physically observable quantities like the observed electron mass stay fixed, which means that the constants in the Lagrangian defining the theory depend on the spacing. By allowing the constants to vary with the lattice spacing, all the results at long distances become insensitive to the lattice, defining a continuum limit.

The renormalization procedure only works for a certain limited class of quantum field theories, called *renormalizable quantum field theories*. A theory is *perturbatively renormalizable* when the constants in the Lagrangian only diverge at worst as *logarithms* of the lattice spacing for very short spacings. The continuum limit is then well defined in perturbation theory, and even if it is not fully well defined non-perturbatively, the problems only show up at distance scales that are exponentially small in the inverse coupling for weak couplings. The Standard Model of particle physics is perturbatively renormalizable, and so are its component theories (quantum electrodynamics/electroweak theory and quantum chromodynamics). Of the three components, quantum electrodynamics is believed to not have a continuum limit by itself, while the asymptotically free $SU(2)$ and $SU(3)$ weak and strong color interactions are nonperturbatively well defined.

The renormalization group as developed along Wilson's breakthrough insights relates effective field theories at a given scale to such at contiguous scales. It thus describes how renormalizable theories emerge as the long distance low-energy effective field theory for any given high-energy theory. As a consequence, renormalizable theories are insensitive to the precise nature of the underlying high-energy short-distance phenomena (the macroscopic physics is dominated by only a few *"relevant" observables*). This is a blessing in practical terms, because it allows physicists to formulate low energy theories without detailed knowledge of high-energy phenomena. It is also a curse, because once a renormalizable theory such as the standard model is found to work, it provides very few clues to higher-energy processes.

The only way high-energy processes can be seen in the standard model is when they allow otherwise forbidden events, or else if they reveal predicted compelling quantitative relations among the coupling constants of the theories or models.

On account of renormalization, the couplings of QFT vary with scale, thereby confining quarks into hadrons, allowing the study of weakly-coupled quarks inside hadrons, and enabling speculation on ultra-high energy behavior.

Gauge Freedom

A gauge theory is a theory that admits a symmetry with a local parameter. For example, in every quantum theory, the global phase of the wave function is arbitrary and does not represent something physical. Consequently, the theory is invariant under a global change of phases (adding a constant to the phase of all wave functions, everywhere); this is a global symmetry. In quantum electrodynamics, the theory is also invariant under a *local* change of phase, that is – one may shift the phase of all wave functions so that the shift may be different at every point in space-time. This is a *local* symmetry. However, in order for a well-defined derivative operator to exist, one must introduce a new field, the gauge field, which also transforms in order for the local change of variables (the phase in our example) not to affect the derivative. In quantum electrodynamics, this gauge field is the electromagnetic field. The change of local gauge of variables is termed gauge transformation.

By Noether's theorem, for every such symmetry there exists an associated conserved current. The aforementioned symmetry of the wavefunction under global phase changes implies the conservation of electric charge. Since the excitations of fields represent particles, the particle associated with excitations of the gauge field is the gauge boson, e.g., the photon in the case of quantum electrodynamics.

The degrees of freedom in quantum field theory are local fluctuations of the fields. The existence of a gauge symmetry reduces the number of degrees of freedom, simply because some fluctuations of the fields can be transformed to zero by gauge transformations, so they are equivalent to having no fluctuations at all, and they, therefore, have no physical meaning. Such fluctuations are usually called "non-physical degrees of freedom" or *gauge artifacts*; usually, some of them have a negative norm, making them inadequate for a consistent theory. Therefore, if a classical field theory has a gauge symmetry, then its quantized version (the corresponding quantum field theory) will have this symmetry as well. In other words, a gauge symmetry cannot have a quantum anomaly.

In general, the gauge transformations of a theory consist of several different transformations, which may not be commutative. These transformations are combine into the framework of a gauge group; infinitesimal gauge transformations are the gauge group generators. Thus, the number of gauge bosons is the group dimension (i.e., the number of generators forming the basis of the corresponding Lie algebra).

All the known fundamental interactions in nature are described by gauge theories (possibly barring the Higgs multiplet couplings, if considered in isolation). These are:

- Quantum chromodynamics, whose gauge group is $SU(3)$. The gauge bosons are eight gluons.

- The electroweak theory, whose gauge group is $U(1) \times SU(2)$, (a direct product of $U(1)$ and $SU(2)$). The gauge bosons are the photon and the massive $W^{\pm}$ and Z^0 bosons.

- Gravity, whose classical theory is general relativity, relies on the equivalence principle, which is essentially a form of gauge symmetry. Its action may also be written as a gauge theory of the Lorentz group on tangent space.

Supersymmetry

Supersymmetry assumes that every fundamental fermion has a superpartner that is a boson and vice versa. Its gauge theory, Supergravity, is an extension of general relativity. Supersymmetry is a key ingredient for the consistency of string theory.

It was utilized in order to solve the so-called Hierarchy Problem of the standard model, that is, to explain why particles not protected by any symmetry (like the Higgs boson) do not receive radiative corrections to their mass, driving it to the larger scales such as that of GUTs, or the Planck mass of gravity. The way supersymmetry protects scale hierarchies is the following: since for every particle there is a superpartner with the same mass but different statistics, any loop in a radiative correction is cancelled by the loop corresponding to its superpartner, rendering the theory more UV finite.

Since, however, no super partners have been observed, if supersymmetry existed it should be broken severely (through a so-called soft term, which breaks supersymmetry without ruining its helpful features). The simplest models of this breaking require that the energy of the superpartners not be too high; in these cases, supersymmetry could be observed by experiments at the Large Hadron Collider. However, to date, after the observation of the Higgs boson there, no such superparticles have been discovered.

Axiomatic Approaches

The preceding description of quantum field theory follows the spirit in which most physicists approach the subject. However, it is not mathematically rigorous. Over the past several decades, there have been many attempts to put quantum field theory on a firm mathematical footing by formulating a set of axioms for it. These attempts fall into two broad classes.

The first class of axioms, first proposed during the 1950s, include the Wightman, Osterwalder–Schrader, and Haag–Kastler systems. They attempted to formalize the physicists' notion of an "operator-valued field" within the context of functional analysis and enjoyed limited success. It was possible to prove that any quantum field theory satisfying these axioms satisfied certain general theorems, such as the spin-statistics theorem and the CPT theorem. Unfortunately, it proved extraordinarily difficult to show that any realistic field theory, including the Standard Model, satisfied these axioms. Most of the theories that could be treated with these analytic axioms were physically trivial, being restricted to low-dimensions and lacking interesting dynamics. The construction of theories satisfying one of these sets of axioms falls in the field of constructive quantum field theory. Important work was done in this area in the 1970s by Segal, Glimm, Jaffe and others.

During the 1980s, the second set of axioms based on geometric ideas was proposed. This line of investigation, which restricts its attention to a particular class of quantum field theories known as topological quantum field theories, is associated most closely with Michael Atiyah and Graeme Segal, and was notably expanded upon by Edward Witten, Richard Borcherds, and Maxim Kontsevich. However, most of the physically relevant quantum field theories, such as the Standard Model, are not topological quantum field theories; the quantum field theory of the fractional quantum Hall effect is a notable exception. The main impact of axiomatic topological quantum field theory has been on mathematics, with important applications in representation theory, algebraic topology, and differential geometry.

Finding the proper axioms for quantum field theory is still an open and difficult problem in mathematics. One of the Millennium Prize Problems—proving the existence of a mass gap in Yang–Mills theory—is linked to this issue.

Haag's Theorem

From a mathematically rigorous perspective, there exists no interaction picture in a Lorentz-covariant quantum field theory. This implies that the perturbative approach of Feynman diagrams in QFT is not strictly justified, despite producing vastly precise predictions validated by experiment. This is called Haag's theorem, but most particle physicists relying on QFT largely shrug it off, as not really limiting the power of the theory.

Common Integrals in Quantum Field Theory

There are common integrals in quantum field theory that appear repeatedly. These integrals are all variations and generalizations of gaussian integrals to the complex plane and to multiple dimensions. Other integrals can be approximated by versions of the gaussian integral. Fourier integrals are also considered.

Variations on a Simple Gaussian Integral

Gaussian Integral

The first integral, with broad application outside of quantum field theory, is the gaussian integral.

$$G \equiv \int_{-\infty}^{\infty} e^{-\frac{1}{2}x^2} \, dx$$

In physics the factor of 1/2 in the argument of the exponential is common.

Note:

$$G^2 = \left(\int_{-\infty}^{\infty} e^{-\frac{1}{2}x^2} \, dx \right) \cdot \left(\int_{-\infty}^{\infty} e^{-\frac{1}{2}y^2} \, dy \right) = 2\pi \int_{0}^{\infty} re^{-\frac{1}{2}r^2} \, dr = 2\pi \int_{0}^{\infty} e^{-w} \, dw = 2\pi . G^2 = \left(\int_{-\infty}^{\infty} e^{-\frac{1}{2}x^2} \, dx \right) \cdot \left(\int_{-\infty}^{\infty} e^{-\frac{1}{2}y^2} \, dy \right) = 2\pi \int_{0}^{\infty} re^{-\frac{1}{2}r^2} \, dr = 2\pi \int_{0}^{\infty} e^{-w} \, dw = 2\pi .$$

Thus we obtain

$$\int_{-\infty}^{\infty} e^{-\frac{1}{2}x^2}\, dx = \sqrt{2\pi}.$$

Slight Generalization of the Gaussian Integral

$$\int_{-\infty}^{\infty} e^{-\frac{1}{2}ax^2}\, dx = \sqrt{\frac{2\pi}{a}}$$

where we have scaled

$$x \to \frac{x}{\sqrt{a}}.$$

Integrals of Exponents and Even Powers of x

$$\int_{-\infty}^{\infty} x^2 e^{-\frac{1}{2}ax^2}\, dx = -2\frac{d}{da}\int_{-\infty}^{\infty} e^{-\frac{1}{2}ax^2}\, dx = -2\frac{d}{da}\left(\frac{2\pi}{a}\right)^{\frac{1}{2}} = \left(\frac{2\pi}{a}\right)^{\frac{1}{2}}\frac{1}{a}$$

and

$$\int_{-\infty}^{\infty} x^4 e^{-\frac{1}{2}ax^2}\, dx = \left(-2\frac{d}{da}\right)\left(-2\frac{d}{da}\right)\int_{-\infty}^{\infty} e^{-\frac{1}{2}ax^2}\, dx = \left(-2\frac{d}{da}\right)\left(-2\frac{d}{da}\right)\left(\frac{2\pi}{a}\right)^{\frac{1}{2}} = \left(\frac{2\pi}{a}\right)^{\frac{1}{2}}\frac{3}{a^2}$$

In general

$$\int_{-\infty}^{\infty} x^{2n} e^{-\frac{1}{2}ax^2}\, dx = \left(\frac{2\pi}{a}\right)^{\frac{1}{2}}\frac{1}{a^n}(2n-1)(2n-3)\cdots 5\cdot 3\cdot 1 = \left(\frac{2\pi}{a}\right)^{\frac{1}{2}}\frac{1}{a^n}(2n-1)!!$$

Note that the integrals of exponents and odd powers of x are 0, due to odd symmetry.

Integrals with a Linear Term in the Argument of the Exponent

$$\int_{-\infty}^{\infty} \exp\left(-\frac{1}{2}ax^2 + Jx\right) dx$$

This integral can be performed by completing the square:

$$\left(-\frac{1}{2}ax^2 + Jx\right) = -\frac{1}{2}a\left(x^2 - \frac{2Jx}{a} + \frac{J^2}{a^2} - \frac{J^2}{a^2}\right) = -\frac{1}{2}a\left(x - \frac{J}{a}\right)^2 + \frac{J^2}{2a}$$

Therefore:

$$\int_{-\infty}^{\infty} \exp\left(-\frac{1}{2}ax^2 + Jx\right)dx = \exp\left(\frac{J^2}{2a}\right)\int_{-\infty}^{\infty} \exp\left[-\frac{1}{2}a\left(x-\frac{J}{a}\right)^2\right]dx$$

$$= \exp\left(\frac{J^2}{2a}\right)\int_{-\infty}^{\infty} \exp\left(-\frac{1}{2}ax^2\right)dx$$

$$= \left(\frac{2\pi}{a}\right)^{\frac{1}{2}} \exp\left(\frac{J^2}{2a}\right)$$

Integrals with an Imaginary Linear Term in the Argument of the Exponent

The integral

$$\int_{-\infty}^{\infty} \exp\left(-\frac{1}{2}ax^2 + iJx\right)dx = \left(\frac{2\pi}{a}\right)^{\frac{1}{2}} \exp\left(-\frac{J^2}{2a}\right)$$

is proportional to the Fourier transform of the gaussian where J is the conjugate variable of x.

By again completing the square we see that the Fourier transform of a gaussian is also a gaussian, but in the conjugate variable. The larger a is, the narrower the gaussian in x and the wider the gaussian in J. This is a demonstration of the uncertainty principle.

This integral is also known as the Hubbard-Stratonovich transformation used in field theory.

Integrals with a Complex Argument of the Exponent

The integral of interest is

$$\int_{-\infty}^{\infty} \exp\left(\frac{1}{2}iax^2 + iJx\right)dx.$$

We now assume that a and J may be complex.

Completing the square

$$\left(\frac{1}{2}iax^2 + iJx\right) = \frac{1}{2}ia\left(x^2 + \frac{2Jx}{a} + \left(\frac{J}{a}\right)^2 - \left(\frac{J}{a}\right)^2\right) = -\frac{1}{2}\frac{a}{i}\left(x+\frac{J}{a}\right)^2 - \frac{iJ^2}{2a}.$$

By analogy with the previous integrals

$$\int_{-\infty}^{\infty} \exp\left(\frac{1}{2}iax^2 + iJx\right)dx = \left(\frac{2\pi i}{a}\right)^{\frac{1}{2}} \exp\left(\frac{-iJ^2}{2a}\right).$$

This result is valid as an integration in the complex plane as long as a has a positive imaginary part.

Gaussian Integrals in Higher Dimensions

The one-dimensional integrals can be generalized to multiple dimensions.

$$\int \exp\left(-\frac{1}{2}x \cdot A \cdot x + J \cdot x\right) d^n x = \sqrt{\frac{(2\pi)^n}{\det A}} \exp\left(\frac{1}{2}J \cdot A^{-1} \cdot J\right)$$

Here A is a real symmetric matrix.

This integral is performed by diagonalization of A with an orthogonal transformation

$$D = O^{-1}AO = O^T AO$$

where D is a diagonal matrix and O is an orthogonal matrix. This decouples the variables and allows the integration to be performed as n one-dimensional integrations.

This is best illustrated with a two-dimensional example.

Example: Simple Gaussian Integration in Two Dimensions

The gaussian integral in two dimensions is

$$\int \exp\left(-\frac{1}{2}A_{ij}x^i x^j\right) d^2 x = \sqrt{\frac{(2\pi)^2}{\det A}}$$

where A is a two-dimensional symmetric matrix with components specified as

$$A = \begin{bmatrix} a & c \\ c & b \end{bmatrix}$$

and we have used the Einstein summation convention.

Diagonalize The Matrix

The first step is to diagonalize the matrix. Note that

$$A_{ij}x^i x^j \equiv x^T A x = x^T \left(OO^T\right) A \left(OO^T\right) x = \left(x^T O\right)\left(O^T AO\right)\left(O^T x\right)$$

where, since A is a real symmetric matrix, we can choose O to be orthogonal, and hence also a unitary matrix. O can be obtained from the eigenvectors of A. We choose O such that: $D \equiv O^T AO$ is diagonal.

Eigenvalues of A

To find the eigenvectors of A one first finds the eigenvalues λ of A given by

$$\begin{bmatrix} a & c \\ c & b \end{bmatrix}\begin{bmatrix} u \\ v \end{bmatrix} = \lambda \begin{bmatrix} u \\ v \end{bmatrix}.$$

The eigenvalues are solutions of the characteristic polynomial

$$(a-\lambda)(b-\lambda) - c^2 = 0$$

$$\lambda^2 - \lambda(a+b) + ab - c^2 = 0$$

which are found using the quadratic equation:

$$\lambda_{\pm} = \frac{1}{2}(a+b) \pm \frac{1}{2}\sqrt{(a+b)^2 - 4(ab - c^2)}.$$

$$\lambda_{\pm} = \frac{1}{2}(a+b) \pm \frac{1}{2}\sqrt{a^2 + 2ab + b^2 - 4ab + 4c^2}.$$

$$\lambda_{\pm} = \frac{1}{2}(a+b) \pm \frac{1}{2}\sqrt{(a-b)^2 + 4c^2}.$$

Eigenvectors of A

Substitution of the eigenvalues back into the eigenvector equation yields

$$v = -\frac{(a-\lambda_{\pm})u}{c}, \qquad v = -\frac{cu}{(b-\lambda_{\pm})}.$$

From the characteristic equation we know

$$\frac{a-\lambda_{\pm}}{c} = \frac{c}{b-\lambda_{\pm}}.$$

Also note

$$\frac{a-\lambda_{\pm}}{c} = -\frac{b-\lambda_{\mp}}{c}.$$

The eigenvectors can be written as:

$$\begin{bmatrix} \dfrac{1}{\eta} \\ -\dfrac{a-\lambda_{-}}{c\eta} \end{bmatrix}, \qquad \begin{bmatrix} -\dfrac{b-\lambda_{+}}{c\eta} \\ \dfrac{1}{\eta} \end{bmatrix}$$

for the two eigenvectors. Here η is a normalizing factor given by

$$\eta = \sqrt{1 + \left(\frac{a - \lambda_-}{c}\right)^2} = \sqrt{1 + \left(\frac{b - \lambda_+}{c}\right)^2}.$$

It is easily verified that the two eigenvectors are orthogonal to each other.

Construction of the Orthogonal Matrix

The orthogonal matrix is constructed by assigning the normalized eigenvectors as columns in the orthogonal matrix

$$O = \begin{bmatrix} \dfrac{1}{\eta} & -\dfrac{b - \lambda_+}{c\eta} \\ -\dfrac{a - \lambda_-}{c\eta} & \dfrac{1}{\eta} \end{bmatrix}.$$

Note that $\det(O) = 1$.

If we define

$$\sin(\theta) = -\frac{a - \lambda_-}{c\eta}$$

then the orthogonal matrix can be written

$$O = \begin{bmatrix} \cos(\theta) & -\sin(\theta) \\ \sin(\theta) & \cos(\theta) \end{bmatrix}$$

which is simply a rotation of the eigenvectors with the inverse:

$$O^{-1} = O^T = \begin{bmatrix} \cos(\theta) & \sin(\theta) \\ -\sin(\theta) & \cos(\theta) \end{bmatrix}.$$

Diagonal Matrix

The diagonal matrix becomes

$$D = O^T A O = \begin{bmatrix} \lambda_- & 0 \\ 0 & \lambda_+ \end{bmatrix}$$

with eigenvectors

$$\begin{bmatrix} 1 \\ 0 \end{bmatrix}, \quad \begin{bmatrix} 0 \\ 1 \end{bmatrix}$$

Numerical Example

$$A = \begin{bmatrix} 2 & 1 \\ 1 & 1 \end{bmatrix}$$

The eigenvalues are

$$\lambda_\pm = \frac{3}{2} \pm \frac{\sqrt{5}}{2}.$$

The eigenvectors are

$$\frac{1}{\eta} \begin{bmatrix} 1 \\ -\frac{1}{2} - \frac{\sqrt{5}}{2} \end{bmatrix}, \quad \frac{1}{\eta} \begin{bmatrix} \frac{1}{2} + \frac{\sqrt{5}}{2} \\ 1 \end{bmatrix}$$

where

$$\eta = \sqrt{\frac{5}{2} + \frac{\sqrt{5}}{2}}.$$

Then

$$O = \begin{bmatrix} \dfrac{1}{\eta} & \dfrac{1}{\eta}\left(\dfrac{1}{2} + \dfrac{\sqrt{5}}{2}\right) \\ \dfrac{1}{\eta}\left(-\dfrac{1}{2} - \dfrac{\sqrt{5}}{2}\right) & \dfrac{1}{\eta} \end{bmatrix}$$

$$O^{-1} = \begin{bmatrix} \dfrac{1}{\eta} & \dfrac{1}{\eta}\left(-\dfrac{1}{2} - \dfrac{\sqrt{5}}{2}\right) \\ \dfrac{1}{\eta}\left(\dfrac{1}{2} + \dfrac{\sqrt{5}}{2}\right) & \dfrac{1}{\eta} \end{bmatrix}$$

The diagonal matrix becomes

$$D = O^T A O = \begin{bmatrix} \lambda_- & 0 \\ 0 & \lambda_+ \end{bmatrix} = \begin{bmatrix} \dfrac{3}{2} - \dfrac{\sqrt{5}}{2} & 0 \\ 0 & \dfrac{3}{2} + \dfrac{\sqrt{5}}{2} \end{bmatrix}$$

with eigenvectors

$$\begin{bmatrix} 1 \\ 0 \end{bmatrix}, \quad \begin{bmatrix} 0 \\ 1 \end{bmatrix}$$

Rescale the Variables and Integrate

With the diagonalization the integral can be written

$$\int \exp\left(-\frac{1}{2}x^T A x\right) d^2 x = \int \exp\left(-\frac{1}{2}\sum_{j=1}^{2} \lambda_j y_j^2\right) d^2 y$$

where

$$y = O^T x.$$

Since the coordinate transformation is simply a rotation of coordinates the Jacobian determinant of the transformation is one yielding

$$dy^2 = dx^2$$

The integrations can now be performed.

$$\int \exp\left(-\frac{1}{2}x^T A x\right) d^2 x \quad = \int \exp\left(-\frac{1}{2}\sum_{j=1}^{2} \lambda_j y_j^2\right) d^2 y$$

$$= \prod_{j=1}^{2}\left(\frac{2\pi}{\lambda_j}\right)^{\frac{1}{2}}$$

$$= \left(\frac{(2\pi)^2}{\prod_{j=1}^{2} \lambda_j}\right)^{\frac{1}{2}}$$

$$= \left(\frac{(2\pi)^2}{\det\left(O^{-1}AO\right)}\right)^{\frac{1}{2}}$$

$$= \left(\frac{(2\pi)^2}{\det\left(A\right)}\right)^{\frac{1}{2}}$$

which is the advertised solution.

Integrals with Complex and Linear Terms in Multiple Dimensions

With the two-dimensional example it is now easy to see the generalization to the complex plane and to multiple dimensions.

Integrals with a Linear Term in the Argument

$$\int \exp\left(-\frac{1}{2}x \cdot A \cdot x + J \cdot x\right) d^n x = \sqrt{\frac{(2\pi)^n}{\det A}}\, \exp\left(\frac{1}{2}J \cdot A^{-1} \cdot J\right)$$

Integrals with an Imaginary Linear Term

$$\int \exp\left(-\frac{1}{2}x \cdot A \cdot x + iJ \cdot x\right) d^n x = \sqrt{\frac{(2\pi)^n}{\det A}}\, \exp\left(-\frac{1}{2}J \cdot A^{-1} \cdot J\right)$$

Integrals with a Complex Quadratic Term

$$\int \exp\left(\frac{i}{2}x \cdot A \cdot x + iJ \cdot x\right) d^n x = \sqrt{\frac{(2\pi i)^n}{\det A}}\, \exp\left(-\frac{i}{2}J \cdot A^{-1} \cdot J\right)$$

Integrals with Differential Operators in the Argument

As an example consider the integral

$$\int \exp\left[\int d^4 x \left(-\frac{1}{2}\varphi \hat{A}\varphi + J\varphi\right)\right] D\varphi$$

where $\hat{A}$ is a differential operator with φ and J functions of spacetime, and $D\varphi$ indicates integration over all possible paths. In analogy with the matrix version of this integral the solution is

$$\int \exp\left[\int d^4 x \left(-\frac{1}{2}\varphi \hat{A}\varphi + J\varphi\right)\right] D\varphi \propto \exp\left(\frac{1}{2}\int d^4 x\, d^4 y J(x)D(x-y)J(y)\right)$$

where

$$\hat{A}D(x-y) = \delta^4(x-y)$$

and $D(x-y)$, called the propagator, is the inverse of $\hat{A}$, and $\delta^4(x-y)$ is the Dirac delta function.

Similar arguments yield

$$\int \exp\left[\int d^4 x \left(-\frac{1}{2}\varphi \hat{A}\varphi + iJ\varphi\right)\right] D\varphi \propto \exp\left(-\frac{1}{2}\int d^4 x\, d^4 y J(x)D(x-y)J(y)\right),$$

and

$$\int \exp\left[i\int d^4 x \left(\frac{1}{2}\varphi \hat{A}\varphi + J\varphi\right)\right] D\varphi \propto \exp\left(\frac{i}{2}\int d^4 x\, d^4 y J(x)D(x-y)J(y)\right).$$

Integrals that can be Approximated by the Method of Steepest Descent

In quantum field theory n-dimensional integrals of the form

$$\int_{-\infty}^{\infty} \exp\left(-\frac{1}{\hbar} f(q)\right) d^n q$$

appear often. Here $\hbar$ is the reduced Planck's constant and f is a function with a positive minimum at $q = q_0$. These integrals can be approximated by the method of steepest descent.

For small values of Planck's constant, f can be expanded about its minimum

$$\int_{-\infty}^{\infty} \exp\left[-\frac{1}{\hbar}\left(f(q_0) + \frac{1}{2}(q - q_0)^2 f''(q - q_0) + \cdots\right)\right] d^n q .$$

Here f'' is the n by n matrix of second derivatives evaluated at the minimum of the function.

If we neglect higher order terms this integral can be integrated explicitly.

$$\int_{-\infty}^{\infty} \exp\left[-\frac{1}{\hbar}(f(q))\right] d^n q \approx \exp\left[-\frac{1}{\hbar}(f(q_0))\right]\sqrt{\frac{(2\pi\hbar)^n}{\det f''}} .$$

Integrals that can be Approximated by the Method of Stationary Phase

A common integral is a path integral of the form

$$\int \exp\left(\frac{i}{\hbar} S(q, \dot{q})\right) Dq$$

where $S(q, \dot{q})$ is the classical action and the integral is over all possible paths that a particle may take. In the limit of small $\hbar$ the integral can be evaluated in the stationary phase approximation. In this approximation the integral is over the path in which the action is a minimum. Therefore, this approximation recovers the classical limit of mechanics.

Fourier Integrals

Dirac Delta Distribution

The Dirac delta distribution in spacetime can be written as a Fourier transform

$$\int \frac{d^4 k}{(2\pi)^4} \exp(ik(x - y)) = \delta^4(x - y).$$

In general, for any dimension N

$$\int \frac{d^N k}{(2\pi)^N} \exp(ik(x - y)) = \delta^N(x - y).$$

Fourier Integrals of Forms of the Coulomb Potential

Laplacian of 1/r

While not an integral, the identity in three-dimensional Euclidean space

$$-\frac{1}{4\pi}\nabla^2\left(\frac{1}{r}\right) = \delta(\mathbf{r})$$

where

$$= \mathbf{r}\cdot\mathbf{r}$$

is a consequence of Gauss's theorem and can be used to derive integral identities.

This identity implies that the Fourier integral representation of 1/r is

$$\int\frac{d^3k}{(2\pi)^3}\frac{\exp(i\mathbf{k}\cdot\mathbf{r})}{k^2} = \frac{1}{4\pi r}.$$

Yukawa Potential: The Coulomb Potential with Mass

The Yukawa potential in three dimensions can be represented as an integral over a Fourier transform

$$\int\frac{d^3k}{(2\pi)^3}\frac{\exp(i\mathbf{k}\cdot\mathbf{r})}{k^2+m^2} = \frac{e^{-mr}}{4\pi r}$$

where

$$r^2 = \mathbf{r}\cdot\mathbf{r}, \qquad k^2 = \mathbf{k}\cdot\mathbf{k}.$$

See Static forces and virtual-particle exchange for an application of this integral.

In the small m limit the integral reduces to $1/4\pi r$.

To derive this result note:

$$\begin{aligned}
\int\frac{d^3k}{(2\pi)^3}\frac{\exp(i\mathbf{k}\cdot\mathbf{r})}{k^2+m^2} &= \int_0^\infty\frac{k^2 dk}{(2\pi)^2}\int_{-1}^1 du\,\frac{e^{ikru}}{k^2+m^2}\\
&= \frac{2}{r}\int_0^\infty\frac{kdk}{(2\pi)^2}\frac{\sin(kr)}{k^2+m^2}\\
&= \frac{1}{ir}\int_{-\infty}^\infty\frac{kdk}{(2\pi)^2}\frac{e^{ikr}}{k^2+m^2}\\
&= \frac{1}{ir}\int_{-\infty}^\infty\frac{kdk}{(2\pi)^2}\frac{e^{ikr}}{(k+im)(k-im)}\\
&= \frac{1}{ir}\frac{2\pi i}{(2\pi)^2}\frac{im}{2im}e^{-mr}\\
&= \frac{1}{4\pi r}e^{-mr}
\end{aligned}$$

Modified Coulomb Potential with Mass

$$\int \frac{d^3k}{(2\pi)^3} \left(\hat{\mathbf{k}} \cdot \hat{\mathbf{r}}\right)^2 \frac{\exp\left(i\mathbf{k} \cdot \mathbf{r}\right)}{k^2 + m^2} = \frac{e^{-mr}}{4\pi r} \left\{ 1 + \frac{2}{mr} - \frac{2}{(mr)^2}\left(e^{mr} - 1\right) \right\}$$

where the hat indicates a unit vector in three dimensional space. The derivation of this result is as follows:

$$\int \frac{d^3k}{(2\pi)^3} \left(\hat{\mathbf{k}} \cdot \hat{\mathbf{r}}\right)^2 \frac{\exp\left(i\mathbf{k} \cdot \mathbf{r}\right)}{k^2 + m^2} = \int_0^\infty \frac{k^2 dk}{(2\pi)^2} \int_{-1}^1 du\, u^2 \frac{e^{ikru}}{k^2 + m^2}$$

$$= 2\int_0^\infty \frac{k^2 dk}{(2\pi)^2} \frac{1}{k^2 + m^2} \left\{ \frac{1}{kr}\sin(kr) + \frac{2}{(kr)^2}\cos(kr) - \frac{2}{(kr)^3}\sin(kr) \right\}$$

$$= \frac{e^{-mr}}{4\pi r} \left\{ 1 + \frac{2}{mr} - \frac{2}{(mr)^2}\left(e^{mr} - 1\right) \right\}$$

Note that in the small m limit the integral goes to the result for the Coulomb potential since the term in the brackets goes to 1.

Longitudinal Potential with Mass

$$= \frac{1}{2} \frac{e^{-mr}}{4\pi r} \left([1 - \hat{\mathbf{r}}\hat{\mathbf{r}}] + \left\{ 1 + \frac{2}{mr} - \frac{2}{(mr)^2}\left(e^{mr} - 1\right) \right\}[1 + \hat{\mathbf{r}}\hat{\mathbf{r}}] \right)$$

where the hat indicates a unit vector in three dimensional space. The derivation for this result is as follows:

$$\int \frac{d^3k}{(2\pi)^3} \hat{\mathbf{k}}\hat{\mathbf{k}} \frac{\exp\left(i\mathbf{k} \cdot \mathbf{r}\right)}{k^2 + m^2} = \int \frac{d^3k}{(2\pi)^3} \left[\left(\hat{\mathbf{k}} \cdot \hat{\mathbf{r}}\right)^2 \hat{\mathbf{r}}\hat{\mathbf{r}} + \left(\hat{\mathbf{k}} \cdot \hat{\ }\right)^2 \hat{\ }\hat{\ } + \left(\hat{\ } \cdot \hat{\phi}\right)^2 \hat{\phi}\hat{\phi} \right] \frac{\exp\left(i\mathbf{k} \cdot \mathbf{r}\right)}{k^2 + m^2}$$

$$= \frac{e^{-mr}}{4\pi r} \left\{ 1 + \frac{2}{mr} - \frac{2}{(mr)^2}\left(e^{mr} - 1\right) \right\} \left\{ 1 - \frac{1}{2}[1 - \hat{\mathbf{r}}\hat{\mathbf{r}}] \right\} + \int_0^\infty \frac{k^2 dk}{(2\pi)^2} \int_{-1}^1 du\, \frac{e^{ikru}}{k^2 + m^2} \frac{1}{2}[1 - \hat{\mathbf{r}}\hat{\mathbf{r}}]$$

$$= \frac{1}{2} \frac{e^{-mr}}{4\pi r}[1 - \hat{\mathbf{r}}\hat{\mathbf{r}}] + \frac{e^{-mr}}{4\pi r} \left\{ 1 + \frac{2}{mr} - \frac{2}{(mr)^2}\left(e^{mr} - 1\right) \right\} \left\{ \frac{1}{2}[1 + \hat{\mathbf{r}}\hat{\mathbf{r}}] \right\}$$

$$= \frac{1}{2} \frac{e^{-mr}}{4\pi r} \left([1 - \hat{\mathbf{r}}\hat{\mathbf{r}}] + \left\{ 1 + \frac{2}{mr} - \frac{2}{(mr)^2}\left(e^{mr} - 1\right) \right\}[1 + \hat{\mathbf{r}}\hat{\mathbf{r}}] \right)$$

Note that in the small m limit the integral reduces to

$$\frac{1}{2} \frac{1}{4\pi r}[1 - \hat{\mathbf{r}}\hat{\mathbf{r}}].$$

Transverse Potential with Mass

$$\int \frac{d^3k}{(2\pi)^3} \left[1 - \hat{\mathbf{k}}\hat{\mathbf{k}}\right] \frac{\exp\left(i\mathbf{k} \cdot \mathbf{r}\right)}{k^2 + m^2} = \frac{1}{2} \frac{e^{-mr}}{4\pi r} \left\{ \frac{2}{(mr)^2}\left(e^{mr} - 1\right) - \frac{2}{mr} \right\}[1 + \hat{\mathbf{r}}\hat{\mathbf{r}}]$$

In the small mr limit the integral goes to

$$\frac{1}{2}\frac{1}{4\pi r}\left[1+\hat{\mathbf{r}}\hat{\mathbf{r}}\right].$$

For large distance, the integral falls off as the inverse cube of r

$$\frac{1}{4\pi m^2 r^3}\left[1+\hat{\mathbf{r}}\hat{\mathbf{r}}\right].$$

Angular Integration in Cylindrical Coordinates

There are two important integrals. The angular integration of an exponential in cylindrical coordinates can be written in terms of Bessel functions of the first kind

$$\int_0^{2\pi}\frac{d\varphi}{2\pi}\exp\left(ip\cos(\varphi)\right)=J_0(p)$$

and

$$\int_0^{2\pi}\frac{d\varphi}{2\pi}\cos(\varphi)\exp\left(ip\cos(\varphi)\right)=iJ_1(p).$$

Bessel Functions

Integration of the Cylindrical Propagator with Mass

First Power of a Bessel Function

$$\int_0^\infty\frac{k\,dk}{k^2+m^2}J_0\left(kr\right)=K_0(mr).$$

For $mr \ll 1$, we have

$$K_0(mr)\rightarrow-\ln\left(\frac{mr}{2}\right)+0.5772.$$

Squares of Bessel Functions

The integration of the propagator in cylindrical coordinates is

$$\int_0^\infty\frac{k\,dk}{k^2+m^2}J_1^2(kr)=I_1(mr)K_1(mr).$$

For small mr the integral becomes

$$\int_o^\infty\frac{k\,dk}{k^2+m^2}J_1^2(kr)\rightarrow\frac{1}{2}\left[1-\frac{1}{8}(mr)^2\right].$$

For large mr the integral becomes

$$\int_0^\infty \frac{k\,dk}{k^2+m^2}\,J_1^2(kr) \to \frac{1}{2}\left(\frac{1}{mr}\right).$$

In general

$$\int_0^\infty \frac{k\,dk}{k^2+m^2}J_\nu^2(kr) = I_\nu(mr)K_\nu(mr) \qquad \Re(\nu) > -1.$$

Integration over a Magnetic Wave Function

The two-dimensional integral over a magnetic wave function is

$$\frac{2a^{2n+2}}{n!}\int_0^\infty dr\, r^{2n+1}\exp\left(-a^2r^2\right)J_0(kr) = M\left(n+1,1,-\frac{k^2}{4a^2}\right).$$

Here, M is a confluent hypergeometric function. For an application of this integral see Charge density spread over a wave function.

First Quantization

A first quantization of a physical system is a semi-classical treatment of quantum mechanics, in which particles or physical objects are treated using quantum wave functions but the surrounding environment (for example a potential well or a bulk electromagnetic field or gravitational field) is treated classically. First quantization is appropriate for studying a single quantum-mechanical system being controlled by a laboratory apparatus that is itself large enough that classical mechanics is applicable to most of the apparatus.

One-particle Systems

In general, the one-particle state could be described by a complete set of quantum numbers denoted by ν. For example, the three quantum numbers n,l,m associated to an electron in a coulomb potential, like the hydrogen atom, form a complete set (ignoring spin). Hence, the state is called $|\nu\rangle$ and is an eigenvector of the Hamiltonian operator. One can obtain a state function representation of the state using $\psi_\nu(\mathbf{r}) = \langle \mathbf{r}|\nu\rangle$..All eigenvectors of a Hermitian operator form a complete basis, so one can construct any state $|\psi\rangle = \sum_\nu |\nu\rangle\langle\nu|\psi\rangle$ obtaining the completeness relation:

$$\sum_\nu |\nu\rangle\langle\nu| = \hat{\mathbf{1}}$$

All the properties of the particle could be known using this vector basis.

Many-particle Systems

When turning to N-particle systems, i.e., systems containing N identical particles i.e. particles characterized by the same physical parameters such as mass, charge and spin, is necessary an extension of single-particle state function $\psi(\mathbf{r})$ to the N-particle state function $\psi(\mathbf{r}_1, \mathbf{r}_2, ..., \mathbf{r}_N)$. A fundamental difference between classical and quantum mechanics concerns the concept of indistinguishability of identical particles. Only two species of particles are thus possible in quantum physics, the so-called bosons and fermions which obey the rules:

$$\psi(\mathbf{r}_1, ..., \mathbf{r}_j, ..., \mathbf{r}_k, ..., \mathbf{r}_N) = +\psi(\mathbf{r}_1, ..., \mathbf{r}_k, ..., \mathbf{r}_j, ..., \mathbf{r}_N) \text{ (bosons)},$$

$$\psi(\mathbf{r}_1, ..., \mathbf{r}_j, ..., \mathbf{r}_k, ..., \mathbf{r}_N) = -\psi(\mathbf{r}_1, ..., \mathbf{r}_k, ..., \mathbf{r}_j, ..., \mathbf{r}_N) \text{ (fermions)}.$$

Where we have interchanged two coordinates $(\mathbf{r}_j, \mathbf{r}_k)$ of the state function. The usual wave function is obtained using the slater determinant and the identical particles theory. Using this basis, it is possible to solve any many-particle problem.

Second Quantization

Second quantization is a formalism used to describe and analyze quantum many-body systems. It is also known as canonical quantization in quantum field theory, in which the fields (typically as the wave functions of matter) are thought of as field operators, in a manner similar to how the physical quantities (position, momentum, etc.) are thought of as operators in first quantization. The key ideas of this method were introduced in 1927 by Dirac, and were developed, most notably, by Fock and Jordan later.

In this approach, the quantum many-body states are represented in the Fock state basis, which are constructed by filling up each single-particle state with a certain number of identical particles. The second quantization formalism introduces the creation and annihilation operators to construct and handle the Fock states, providing useful tools to the study of the quantum many-body theory.

Quantum Many-body States

The starting point of the second quantization formalism is the notion of indistinguishability of particles in quantum mechanics. Unlike in classical mechanics, where each particle is labeled by a distinct position vector $\mathbf{r}_i$ and different configurations of the set of $\mathbf{r}_i$'s correspond to different many-body states, *in quantum mechanics, the particles are identical, such that exchanging two particles, i.e. $\mathbf{r}_i \leftrightarrow \mathbf{r}_j$, does not lead to a different many-body quantum state*. This implies that the quantum many-body wave function must be invariant (up to a phase factor) under the exchange of two particles. According to the statistics of the particles, the many-body wave function can either be symmetric or antisymmetric under the particle exchange:

$$\Psi_B(\cdots, \mathbf{r}_i, \cdots, \mathbf{r}_j, \cdots) = +\Psi_B(\cdots, \mathbf{r}_j, \cdots, \mathbf{r}_i, \cdots) \text{ if the particles are bosons,}$$

$\Psi_{\mathrm{F}}(\cdots,\mathbf{r}_i,\cdots,\mathbf{r}_j,\cdots) = -\Psi_{\mathrm{F}}(\cdots,\mathbf{r}_j,\cdots,\mathbf{r}_i,\cdots)$ if the particles are fermions.

This exchange symmetry property imposes a constraint on the many-body wave function. Each time a particle is added or removed from the many-body system, the wave function must be properly symmetrized or anti-symmetrized to satisfy the symmetry constraint. In the first quantization formalism, this constraint is guaranteed by representing the wave function as linear combination of permanents (for bosons) or determinants (for fermions) of single-particle states. In the second quantization formalism, the issue of symmetrization is automatically taken care of by the creation and annihilation operators, such that its notation can be much simpler.

First-quantized Many-body Wave Function

Consider a complete set of single-particle wave functions $\psi_\alpha(\mathbf{r})$ labeled by α (which may be a combined index of a number of quantum numbers). The following wave function

$$\Psi[\mathbf{r}_i] = \prod_{i=1}^{N} \psi_{\alpha_i}(\mathbf{r}_i) \equiv \psi_{\alpha_1} \otimes \psi_{\alpha_2} \otimes \cdots \otimes \psi_{\alpha_N}$$

represents an N-particle state with the ith particle occupying the single-particle state $|\alpha_i\rangle$. In the shorthanded notation, the position argument of the wave function may be omitted, and it is assumed that the ith single-particle wave function describes the state of the ith particle. The wave function Ψ has not been symmetrized or anti-symmetrized, thus in general not qualified as a many-body wave function for identical particles. However, it can be brought to the symmetrized (anti-symmetrized) form by the symmetrization (anti-symmetrization) operators, denoted $\mathcal{S}$ ($\mathcal{A}$).

For bosons, the many-body wave function must be symmetrized,

$$\Psi_{\mathrm{B}}[\mathbf{r}_i] = \mathcal{N}\mathcal{S}\Psi[\mathbf{r}_i] = \mathcal{N}\sum_{\pi\in S_N}\prod_{i=1}^{N}\psi_{\alpha_{\pi(i)}}(\mathbf{r}_i) = \mathcal{N}\sum_{\pi\in S_N}\psi_{\alpha_{\pi(1)}}\otimes\psi_{\alpha_{\pi(2)}}\otimes\cdots\otimes\psi_{\alpha_{\pi(N)}};$$

while for fermions, the many-body wave function must be anti-symmetrized,

$$\Psi_{F}[\mathbf{r}_i] = \mathcal{N}\mathcal{A}\Psi[\mathbf{r}_i] = \mathcal{N}\sum_{\pi\in S_N}(-1)^{\pi}\prod_{i=1}^{N}\psi_{\alpha_{\pi(i)}}(\mathbf{r}_i) = \mathcal{N}\sum_{\pi\in S_N}(-1)^{\pi}\psi_{\alpha_{\pi(1)}}\otimes\psi_{\alpha_{\pi(2)}}\otimes\cdots\otimes\psi_{\alpha_{\pi(N)}}.$$

Here π is an element in the N-body permutation group (or symmetric group) S_N, which performs a permutation among the state labels α_i, and $(-1)^\pi$ denotes the corresponding permutation sign. $\mathcal{N}$ is the normalization operator that normalizes the wave function. (It is the operator that applies a suitable numerical normalization factor to the symmetrized tensors of degree n.)

If one arranges the single-particle wave functions in a matrix U, such that the row-i column-j matrix element is $U_{ij} = \psi_{\alpha_j}(\mathbf{r}_i) \equiv \langle\mathbf{r}_i|\alpha_j\rangle$, then the boson many-body wave function can be simply written as a permanent $\Psi_B = \mathcal{N}\,\mathrm{perm}\,U$, and the fermion many-body wave function as a determinant $\Psi_F = \mathcal{N}\det U$ (also known as the Slater determinant).

Second-quantized Fock States

First quantized wave functions involve complicated symmetrization procedures to describe phys-ically realizable many-body states because the language of first quantization is redundant for in-distinguishable particles. In the first quantization language, the many-body state is described by answering a series of questions like *"which particle is on which state"*. However these are not physical questions, because the particles are identical, and it is impossible to tell which particle is which in the first place. The seemingly different states $\psi_1 \otimes \psi_2$ and $\psi_2 \otimes \psi_1$ are actually redundant names of the same quantum many-body state. So the symmetrization (or anti-symmetrization) must be introduced to eliminate this redundancy in the first quantization description.

In the second quantization language, instead of asking "each particle on which state", one asks *"how many particles are there on each state"*. Because this description does not refer to the la-beling of particles, it contains no redundant information, and hence leads to a precise and simpler description of the quantum many-body state. In this approach, the many-body state is represented in the occupation number basis, and the basis state is labeled by the set of occupation numbers, denoted

$$| [n_\alpha] \rangle \equiv | n_1, n_2, \cdots, n_\alpha, \cdots \rangle,$$

meaning that there are n_α particles in the single-particle state $| \alpha \rangle$ (or as ψ_α). The occupation numbers sum up to the total number of particles, i.e. $\sum n_\alpha = N$. For fermions, the occupation number n_α can only be 0 or 1, due to the Pauli exclusion principle; while for bosons it can be any non negative integer

$$n_\alpha = \begin{cases} 0,1 & \text{fermions,} \\ 0,1,2,3,\ldots & \text{bosons.} \end{cases}$$

The occupation number states $| [n_\alpha] \rangle$ are also known as the Fock states. All the Fock states form a complete set of basis of the many-body Hilbert space, or the Fock space. Any generic quantum many-body state can be expressed as a linear combination of Fock states.

Note that besides providing a more efficient language, Fock space allows for a variable number of particles. As a Hilbert space, it is isomorphic to the sum of the n-particle bosonic or fermionic tensor spaces described in the previous section, including a one-dimensional zero-particle space C.

The Fock state with all occupation numbers equal to zero is called the vacuum state, denoted $| 0 \rangle \equiv | \cdots, 0_\alpha, \cdots \rangle$. . The Fock state with only one non-zero occupation number is a single-mode Fock state, denoted $| n_\alpha \rangle \equiv | \cdots, 0, n_\alpha, 0, \cdots \rangle$. In terms of the first quantized wave function, the vacuum state is the unit of tensor product, and can be denoted as $| 0 \rangle = 1$. The single-particle state is reduced to its wave function $| 1_\alpha \rangle = \psi_\alpha$. Other single-mode many-body (boson) state are just the tensor product of the wave function of that mode, such as $| 2_\alpha \rangle = \psi_\alpha \otimes \psi_\alpha$ and $_\alpha \rangle = \overset{\otimes}{_\alpha}$. For multi-mode Fock states (meaning more than one single-particle state $| \alpha \rangle$ is involved), the corre-sponding first-quantized wave function will require proper symmetrization according to the particle statistics, e.g. $|1_1, 1_2\rangle = (\psi_1 \psi_2 + \psi_2 \psi_1)/\sqrt{2}$ for a boson state, and $|1_1, 1_2\rangle = (\psi_1 \psi_2 - \psi_2 \psi_1)/\sqrt{2}$ for a fermion state (the symbol $\otimes$ between ψ_1 and ψ_2 is omitted for simplicity). In general, the normalization is found to be $\sqrt{\dfrac{\prod_\alpha n_\alpha!}{N!}}$, where N is the total number of particles. For fermion, this expression reduces to $\dfrac{1}{\sqrt{}}$ as n_α can only be either zero or one. So the first-quantized wave function corresponding to the Fock state reads

$$|[n_\alpha]\rangle_B = \left(\frac{\prod\limits_\alpha n_\alpha!}{N!} \right)^{1/2} \mathcal{S} \bigotimes_\alpha \psi_\alpha^{\otimes n_\alpha}$$

for bosons and

$$|[n_\alpha]\rangle_F = \frac{1}{\sqrt{N!}} \mathcal{A} \bigotimes_\alpha \psi_\alpha^{\otimes n_\alpha}$$

for fermions. Note that for fermions, $n_\alpha = 0, 1$ only, so the tensor product above is effectively just a product over all occupied single-particle states.

Creation and Annihilation Operators

The creation and annihilation operators are introduced to add or remove a particle from the many-body system. These operators lie at the core of the second quantization formalism, bridging the gap between the first- and the second-quantized states. Applying the creation (annihilation) operator to a first-quantized many-body wave function will insert (delete) a single-particle state from the wave function in a symmetrized way depending on the particle statistics. On the other hand, all the second-quantized Fock states can be constructed by applying the creation operators to the vacuum state repeatedly.

The creation and annihilation operators (for bosons) are originally constructed in the context of the quantum harmonic oscillator as the raising and lowering operators, which are then generalized to the field operators in the quantum field theory. They are fundamental to the quantum many-body theory, in the sense that every many-body operator (including the Hamiltonian of the many-body system and all the physical observables) can be expressed in terms of them.

Insertion and Deletion Operation

The creation and annihilation of a particle is implemented by the insertion and deletion of the single-particle state from the first quantized wave function in an either symmetric or anti-symmetric manner. Let ψ_α be a single-particle state, let 1 be the tensor identity (it is the generator of the zero-particle space C and satisfies $\psi_\alpha \equiv 1 \otimes \psi_\alpha \equiv \psi_\alpha \otimes 1$ in the tensor algebra over the fundamental Hilbert space), and let $\varnothing = \psi_{\alpha_1} \otimes \psi_{\alpha_2} \otimes \cdots$ be a generic tensor product state. The insertion $\otimes_\pm$ and the deletion $\oslash_\pm$ operators are linear operators defined by the following recursive equations

$$\psi_\alpha \otimes_\pm 1 = \psi_\alpha, \quad \psi_\alpha \otimes_\pm (\psi_\beta \otimes \Psi) = \psi_\alpha \otimes \psi_\beta \otimes \Psi \pm \psi_\beta \otimes (\psi_\alpha \otimes_\pm \Psi);$$

$$\psi_\alpha \oslash_\pm 1 = 0, \quad \psi_\alpha \oslash_\pm (\psi_\beta \otimes \Psi) = \delta_{\alpha\beta} \Psi \pm \psi_\beta \otimes (\psi_\alpha \oslash_\pm \Psi).$$

Here $\delta_{\alpha\beta}$ is the Kronecker delta symbol, which gives 1 if $\alpha = \beta$, and 0 otherwise. The subscript $\pm$ of the insertion or deletion operators indicates whether symmetrization (for bosons) or anti-symmetrization (for fermions) is implemented.

Boson Creation and Annihilation Operators

The boson creation (resp. annihilation) operator is usually denoted as $b_\alpha^\dagger$ (resp. b_α). The creation operator $b_\alpha^\dagger$ adds a boson to the single-particle state $|\alpha\rangle$, and the annihilation operator b_α removes a boson from the single-particle state $|\ \rangle$..The creation and annihilation operators are Hermitian conjugate to each other, but neither of them are Hermitian operators ($b_\alpha \neq b_\alpha^\dagger$).

Definition

The boson creation (annihilation) operator is a linear operator, whose action on a N-particle first-quantized wave function $\emptyset$ is defined as

$$b_\alpha^\dagger \Psi = \frac{1}{\sqrt{N+1}} \psi_\alpha \otimes_+ \Psi,$$

$$b_\alpha \Psi = \frac{1}{\sqrt{N}} \psi_\alpha \oslash_+ \Psi,$$

where $\psi_\alpha \otimes_+$ inserts the single-particle state ψ_α in $N+1$ possible insertion positions symmetrically, and $\psi_\alpha \oslash_+$ deletes the single-particle state ψ_α from N possible deletion positions symmetrically.

Examples

Hereinafter the tensor symbol $\otimes$ between single-particle states is omitted for simplicity. Take the state $|1_1,1_2\rangle = (\psi_1\psi_2 + \psi_2\psi_1)/\sqrt{2}$,, create one more boson on the state ψ_1,

$$b_1^\dagger |1_1,1_2\rangle = \frac{1}{\sqrt{2}}(b_1^\dagger \psi_1\psi_2 + b_1^\dagger \psi_2\psi_1)$$

$$= \frac{1}{\sqrt{2}}\left(\frac{1}{\sqrt{3}}\psi_1 \otimes_+ \psi_1\psi_2 + \frac{1}{\sqrt{3}}\psi_1 \otimes_+ \psi_2\psi_1\right)$$

$$= \frac{1}{\sqrt{2}}\left(\frac{1}{\sqrt{3}}(\psi_1\psi_1\psi_2 + \psi_1\psi_1\psi_2 + \psi_1\psi_2\psi_1) + \frac{1}{\sqrt{3}}(\psi_1\psi_2\psi_1 + \psi_2\psi_1\psi_1 + \psi_2\psi_1\psi_1)\right)$$

$$= \frac{\sqrt{2}}{\sqrt{3}}(\psi_1\psi_1\psi_2 + \psi_1\psi_2\psi_1 + \psi_2\psi_1\psi_1)$$

$$= \sqrt{2}|2_1,1_2\rangle.$$

Then annihilate one boson from the state ψ_1,

$$b_1 \mid 2_1, 1_2 \rangle = \frac{1}{\sqrt{3}} (b_1 \psi_1 \psi_1 \psi_2 + b_1 \psi_1 \psi_2 \psi_1 + b_1 \psi_2 \psi_1 \psi_1)$$

$$= \frac{1}{\sqrt{3}} \left(\frac{1}{\sqrt{3}} \psi_1 \oslash_+ \psi_1 \psi_1 \psi_2 + \frac{1}{\sqrt{3}} \psi_1 \oslash_+ \psi_1 \psi_2 \psi_1 + \frac{1}{\sqrt{3}} \psi_1 \oslash_+ \psi_2 \psi_1 \psi_1 \right)$$

$$= \frac{1}{\sqrt{3}} \left(\frac{1}{\sqrt{3}} (\psi_1 \psi_2 + \psi_1 \psi_2 + 0) + \frac{1}{\sqrt{3}} (\psi_2 \psi_1 + 0 + \psi_1 \psi_2) + \frac{1}{\sqrt{3}} (0 + \psi_2 \psi_1 + \psi_2 \psi_1) \right)$$

$$= \psi_1 \psi_2 + \psi_2 \psi_1$$

$$= \sqrt{2} \mid 1_1, 1_2 \rangle.$$

Action on Fock States

Starting from the single-mode vacuum state $\mid 0_\alpha \rangle = 1$, applying the creation operator $b_\alpha^\dagger$ repeatedly, one finds

$$b_\alpha^\dagger \mid 0_\alpha \rangle = \psi_\alpha \otimes_+ 1 = \psi_\alpha = \mid 1_\alpha \rangle,$$

$$b_\alpha^\dagger \mid n_\alpha \rangle = \frac{1}{\sqrt{n_\alpha + 1}} \psi_\alpha \otimes_+ \psi_\alpha^{\otimes n_\alpha} = \sqrt{n_\alpha + 1} \psi_\alpha^{\otimes(n_\alpha + 1)} = \sqrt{n_\alpha + 1} \mid n_\alpha + 1 \rangle.$$

The creation operator raises the boson occupation number by 1. Therefore, all the occupation number states can be constructed by the boson creation operator from the vacuum state

$$\mid n_\alpha \rangle = \frac{1}{\sqrt{n_\alpha!}} (b_\alpha^\dagger)^{n_\alpha} \mid 0_\alpha \rangle.$$

On the other hand, the annihilation operator b_α lowers the boson occupation number by 1

$$b_\alpha \mid n_\alpha \rangle = \frac{1}{\sqrt{n_\alpha}} \psi_\alpha \oslash_+ \psi_\alpha^{\otimes n_\alpha} = \sqrt{n_\alpha} \psi_\alpha^{\otimes(n_\alpha - 1)} = \sqrt{n_\alpha} \mid n_\alpha - 1 \rangle.$$

It will also quench the vacuum state $b_\alpha \mid 0_\alpha \rangle = 0$ as there has been no boson left in the vacuum state to be annihilated. Using the above formulae, it can be shown that

$$b_\alpha^\dagger b_\alpha \mid n_\alpha \rangle = n_\alpha \mid n_\alpha \rangle,$$

meaning that $\hat{n}_\alpha = b_\alpha^\dagger b_\alpha$ defines the boson number operator.

The above result can be generalized to any Fock state of bosons.

$$b_{\dot{a}}^\dagger \mid \cdots, n_{\grave{a}}, n_{\dot{a}}, n_{\tilde{a}}, \ldots \rangle = \sqrt{n_{\dot{a}} + 1} \mid \cdots, n_{\grave{a}}, n_{\dot{a}} + 1, n_{\tilde{a}}, \ldots \rangle.$$

$$b_\alpha \mid \cdots, n_\beta, n_\alpha, n_\gamma, \cdots \rangle = \sqrt{n_\alpha} \mid \cdots, n_\beta, n_\alpha - 1, n_\gamma, \cdots \rangle.$$

These two equations can be considered as the defining properties of boson creation and annihilation operators in the second-quantization formalism. The complicated symmetrization of the underlying first-quantized wave function is automatically taken care of by the creation and annihilation operators (when acting on the first-quantized wave function), so that the complexity is not revealed on the second-quantized level, and the second-quantization formulae are simple and neat.

Operator Identities

The following operator identities follow from the action of the boson creation and annihilation operators on the Fock state,

$$[b_\alpha^\dagger, b_\beta^\dagger] = [b_\alpha, b_\beta] = 0, \quad [b_\alpha, b_\beta^\dagger] = \delta_{\alpha\beta}.$$

These commutation relations can be considered as the algebraic definition of the boson creation and annihilation operators. The fact that the boson many-body wave function is symmetric under particle exchange is also manifested by the commutation of the boson operators.

The raising and lowering operators of the quantum harmonic oscillator also satisfies the same set of commutation relations, implying that the bosons can be interpreted as the energy quanta (phonons) of an oscillator. This is indeed the idea of quantum field theory, which considers each mode of the matter field as an oscillator subject to quantum fluctuations, and the bosons are treated as the excitations (or energy quanta) of the field.

Fermion Creation and Annihilation Operators

The fermion creation (annihilation) operator is usually denoted as $c_\alpha^\dagger$ (c_α). The creation operator $c_\alpha^\dagger$ adds a fermion to the single-particle state $|\alpha\rangle$, and the annihilation operator c_α removes a fermion from the single-particle state $|\alpha\rangle$. The creation and annihilation operators are Hermitian conjugate to each other, but neither of them are Hermitian operators ($c_\alpha \neq c_\alpha^\dagger$). The Hermitian combination of the fermion creation and annihilation operators

$$\chi_{\alpha,\text{Re}} = (c_\alpha + c_\alpha^\dagger)/2, \quad \chi_{\alpha,\text{Im}} = (c_\alpha - c_\alpha^\dagger)/(2i),$$

are called Majorana fermion operators.

Definition

The fermion creation (annihilation) operator is a linear operator, whose action on a N-particle first-quantized wave function $\emptyset$ is defined as

$$c_\alpha^\dagger \Psi = \frac{1}{\sqrt{N+1}} \psi_\alpha \otimes_- \Psi,$$

$$c_\alpha \Psi = \frac{1}{\sqrt{N}} \psi_\alpha \oslash_- \Psi,$$

where $\psi_\alpha \otimes_-$ inserts the single-particle state ψ_α in $N+1$ possible insertion positions anti-symmetrically, and $\psi_\alpha \oslash_-$ deletes the single-particle state ψ_α from N possible deletion positions anti-symmetrically.

Examples

Hereinafter the tensor symbol $\otimes$ between single-particle states is omitted for simplicity. Take the state $|1_1, 1_2\rangle = (\psi_1\psi_2 - \psi_2\psi_1)/\sqrt{2}$, attempt to create one more fermion on the occupied ψ_1 state will quench the whole many-body wave function,

$$c_1^\dagger |1_1, 1_2\rangle = \qquad \frac{1}{\sqrt{2}}(c_1^\dagger \psi_1\psi_2 - c_1^\dagger \psi_2\psi_1)$$

$$= \qquad \frac{1}{\sqrt{2}}\left(\frac{1}{\sqrt{3}}\psi_1 \otimes_- \psi_1\psi_2 - \frac{1}{\sqrt{3}}\psi_1 \otimes_- \psi_2\psi_1\right)$$

$$= \qquad \frac{1}{\sqrt{2}}\left(\frac{1}{\sqrt{3}}(\psi_1\psi_1\psi_2 - \psi_1\psi_1\psi_2 + \psi_1\psi_2\psi_1) - \frac{1}{\sqrt{3}}(\psi_1\psi_2\psi_1 - \psi_2\psi_1\psi_1 + \psi_2\psi_1\psi_1)\right)$$

$$= \qquad 0.$$

Annihilate a fermion on the ψ_2 state, take the state $|1_1, 1_2\rangle = (\psi_1\psi_2 - \psi_2\psi_1)/\sqrt{2}$,

$$c_2 |1_1, 1_2\rangle = \qquad \frac{1}{\sqrt{2}}(c_2\psi_1\psi_2 - c_2\psi_2\psi_1)$$

$$= \qquad \frac{1}{\sqrt{2}}\left(\frac{1}{\sqrt{2}}\psi_2 \oslash_- \psi_1\psi_2 - \frac{1}{\sqrt{2}}\psi_2 \oslash_- \psi_2\psi_1\right)$$

$$= \qquad \frac{1}{\sqrt{2}}\left(\frac{1}{\sqrt{2}}(0 - \psi_1) - \frac{1}{\sqrt{2}}(\psi_1 - 0)\right)$$

$$= \qquad -\psi_1$$

$$= \qquad -|1_1, 0_2\rangle.$$

The minus sign (known as the fermion sign) appears due to the anti-symmetric property of the fermion wave function.

Action on Fock States

Starting from the single-mode vacuum state $|0_\alpha\rangle = 1$, applying the fermion creation operator $c_\alpha^\dagger$,

$$c_\alpha^\dagger |0_\alpha\rangle = \psi_\alpha \otimes_- 1 = \psi_\alpha = |1_\alpha\rangle,$$

$$c_\alpha^\dagger |1_\alpha\rangle = \frac{1}{\sqrt{2}}\psi_\alpha \otimes_- \psi_\alpha = 0.$$

If the single-particle state $|\alpha\rangle$ is empty, the creation operator will fill the state with a fermion. However, if the state is already occupied by a fermion, further application of the creation operator will quench the state, demonstrating the Pauli exclusion principle that two identical fermions can not occupy the same state simultaneously. Nevertheless, the fermion can be removed from the occupied state by the fermion annihilation operator c_α,

$$c_\alpha \left| 1_\alpha \right\rangle = \psi_\alpha \oslash_- \psi_\alpha = 1 = \left| 0_\alpha \right\rangle,$$

$$c_\alpha \left| 0_\alpha \right\rangle = 0.$$

The vacuum state is quenched by the action of the annihilation operator.

Similar to the boson case, the fermion Fock state can be constructed from the vacuum state using the fermion creation operator

$$\left| n_\alpha \right\rangle = (c_\alpha^\dagger)^{n_\alpha} \left| 0_\alpha \right\rangle.$$

It is easy to check (by enumeration) that

$$c_\alpha^\dagger c_\alpha \left| n_\alpha \right\rangle = n_\alpha \left| n_\alpha \right\rangle,$$

meaning that $\hat{n}_\alpha = c_\alpha^\dagger c_\alpha$ defines the fermion number operator.

The above result can be generalized to any Fock state of fermions.

$$c_\alpha^\dagger \left| \cdots, n_\beta, n_\alpha, n_\gamma, \cdots \right\rangle = (-1)^{\sum\limits_{\beta < \alpha} n_\beta} (1 - n_\alpha) \left| \cdots, n_\beta, 1 - n_\alpha, n_\gamma, \cdots \right\rangle.$$

$$c_\alpha \left| \cdots, n_\beta, n_\alpha, n_\gamma, \cdots \right\rangle = (-1)^{\sum\limits_{\beta < \alpha} n_\beta} n_\alpha \left| \cdots, n_\beta, 1 - n_\alpha, n_\gamma, \cdots \right\rangle.$$

Recall that the occupation number n_α can only take 0 or 1 for fermions. These two equations can be considered as the defining properties of fermion creation and annihilation operators in the second quantization formalism. Note that the fermion sign structure $(-1)^{\sum\limits_{\beta < \alpha} n_\beta}$, also known as the Jordan-Wigner string, requires there to exist a predefined ordering of the single-particle states (the spin structure) and involves a counting of the fermion occupation numbers of all the preceding states; therefore the fermion creation and annihilation operators are considered non-local in some sense. This observation leads to the idea that fermions are emergent particles in the long-range entangled local qubit system.

Operator Identities

The following operator identities follow from the action of the fermion creation and annihilation operators on the Fock state,

$$\{c_\alpha^\dagger, c_\beta^\dagger\} = \{c_\alpha, c_\beta\} = 0, \quad \{c_\alpha, c_\beta^\dagger\} = \delta_{\alpha\beta}.$$

These anti-commutation relations can be considered as the algebraic definition of the fermion creation and annihilation operators. The fact that the fermion many-body wave function is anti-symmetric under particle exchange is also manifested by the anti-commutation of the fermion operators.

Quantum Field Operators

Defining $a_\nu^\dagger$ as a general annihilation(creation) operator for a single-particle state ν that could be either fermionic $(c_\nu^\dagger)$ or bosonic $(b_\nu^\dagger)$, the real space representation of the operators defines the quantum field operators $\Psi(\mathbf{r})$ and $\Psi^\dagger(\mathbf{r})$ by

$$\Psi(\mathbf{r}) = \sum_\nu \psi_\nu(\mathbf{r}) a_\nu$$

$$\Psi^\dagger(\mathbf{r}) = \sum_\nu \psi_\nu^*(\mathbf{r}) a_\nu^\dagger$$

These are second quantization operators, with coefficients $\psi_\nu(\mathbf{r})$ and $\psi_\nu^*(\mathbf{r})$ that are ordinary first-quantization wavefunctions. Thus, for example, any expectation values will be ordinary first-quantization wavefunctions. Loosely speaking, $\Psi^\dagger(\mathbf{r})$ is the sum of all possible ways to add a particle to the system at position r through any of the basis states $\psi_\nu(\mathbf{r})$.

Since $\Psi(\mathbf{r})$ and $\Psi^\dagger(\mathbf{r})$ are second quantization operators defined in every point in space they are called quantum field operators. They obey the following fundamental commutator and anti-commutator relations,

$$\left[\Psi(\mathbf{r}_1), \Psi^\dagger(\mathbf{r}_2)\right] = \delta(\mathbf{r}_1 - \mathbf{r}_2) \text{ boson fields,}$$

$$\{\Psi(\mathbf{r}_1), \Psi^\dagger(\mathbf{r}_2)\} = \delta(\mathbf{r}_1 - \mathbf{r}_2) \text{ fermion fields.}$$

In homogeneous systems it is often desirable to transform between real space and the momentum representations, hence, the quantum fields operators in Fourier basis yields:

$$\Psi(\mathbf{r}) = \frac{1}{\sqrt{V}} \sum_\mathbf{k} e^{i\mathbf{k}\cdot\mathbf{r}} a_\mathbf{k}$$

$$\Psi^\dagger(\mathbf{r}) = \frac{1}{\sqrt{V}} \sum_\mathbf{k} e^{-i\mathbf{k}\cdot\mathbf{r}} a^\dagger_\mathbf{k}$$

Comment on Nomenclature

The term "second quantization" is a misnomer that has persisted for historical reasons. One is not quantizing "again", as the term "second" might suggest; the field that is being quantized is not a Schrödinger wave function that was produced as the result of quantizing a particle, but is a classical field (such as the electromagnetic field or Dirac spinor field) that was not previously quantized. One is merely shifting from a semiclassical treatment of the system to a fully quantum-mechanical one.

References

- Pais, A. (1994) [1986]. Inward Bound: Of Matter and Forces in the Physical World (reprint ed.). Oxford, New York, Toronto: Oxford University Press. ISBN 978-0198519973.

- Schweber, S. S. (1994). QED and the Men Who Made It: Dyson, Feynman, Schwinger, and Tomonaga. Princeton University Press. ISBN 9780691033273.

- Newton, T. D.; Wigner, E.P. (1949). "Localized states for elementary systems". Rev. Mod. Phys. APS. 21 (3). Bibcode:1949RvMP...21..400N. doi:10.1103/RevModPhys.21.400. ISSN 0034-6861.

- Scharf, Günter (2014) [1989]. Finite Quantum Electrodynamics: The Causal Approach (third ed.). Dover Publications. ISBN 978-0486492735.

- Zee, Anthony (2010). Quantum Field Theory in a Nutshell (2nd ed.). Princeton University Press. ISBN 978-0691140346

Permissions

Index